The Future of Energy Systems: Integration of Energy Technologies in Distribution Grids

The Future of Energy Systems: Integration of Energy Technologies in Distribution Grids

Guest Editor

Mahmoud Ghofrani

MDPI

Basel • Beijing • Wuhan • Barcelona • Belgrade • Novi Sad • Cluj • Manchester

Guest Editor
Mahmoud Ghofrani
School of Science, Technology,
Engineering and Mathematics (STEM)
University of Washington
Bothell, WA
USA

Editorial Office
MDPI AG
Grosspeteranlage 5
4052 Basel, Switzerland

This is a reprint of the Special Issue, published open access by the journal *Energies* (ISSN 1996-1073), freely accessible at: https://www.mdpi.com/journal/energies/special_issues/3L3SE4Q7SW.

For citation purposes, cite each article independently as indicated on the article page online and as indicated below:

Lastname, A.A.; Lastname, B.B. Article Title. *Journal Name* **Year**, *Volume Number*, Page Range.

ISBN 978-3-7258-7070-7 (Hbk)
ISBN 978-3-7258-7071-4 (PDF)
https://doi.org/10.3390/books978-3-7258-7071-4

Contents

About the Editor

Mahmoud Ghofrani

Mahmoud Ghofrani is currently an Associate Professor affiliated with the School of Science, Technology, Engineering and Mathematics at University of Washington, Bothell. His research interests include power systems operation and planning, renewable energy systems, smart grids, electric vehicles and the electricity market.

Preface

This Reprint is dedicated to advancing knowledge on the modelling, optimisation, digitalisation, and secure operation of modern power systems in the context of accelerating renewable integration and electrification. As renewable energy sources, electric vehicles, distributed energy resources, and advanced digital technologies expand, power systems are experiencing structural and operational transformation at both the transmission and distribution levels. Challenges such as high prosumer penetration, network reconfiguration, load forecasting uncertainty, flexible demand management, and growing cybersecurity exposure demand integrated and data-driven solutions.

The scope of this Reprint spans long-term system modelling focusing on net-zero energy systems, short-term load forecasting using machine learning and deep learning approaches, optimisation of distribution network operation, flexible demand side management in smart cities, and technical frameworks for distribution system operators integrating distributed resources. Several contributions address cybersecurity and resilience, including analyses of cyber-attacks and intrusion detection within electric vehicle charging infrastructure and secure architectures for EV grid integration. The impact of digital technologies on sustainability and smart grid reliability is also examined.

The motivation for this Reprint stems from recognising that the energy transition requires coordinated advances in modelling, optimisation, artificial intelligence, cybersecurity, and regulatory adaptation. Rather than focusing on isolated techniques, the collected works highlight interconnected technical and institutional pathways necessary for reliable, sustainable, and secure power systems.

This Reprint is intended for researchers, engineers, utility professionals, policymakers, and graduate students engaged in smart grids, distributed energy systems, digital energy technologies, and energy security, offering both a reference for current developments and insight into future research directions.

Mahmoud Ghofrani
Guest Editor

energies

MDPI

Systematic Review

Securing the Electrified Future: A Systematic Review of Cyber Attacks, Intrusion and Anomaly Detection, and Authentication in Electric Vehicle Charging Infrastructure

Łukasz Pawlik [1,*], Jacek Łukasz Wilk-Jakubowski [1,2,*], Paweł Tomasz Grabski [3] and Grzegorz Wilk-Jakubowski [2,4]

1 Department of Information Systems, Kielce University of Technology, 7 Tysiąclecia Państwa Polskiego Ave., 25-314 Kielce, Poland
2 Institute of Crisis Management and Computer Modelling, 28-100 Busko-Zdrój, Poland; grzegorzwilkjakubowski@wp.pl
3 Department of Automotive Engineering and Transport, Kielce University of Technology, 7 Tysiąclecia Państwa Polskiego Ave., 25-314 Kielce, Poland; pgrabski@tu.kielce.pl
4 Institute of Internal Security, Old Polish University of Applied Science, 49 Ponurego Piwnika Str., 25-666 Kielce, Poland
* Correspondence: lpawlik@tu.kielce.pl (Ł.P.); jwilk@tu.kielce.pl (J.Ł.W.-J.)

Abstract

The growing popularity of electric vehicles (EVs) necessitates a robust and secure charging infrastructure. This review article analyzes the current state of cybersecurity challenges within EV charging ecosystems, focusing on vulnerabilities affecting charging stations, communication protocols, and grid integration technologies such as vehicle-to-grid (V2G). Based on a systematic literature review, common cyber threats are categorized, including data manipulation, authentication bypass, infrastructure failures, and attacks on backend systems and the power grid. Furthermore, this paper synthesizes existing and emerging risk mitigation strategies, encompassing intrusion detection systems (IDS), secure communication protocols, authentication frameworks, and anomaly detection techniques utilizing machine learning and blockchain technologies. By analyzing literature published between 2017 and 2024, this study aims to provide a comprehensive overview of cybersecurity issues that are critical to ensuring the resilience and reliability of the evolving EV charging infrastructure. The scope of this review is centered specifically on cyber attacks, intrusion and anomaly detection, and authentication mechanisms within EV charging infrastructures, rather than on the broader cybersecurity domain.

Keywords: electric vehicles; charging infrastructure; cybersecurity; cyber attacks; intrusion detection; anomaly detection; power grids; green security; embedded systems

1. Introduction

Electromobility is now a visible pillar of climate policy. In the European Union's roadmap, internal combustion engine vehicles are expected to be phased out of urban traffic by 2050, with electric vehicles (EVs) taking their place [1]. Beyond the well-known environmental gains, deployment at scale hinges on a charging ecosystem that is not only efficient and interoperable but also secure and resilient [2]. In this context, the review does not cover the entirety of cybersecurity research, but instead emphasizes three core domains: cyber attacks, intrusion and anomaly detection, and authentication.

A central challenge is the tight coupling between charging infrastructure and modern power systems. Vehicle-to-grid (V2G) capabilities expand this coupling by enabling bidirectional energy flows, which improves balancing but also broadens the attack surface of the grid–communication stack [2,3]. References often point to complementary energy topics—such as photovoltaic (PV)-assisted charging and self-consumption optimization—as part of infrastructure planning [3,4], and to energy recuperation gains at the vehicle level [3]. Constraints on batteries, their cost and aging, as well as mass-related impacts on energy use, remain relevant background factors for infrastructure planning and user adoption [1,5,6]. These considerations persist, yet the decisive differentiator for dependable scaling is cybersecurity.

Electric vehicle charging stations (EVCSs), back-end platforms (CSMSs) and the protocols that connect vehicles, stations and aggregators are exposed to data manipulation, authentication bypass, denial-of-service and malware propagation, with operational impacts that range from local downtime to distribution-level instability [7–10]. Research responds with several lines of defense: authentication frameworks [8], intrusion detection systems (IDS) and anomaly detection in communication streams [9,11], and safeguards for metering and billing integrity [10]. Recent studies strengthen this picture with machine-learning-based anomaly detection—including generative adversarial network (GAN)-based detectors in EV charging data pipelines [8]—and deep recurrent controllers that mitigate sophisticated, time-varying disturbances in EVCS cyber–physical loops [12]. Work on the ISO 15118 family underscores both its role in secure, interoperable V2G handshakes and concrete vulnerability classes that demand careful engineering and conformance testing [11,13]. Models of large-scale coordinated attacks, including stochastic distributed denial of service (DDoS) traffic, aimed at stations and their networks, quantify service delays and stability risks and motivate robust protective capacity [14]. Parallel efforts explore decentralized authentication and settlement using blockchain to remove single points of failure and enhance privacy in V2G transactions [15–18]. Additionally, studies on price-signal manipulation demonstrate how adversaries could steer charging/discharging behavior and stress distribution feeders if defenses are weak [19]. Emerging EVCS stacks that rely on 5G connectivity further raise the bar for traffic inspection and learning-based IDS to spot covert false data injection (FDI) and DDoS patterns in near real time [20].

Against this backdrop, the aim of this review is to consolidate current knowledge about cybersecurity in EV charging ecosystems with an emphasis on communication protocols, cryptography, anomaly and intrusion detection, and grid integration risks. The review synthesizes literature from 2017–2024 and maps threats and countermeasures across charging points, back-end systems and V2G interfaces. It highlights where defenses are mature and where evidence still rests on simulations rather than field data [2,11]. To keep the discussion grounded in deployment realities, the article also retains the broader infrastructure context referenced above—PV integration, operational constraints and user-level factors—while treating them as boundary conditions for secure design rather than as the primary focus [3,6].

The paper is structured as follows. Section 2 outlines the materials and methods used for the literature selection and classification. Section 3 surveys cybersecurity themes (authentication and cryptography, attacks, intrusion/anomaly detection, blockchain) alongside enabling technologies and infrastructure. Subsequent sections synthesize trends and open gaps, and discuss implications for the secure, scalable integration of EV charging with smart grids [1–4,7–11,14].

2. Materials and Methods

This study employed a systematic review approach focused on analyzing scientific literature published from 2017 to 2024. The selection of publications was based on search results from the Scopus database, with queries restricted to the fields of engineering, computer science, and energy. A total of 102 publications were included in the final analysis. These documents met our defined language, topical, and formal criteria, while literature from non-technical domains was excluded. The documents were classified manually based on their full content, taking into account the type of methodology used, such as experimental studies, literature reviews, case studies, or conceptual modeling.

Additionally, the publications were assigned to four main thematic categories within the domain of cybersecurity and infrastructural technologies for electric vehicles. For qualitative analysis, keywords and a taxonomy were applied based on the occurrence of concepts such as "intrusion detection," "authentication," "charging infrastructure," and "vehicle-to-grid." Special attention was paid to the geographical distribution of the publications and the temporal dynamics in the development of specific algorithmic classes.

In the context of the growing popularity of electric vehicles (EVs) and ongoing advancements in the development in modern charging systems, we conducted a literature analysis to address the following seven key research questions:

- Which anomaly detection techniques are most effective in securing EV charging infrastructure?
- To what extent does blockchain enhance the resilience of EV systems against cyber attacks?
- Which authentication protocols dominate vehicle-to-grid communication, and how do they impact overall system security?
- Does the type of infrastructure (public vs. private) determine the level of cybersecurity protection implemented?
- What are the operational consequences of attacks on charging stations, such as tariff manipulation or grid destabilization?
- To what extent are real-world versus simulated data used in EV cybersecurity research?
- Which regions are leading EV cybersecurity research, and what are the driving factors behind this trend?

2.1. Document Retrieval and Selection

The literature review was conducted using the Scopus database, based on a systematic search methodology inspired by the PRISMA guidelines. Publications from the years 2017–2024 written in English were included in the analysis. These works focused on cybersecurity, deep learning (DL), neural networks (NN), and statistical methods in the context of electric vehicles (EVs). Particular emphasis was placed on studies related to energy consumption forecasting, predictive vehicle control systems, battery resource management, and the integration of electric vehicles with smart power grids.

The selection of literature was carried out based on the thematic relevance of the publications included in the source files, taking into account methodological diversity and the noticeable increase in the number of publications between 2021 and 2024, which confirms the growing global interest in this area. The search was based on a logical combination of keywords related to data analysis methods, electromobility, cybersecurity and charging infrastructure. The following search query was applied:

"TITLE-ABS-KEY(Electric AND Vehicles AND ("Charging" OR "Power Grids" OR "Embedded Systems")) AND PUBYEAR > 2016 AND PUBYEAR < 2025 AND (EXCLUDE (SUBJAREA,"AGRI") OR EXCLUDE (SUBJAREA,"NEUR") OR EXCLUDE (SUBJAREA,"BIOC") OR EXCLUDE (SUBJAREA,"BUSI") OR EXCLUDE (SUBJAREA,"MULT")

OR EXCLUDE (SUBJAREA,"CHEM") OR EXCLUDE (SUBJAREA,"ECON") OR EXCLUDE (SUBJAREA,"MEDI") OR EXCLUDE (SUBJAREA,"CENG") OR EXCLUDE (SUBJAREA,"SOCI") OR EXCLUDE (SUBJAREA,"ARTS") OR EXCLUDE (SUBJAREA,"PHAR") OR EXCLUDE (SUBJAREA,"PSYC") OR EXCLUDE (SUBJAREA,"HEAL") OR EXCLUDE (SUBJAREA,"EART") OR EXCLUDE (SUBJAREA,"ENVI") OR EXCLUDE (SUBJAREA, "MATE") OR EXCLUDE (SUBJAREA,"PHYS") OR EXCLUDE (SUBJAREA,"DECI") OR EXCLUDE (SUBJAREA,"MATH")) AND (EXCLUDE (DOCTYPE,"cr")) AND (LIMIT-TO (LANGUAGE, "English")) AND (LIMIT-TO (EXACTKEYWORD, "Cyber-attacks") OR LIMIT-TO (EXACTKEYWORD, "Cyber Attacks") OR LIMIT-TO (EXACTKEYWORD, "Intrusion Detection") OR LIMIT-TO (EXACTKEYWORD, "Intrusion Detection Systems") OR LIMIT-TO (EXACTKEYWORD, "Intrusion Detection System (IDS)") OR LIMIT-TO (EXACTKEYWORD, "Anomaly Detection") OR LIMIT-TO (EXACTKEYWORD, "Attack Detection") OR LIMIT-TO (EXACTKEYWORD, "Authentication"))"

In the methodology section, a detailed justification for the original search query was added. It was explained that the query was deliberately constructed with precision to select publications specifically related to EV charging in the context of cyber attacks, intrusion and anomaly detection, and authentication.

The next stage of literature selection was based on the results of the query conducted in the Scopus database, which yielded a total of 102 publications meeting the defined criteria. To ensure high relevance, full-text scientific publications were analyzed for the presence of terms related to electric vehicles and charging infrastructure in the context of cybersecurity and embedded systems.

The time range was limited to publications from 2017 to 2024, written in English and classified under subject areas such as engineering, computer science, and energy. The classification process was supported by exclusion criteria targeting non-technical fields (e.g., medicine, psychology, arts), in accordance with the recommendations of previous review studies.

In the methodology section, the justification for selecting specific subject areas was provided. The restriction to key domains was intended to capture the most relevant and directly related publications. This approach minimized the risk of including articles from fields that were not the focus of the analysis, such as social sciences or psychology, and allowed the study to remain concentrated on technical and engineering literature directly addressing the cybersecurity of electric vehicle charging infrastructure.

Based on keyword analysis, the publications were assigned to two primary thematic categories:

(1) Cybersecurity, including authentication, cryptography, intrusion and anomaly detection, and blockchain technologies;
(2) Technologies and infrastructure, focusing on charging systems, integration with the electrical grid (including V2G), and EV onboard systems.

Special attention was also given to research methodology—publications were classified as experimental studies, literature reviews, case studies, or conceptual papers, based on author declarations or content analysis of abstracts.

An additional manual review phase was carried out, covering 196 articles. During this selection, 94 papers were excluded due to not meeting the technical requirements or lacking direct relevance to the research topic, resulting in a final set of 102 publications. In the categorization process, some papers were assigned to more than one group, reflecting the multidimensional nature of the issues addressed in the texts.

The geographical analysis revealed a significant concentration of authors in countries such as China, the United States, India, Germany, and the United Kingdom, with fewer publications from Brazil, Iran, and South Korea. The documents were primarily journal articles or conference papers. Additionally, the analysis incorporated visualizations and data generated using VOSviewer, version 1.6.20 [21], which enabled mapping of keyword relationships and facilitated the identification of key thematic clusters within the analyzed dataset.

A graphical summary illustrating the stages of literature selection, the applied filtering criteria, and the main classification directions is presented in Figure 1, which also depicts the thematic distribution of the analyzed publications.

Figure 1. Data collection and preparation workflow. The blue background indicates the Search in Scopus phase, while the yellow background represents the Defining Categories phase.

2.2. Classification Criteria

A total of 102 scientific publications were selected for analysis based on predefined thematic and formal criteria, covering the years 2017–2024 and written in English. These publications were retrieved from the Scopus database using a query that targeted titles, abstracts, and keywords containing terms related to electric vehicles, power grids, charging, and embedded systems. The final list of documents was narrowed down from an initial set of 196 by excluding items that did not meet the subject area, language, or document type criteria. For further analysis, each document was assigned to five main classification categories: (1) Two thematic groups (cybersecurity and technologies and infrastructure); (2) country of author affiliation; (3) document type; and (4) research methodology.

The classification process was conducted manually based on the analysis of metadata, abstract content, and full-text documents available in English. Where applicable, translated versions were also reviewed.

The first classification group referred to the research focus of the publications, identifying two overarching thematic groups: Cybersecurity and technologies and infrastructure.

Within the cybersecurity group, four specific thematic categories were distinguished, as follows:

- Authentication and cryptography (e.g., authentication, cryptography);
- Cyber attacks (cyber attacks, computer crime, denial of service);
- Intrusion and anomaly detection (intrusion detection systems, anomaly detection);
- Blockchain technologies (blockchain).

The second thematic group addressed technological and infrastructural aspects of electromobility. It included three main subcategories, as follows:

- Electric vehicle charging (electric vehicle charging, charging infrastructure);
- Grid integration (smart grids, V2G, energy transfer);
- Vehicle systems (battery management, embedded systems, vehicular networks, internet of things).

The second classification axis concerned the country of author affiliation. This information was obtained from the bibliographic metadata provided by the Scopus database. Eleven countries were identified as having the highest number of affiliations: United States, China, Canada, Germany, India, Australia, Italy, United Kingdom, Brazil, South Korea, and Turkey. The remaining countries were placed in the "Other" category.

The third classification referred to the document type, which included the following:

- Journal articles (48 studies);
- Conference papers (52 studies);
- Other (e.g., reviews, book chapters; 2 studies).

The fourth classification dimension was based on the research methodology, which was assessed by analyzing the substantive content of the documents and the nature of the conducted studies. Four methodological approaches were identified, as follows:

- Experimental studies, based on original measurements and testing procedures;
- Literature-based analyses, including systematic reviews and meta-analyses;
- Case studies, focusing on implementations and practical applications;
- Conceptual works, involving mathematical models and proposals for new system architectures.

The assignment of categories was conducted according to a predefined classification framework. The data structure was validated using a graphical model representing individual research aspects and their subdivisions. This structure enabled a multi-dimensional and comparative analysis of scientific studies in the field of electromobility, with particular emphasis on cybersecurity threats and integration aspects related to energy infrastructure.

This detailed classification allowed for an in-depth, multi-perspective analysis of the current body of scientific literature concerning energy consumption forecasting and control in electric vehicles.

2.3. Data Processing and Analysis

The bibliographic data, which included 102 scientific publications that met the predefined thematic, linguistic, and temporal criteria, were standardized and prepared for further analysis. First, a manual verification of titles, abstracts, and keywords was performed. Each publication was then assigned to one or more classification categories based on the structure outlined in the source documents. These categories were associated with two primary thematic areas: Cybersecurity and technologies and infrastructure.

For data analysis, the Python 3.12 programming environment was used and the following libraries were applied: Pandas 2.2.2 for tabular data processing, Matplotlib 3.8.4 and

Seaborn 0.13.2 for data visualization, and NumPy 1.26.4 for numerical operations. To streamline data filtering and aggregation, all records were imported into a local PostgreSQL 16.2 relational database. Dedicated tables were created to manage bibliographic information, including authors, affiliations, keywords, year of publication, and DOI. The use of structured SQL queries enabled cross-sectional thematic analyses and comparisons across classification categories.

Publications were classified according to the following criteria:

- Technology of infrastructure (cybersecurity, deep learning, neural networks, statistical methods);
- Application area (battery systems, charging infrastructure, power grids, predictive control);
- Research methodology (experimental study, literature review, case study, conceptual work);
- Country of author affiliation;
- Document type (journal article, conference paper, other).

Particular attention was given to geographical analysis, which made it possible to determine the distribution of publications by country and identify the most active research centers. According to the classification framework, the most frequently represented countries included the United States, China, Canada, Germany, India, Australia, Italy, the United Kingdom, Brazil, South Korea, and Turkey.

A temporal analysis was also conducted, covering the years 2017–2024, allowing for the identification of trends and research dynamics related to data-driven approaches in the context of electromobility and energy management. To visualize relationships between key concepts, network graphs were generated using the VOSviewer tool.

By integrating programming tools, database systems, and manual classification, a deeper understanding of the structure and research directions concerning the security and integration of electric vehicle charging infrastructure with smart energy grids was achieved.

2.4. Review Protocol and Publication Quality

To ensure transparency and reproducibility, the literature review was conducted following a structured protocol inspired by the Preferred Reporting Items for Systematic Reviews and Meta-Analyses (PRISMA) guidelines. The diagram outlines the four main stages of the systematic review process, as follows:

1. Identification: 196 records identified in the Scopus database.
2. Screening: After removing duplicates ($n = 0$), the remaining 196 records were screened, with 94 being excluded by applying an additional keyword filter: "LIMIT-TO (EXACTKEYWORD, "Electric Vehicles")." To ensure the robustness of the screening process, the Scopus query was extended to include a broader set of synonyms related to electric vehicles and charging infrastructure (EVSE, PEV, EVCS, charge point, CSMS), in addition to the term electric vehicles. The extended query took the following form:

   AND (LIMIT-TO (EXACTKEYWORD, "Electric Vehicles") OR LIMIT-TO (EXACTKEYWORD, "EVSE") OR LIMIT-TO (EXACTKEYWORD, "PEV") OR LIMIT-TO (EXACTKEYWORD, "EVCS") OR LIMIT-TO (EXACTKEYWORD, "charge point") OR LIMIT-TO (EXACTKEYWORD, "CSMS")).

   The extended search yielded the same number of publications ($n = 102$) as the original query. This outcome reflects the fact that Scopus does not rely solely on author-supplied keywords, but instead applies its own standardized indexing. As a result, articles in which authors provided alternative terms (e.g., EVSE, PEV, EVCS) were nevertheless indexed under the unified keyword Electric Vehicles. Accordingly, the synonym expansion did not alter the dataset but confirmed the completeness of the initial screening strategy.

3. Eligibility: The full texts of the remaining 102 publications were assessed for eligibility and no further exclusions were made at this stage.
4. Included: A final set of 102 publications was ultimately included in the review.

The flow diagram illustrating this process is presented in Figure 2. It is important to note that no duplicates were identified in the Scopus database after applying the initial filters, as the database itself ensures their absence. However, all records were additionally imported into a PostgreSQL database, where their uniqueness was verified based on DOI identifiers and titles. The exact dates of the searches and data exports were provided, and the Supplementary Materials include the detailed queries used in the process to ensure full reproducibility.

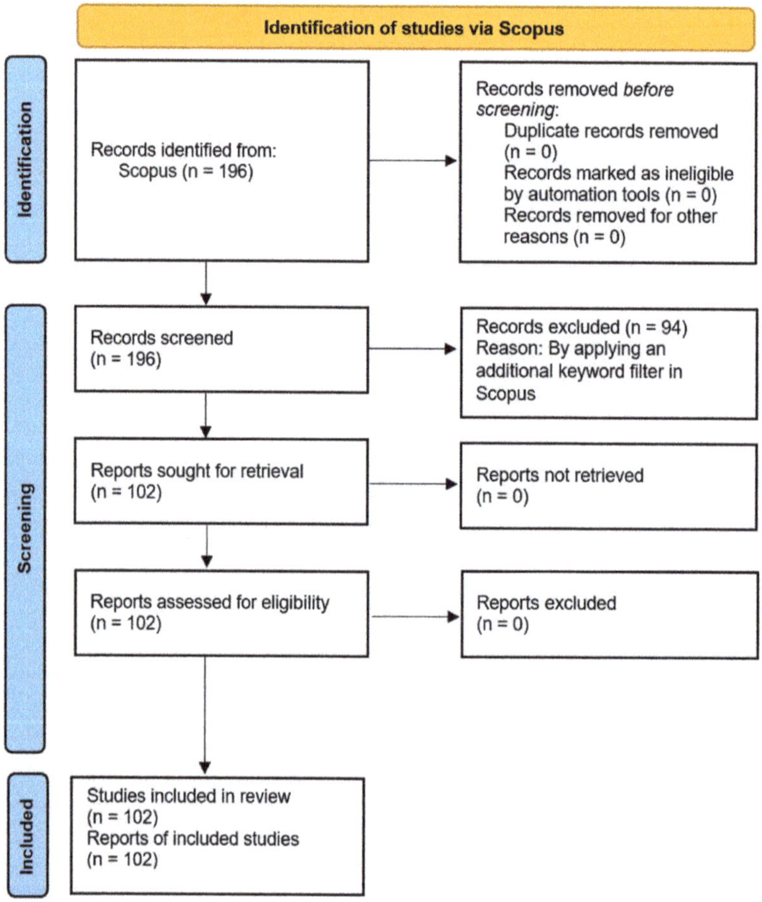

Figure 2. PRISMA flow diagram illustrating the literature selection process, including the stages of identification, screening, eligibility assessment, and final inclusion of publications in the review.

The second classification focused on research topics, which were determined through keyword extraction. Text mining tools, such as VOSviewer, were used to identify dominant terms and their co-occurrence across the reviewed publications. Two primary thematic axes were distinguished: cybersecurity-related issues and electric vehicle technologies and infrastructure.

Within the cybersecurity group, the following thematic categories were identified:

- Authentication and cryptography;
- Cyber attacks;
- Intrusion and anomaly detection;
- Blockchain.

The second thematic group included topics related to the following:

- Vehicle charging;
- Grid integration;
- Vehicle systems, such as battery management systems (BMS), vehicle-to-vehicle (V2V) communication, and vehicular networks.

The geographical classification was performed based on the country of author affiliation. The analyzed dataset was dominated by publications originating from China, India, Germany, the United States, the United Kingdom, Japan, Saudi Arabia, Italy, Canada, and South Korea. Countries with single publication entries were grouped under the category "Other." This approach enabled the identification of geographic centers of research activity in the application of ML/DL/SA methods within electromobility.

Publications were also classified by document type, using a basic division into the following:

- Journal articles (published in peer-reviewed journals);
- Conference papers;
- Other formats.

These classifications were based on metadata provided directly by the Scopus database.

Additionally, the research methodology was evaluated based on abstract content and full-text analysis. Four categories were distinguished, as follows:

- Experimental studies (involving measurements, prototypes, or implementations);
- Literature reviews (focused on surveying existing knowledge);
- Case studies (presenting specific real-world implementations);
- Conceptual papers (theoretical or model-based works proposing new approaches).

For the purposes of this classification, the collected data were manually reviewed by two independent evaluators, who assessed each publication based on predefined criteria. In cases of disagreement, a consensus-based approach was adopted through joint discussion. To quantify inter-rater agreement, Cohen's κ will be computed for the screening and classification stages and reported with a 95% confidence interval. This procedure ensured consistency and enabled a multidimensional analysis of the literature, covering technical, geographical, and methodological aspects. As part of the review protocol, a dedicated quality appraisal step was introduced to evaluate the methodological soundness and relevance of the included studies. A tailored checklist was applied, covering clarity of research objectives, transparency of methodology, robustness of data analysis, and replicability of results. Each publication was assessed independently, and the appraisal scores were used to weight the conclusions drawn from the review. In practice, findings supported by higher-quality studies were given greater interpretative weight, while evidence from lower-quality or less transparent sources was treated with appropriate caution.

The classification categories were determined not only by the presence of relevant terms but also by their frequency and prominence within the keyword co-occurrence structure. This analysis was visualized using a keyword density map (Figure 3). The resulting visualization enabled the identification of the most frequently addressed topics and their interrelationships, highlighting areas of intensive research activity as well as gaps that warrant further exploration.

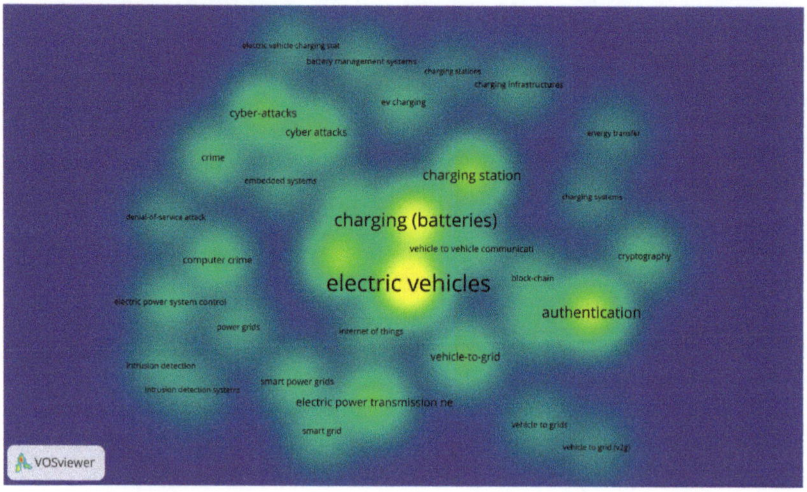

Figure 3. Term density map (VOSviewer). The yellow color indicates the highest density of keywords, signifying the most central and frequently co-occurring topics.

Figure 4 presents a detailed keyword co-occurrence network based on the analyzed publications, forming a thematic map of conceptual linkages within the literature on electric vehicles and related technologies. Each node represents a keyword or concept appearing in the dataset, and its size reflects the frequency of occurrence. The larger the node, the more central its role in the research structure.

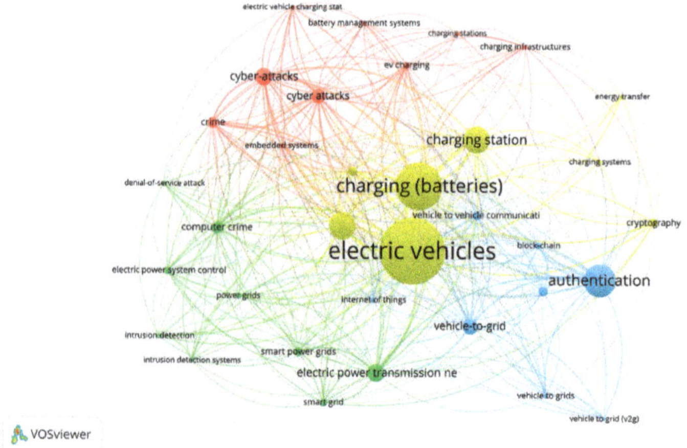

Figure 4. Term co-occurrence network map visualizing interrelationships and thematic clusters identified in the analyzed publications.

Node colors indicate membership in specific thematic clusters, automatically generated using a clustering algorithm. Each cluster groups terms that frequently co-occur across publications, reflecting their logical and topical coherence. For example, the red cluster (top-left area of the map) is focused on cybersecurity issues, including terms such as "cyber-attacks," "cyber attacks," "computer crime," "denial-of-service attack," and "intrusion detection." This terminology indicates a high volume of research dedicated to protecting EV infrastructure from digital threats.

At the center of the map, the largest and most dominant node is "electric vehicles," which serves as the primary anchor for surrounding topics. It is surrounded by terms such as "charging (batteries)," "charging station," "battery management systems," and "vehicle-to-vehicle communication." These are grouped in the yellow cluster, representing a combined focus on infrastructural, technological, and operational aspects of EV energy management and charging systems.

The green cluster, located in the lower section of the map, includes terms related to power grid systems and their integration strategies with EVs. Terms such as "smart power grids," "electric power system control," "transmission networks," "power grids" and "electric power transmission" indicate strong research interest in the integration of electric vehicles with smart grid technologies and bidirectional energy flow concepts, such as vehicle-to-grid (V2G).

On the right-hand side of the map, the blue cluster focuses on data and identity security, featuring terms like "authentication," "blockchain" and "cryptography." Its linkage with other clusters—through terms like "vehicle-to-grid" and "Internet of things"—demonstrates the intersection of security issues with technological and infrastructural dimensions.

The thickness of the edges (links) between nodes indicates the strength of co-occurrence; thicker edges represent higher frequencies of joint appearance. The dense network of connections between clusters suggests a high level of interdisciplinarity, bridging topics across energy systems, computer science, control, and cybersecurity.

This network visualization enables the identification of dominant research areas, as well as gaps and potential directions for further exploration. For instance, the relatively weak linkage between "blockchain" and "battery management systems" may signal an underexplored area in leveraging distributed ledger technologies for EV energy management. Thus, Figure 4 provides a valuable visual perspective on the structure and complexity of the reviewed literature and supports intuitive navigation through the interrelated research domains.

In the following work, literature items ranging from 11 to 112 were analyzed.

3. State of the Art

The rapid growth of electromobility is shaping a complex digital environment where electric vehicles are becoming part of smart grid systems. This new connectivity presents exciting opportunities but also brings notable security risks. It is crucial to protect the integrity and data privacy that moves between vehicles, charging points, and the grid. For this reason, modern research combines cybersecurity with the advancement of electromobility technologies and infrastructure, creating a new field that needs robust protection. This review examines 102 selected publications, sorting them into two primary areas: Cybersecurity (Section 3.1) and technologies and infrastructure (Section 3.2).

3.1. Cybersecurity

This section is structured around the three primary areas that form the focus of this review: authentication, cyber attacks, and intrusion/anomaly detection. The development of electromobility and related digital technologies brings both enormous opportunities and significant challenges related to cybersecurity. Authentication and cryptography mechanisms play a particularly important role, ensuring confidentiality, integrity, and non-repudiation of communication during electric vehicle charging processes and integration with grid infrastructure. As the scale and complexity of energy systems grow, so does the threat of cyber attacks, including denial-of-service attacks, which can destabilize both local charging stations and entire power grids. A key area of research is detecting intruders and anomalies, enabling early response to security breaches in critical systems. Blockchain is also gaining importance, with applications in energy transaction settlement, vehicle-to-grid (V2G) systems, and ensuring data transparency and immutability.

3.1.1. Authentication and Cryptography

The article presents new authentication methods in distributed systems, focusing on minimizing delays in the login process. The authors emphasize the importance of lightweight authentication protocols in the context of devices with limited computing power [8].

The paper analyzes biometric solutions as authentication tools in information systems, highlighting the potential of combining biometrics with classical cryptographic methods [9]. The authors discuss the use of single sign-on (SSO) protocols in large computer networks and propose mechanisms to increase their resilience against spoofing attacks [10]. The article presents the application of post-quantum cryptography in protecting sensitive data, emphasizing the resilience of proposed solutions to future threats posed by quantum computers [11]. The study describes the implementation of homomorphic encryption in cloud systems, pointing out that this technology enables operations on encrypted data without the need for decryption [22]. Finally, research analyzes the use of elliptic curve cryptography in mobile communication systems, indicating that this method ensures a high level of security while requiring minimal resources [15].

3.1.2. Cyber Attacks

The article analyzes the growing scale of cyber attacks targeting critical infrastructure, emphasizing the need for developing global response standards [13]. Within the EV ecosystem, such attacks directly translate into risks for charging stations and control backbones, where manipulation of communication can disrupt open charge point protocol (OCPP) sessions or ISO 15118 handshakes. Advanced persistent threats (APTs), known for persistence and strategic data theft, may target central charging station management systems (CSMSs) to compromise authentication data or disrupt charging coordination [23]. Case studies of ransomware incidents in other critical domains, such as healthcare [14], illustrate how similar techniques could be applied to electric vehicle supply equipment (EVSE) operators. These attacks could encrypt charging management servers and prevent access to billing or session data. Likewise, the evolution of DDoS attacks using IoT botnets [24], demonstrates a clear parallel: large-scale traffic floods can be directed at charging networks, overloading OCPP message queues and causing widespread service denial. Attacks on Supervisory Control and Data Acquisition (SCADA) systems [25], while originally documented in industrial contexts, highlight vulnerabilities relevant to EVSE–grid integration, where tampering with telemetry could destabilize load management. This is a risk particularly highlighted by security standards like IEC 62351-7:2017, which [25] aims to extend. Financial cybercrime [26] and social engineering [27], although often analyzed outside the EV context, point to attack vectors against payment platforms and user accounts linked with EV charging services. Legal harmonization challenges [28] and psychological factors behind cybercrime [29] may be secondary, but they underscore the persistent global nature of these threats. DoS attacks against banking systems [30] and DDoS campaigns in campus networks [31] are valuable by analogy, as they demonstrate the effectiveness of overload strategies that can be repurposed against EVSE authentication servers or CSMS APIs, leading to significant disruption of mobility services.

3.1.3. Intrusion and Anomaly Detection

The article analyzes signature-based intrusion detection systems (IDSs), emphasizing that this method is effective against known attacks but has limited capability to detect new threats [32]. In the EV charging context, this limitation is particularly relevant for OCPP and ISO 15118, where novel manipulation of handshake or tariff messages may bypass legacy IDS signatures.

Hybrid intrusion detection systems combining signature analysis with behavioral methods demonstrate improved detection accuracy and reduced false alarms [33], which is applicable to identifying abnormal CSMS traffic patterns. IDS solutions for industrial networks [34] directly inform EVSE protection, as integration with grid protocols demands anomaly detection that accounts for both power and data flows. Machine learning-based IDS [35] enhance adaptability to new threats, a necessity for dynamic EV environments involving wireless or roaming charging. IDS approaches in cloud environments [20] and IoT networks [36] are highly relevant, as CSMS platforms are increasingly cloud-hosted and connected to heterogeneous IoT devices. Deep learning-based IDS solutions [19] show promise in reducing false positives when applied to large-scale EVSE telemetry. Research on IDS for autonomous vehicles [16] offers transferable insights into vehicular communication patterns, which overlap with EVSE data flows.

Integrated firewall–IDS solutions [37] and flow-based detection [38] provide additional layers of defense suitable for protecting charging stations and backend APIs. Banking sector IDS applications [39], while sector-specific, highlight transferable approaches for fraud detection in EV payment platforms. Telemetry-based solutions [40] are directly applicable to the real-time monitoring of EVSE devices, while phishing detection [41], mobile app security [42], and anomaly detection in industrial [43] or grid systems [44] all highlight methodological tools that can be tailored to EVCS environments. Statistical traffic analysis [45], deep neural networks [46], and IoT anomaly detection [47] further demonstrate techniques that can be adapted to detect misuse of OCPP or ISO 15118 traffic channels. Medical IDS applications [48], although contextually different, underscore the broader utility of anomaly detection in protecting safety-critical systems, reinforcing the case for their application to EVSE networks.

3.1.4. Performance Comparisons and Methodological Limitations

When comparing the families of algorithms listed in Tables 1 and 2, several trade-offs emerge that matter for EVSE/CSMS intrusion and anomaly detection. Linear and polynomial regression provide strong baselines for forecasting (e.g., load and price sensitivity) due to their interpretability and low latency. However, they struggle with non-linear, regime-shifting behaviors typical of charging sessions and are sensitive to multicollinearity. Over-parameterized polynomial models may fit historical dynamics but generalize poorly under novel operating conditions, leading to unstable residual-based alarms.

Support Vector Machines (SVMs) with appropriate kernels often deliver competitive accuracy on medium-sized, high-dimensional feature sets and handle margin maximization well under modest label noise. Their drawbacks include expensive training for large datasets, limited native support for incremental/online updates, and comparatively opaque decision boundaries. Latency on embedded targets may also constrain real-time deployment unless models are carefully pruned.

Tree-based ensembles (Decision Trees, Random Forest, Gradient Boosting) typically offer strong out-of-the-box performance, robustness to heterogeneous features, and useful by-product explanations (feature importance, split logic). Random Forests handle variance well but may underfit fine-grained temporal signatures. Boosting methods capture subtle interactions yet risk overfitting without careful regularization and early stopping. Both families can be brittle under concept drift (e.g., firmware updates, tariff changes), requiring periodic retraining or drift-aware adaptation. Native online learning support is limited in many off-the-shelf implementations.

Table 1. Communication interfaces, protocols, threats, and recommended controls in EV charging infrastructures.

Interface	Protocols	Trust Anchors/Credentials	Common Threats	Recommended Controls
EV ↔ EVSE	ISO 15118-2/ISO 15118-20 (Plug & Charge, Smart Charging, V2G)	EV contract certificates, EVSE station certificates, PKI root CAs	Impersonation of EV/EVSE, MITM during handshake, expired certificates, renegotiation failures	PKI validation with OCSP checks, strict TLS, lifecycle management of certificates, ISO 15118-20 diagnostic/security extensions
EVSE ↔ CSMS	OCPP 2.0.1 (JSON over WebSocket, Device Model, TransactionEvent)	TLS certificates (per OCPP security profiles), signed firmware keys	MITM on OCPP channels, malicious firmware, manipulation of transaction records, key compromise	OCPP Security Profiles, signed firmware validation, Security Event Log, certificate lifecycle management (15,118 workflows)
CSMS ↔ Market/Grid/Roaming	Today: OCPP-based APIs; Emerging: IEC 63110 (charging management), IEC 63119 (roaming)	Operator-issued PKI credentials, roaming contracts, TLS channel certificates	Interoperability failures, weak inter-operator security, inconsistent telemetry, replay attacks	Strong TLS with mutual authentication, harmonization with IEC 63110/63119, monitoring aligned with IEC 62351-7

Table 2. Machine learning algorithms in electromobility.

Algorithm Type	Example Applications	Selected Sources
Linear and polynomial regression	Forecasting energy consumption, optimizing charging schedules	[4,32]
Neural networks (ANN, DNN)	Prediction of energy demand, adaptive control of charging power	[20,36,45]
Decision Trees, Random Forest, Gradient Boosting	Identification of key features affecting energy consumption, detection of unusual charging patterns	[36,42]
Support vector machines (SVM)	Classification of charging patterns, prediction of peak hours, assessment of overload risk	[37,43]
Unsupervised learning (k-means, t-SNE, PCA)	Segmentation of charging station users, dimensionality reduction of telemetry data	[11,32]
Federated learning	Privacy-preserving anomaly detection, distributed predictive models	[16,19]
Generative Adversarial Networks (GANs)	Synthesizing energy data, detecting rare anomalies, testing model robustness	[2,32]
Hybrid systems (SVM + fuzzy logic)	Complex demand modeling, network congestion prediction	[8,14]

Neural networks (ANN/DNN) excel when abundant labeled data are available and can learn non-linear couplings among electrical, protocol, and contextual signals. Recurrent variants (RNN/LSTM/GRU) are particularly effective on session sequences and OCPP/ISO-15118 message streams. Their costs include data hunger, higher latency/energy use, and sensitivity to hyperparameters. Black-box behavior complicates incident response unless paired with post-hoc or inherently explainable mechanisms. Regularization (dropout, weight decay), calibration, and careful thresholding are essential to control false alarms in volatile operating conditions.

Unsupervised methods cover distinct roles. K-means clustering supports quick clustering and coarse segmentation but presumes spherical clusters and is sensitive to scaling and initialization. It may confuse transient load ramps or maintenance modes with attacks. Principal Component Analysis (PCA) reduces dimensionality and can stabilize downstream models, yet information loss may suppress weak but security-relevant signals. T-Distributed Stochastic Neighbor Embedding (t-SNE) is valuable for visualization but is non-parametric and not suited for production inference. In all unsupervised settings, threshold selection (e.g., on reconstruction error or cluster distance) strongly governs the precision–recall trade-off and must be tuned per site and season.

Federated learning addresses data-sharing constraints by training across stations or operators without centralizing raw data, often improving cross-site generalization. Practical limitations include communication overhead, client heterogeneity (non-independent and identically distributed (non-IID) data, intermittent connectivity), and potential privacy leakage through gradients unless combined with secure aggregation and differential privacy. Convergence can degrade when local datasets are small or highly imbalanced.

Generative Adversarial Networks (GANs) can augment rare attack samples and stress-test detectors; they help rebalance training but introduce risks of mode collapse, synthetic–real distribution gaps, and inadvertent bias amplification. Synthetic data should be validated with fidelity metrics and ablation studies, and results reported separately from purely real-world training to avoid overstating operational performance.

Hybrid approaches (e.g., SVM + fuzzy rules) often reduce false positives by encoding domain constraints (grid limits, charger nameplate power) alongside a statistical margin. Their limitations are engineering overhead and rule drift: as tariffs, firmware, or market products evolve, fuzzy rulebases require maintenance to remain aligned with operational realities.

Across all families, reporting should reflect class imbalance (attacks are rare) and the operational cost of errors. Beyond accuracy and ROC-AUC (Receiver Operating Characteristic—Area Under the Curve), precision–recall curves, F1/F_β, Matthews correlation coefficient, and calibrated expected cost are appropriate. Windowed detection latency and throughput (edge vs. cloud execution) should be provided for ISO 15118/OCPP use cases. Because many studies rely on simulated or lab datasets, external validity should be established via cross-site testing, temporal holdout, and, where available, field-grade telemetry. Finally, practitioners should document drift handling (retraining cadence, online adaptation), threshold governance (per-site calibration, seasonal profiles), and explainability provisions so that alarms translate into actionable, auditable responses in real deployments.

3.1.5. Blockchain

Blockchain technologies are increasingly applied to secure EV charging ecosystems, particularly in the context of transaction integrity, decentralized authentication, and V2G energy exchanges. Studies highlight how blockchain can ensure transparency and immutability in charging sessions, preventing tampering with OCPP billing data or ISO 15118 contract certificates [49]. Research on blockchain in supply chains [50] and healthcare [51], while from different sectors, provides transferable models of decentralized trust management that can be adapted to EVSE–CSMS environments.

Smart contracts [52] show promise in automating tariff agreements between EV owners and charging providers, ensuring secure execution without reliance on centralized intermediaries. However, scalability and latency issues [53] remain critical, as charging sessions require near real-time validation to avoid service delays. The integration of blockchain with edge computing [54] and IoT security [55] is particularly relevant, since EVSEs are distributed devices that must securely interact with central management systems. Privacy-preserving techniques in blockchain [56] further support secure handling of user credentials and payment data during charging sessions. While applications in banking [57] and e-government [58] demonstrate broader utility, their adaptation to EVSE/CSMS systems illustrates how blockchain can underpin resilient, transparent, and secure charging infrastructures.

3.2. *Technologies and Infrastructure*

This section dives into the technological backbone and infrastructure of the electromobility ecosystem. While Section 3.1 focused on the cybersecurity aspects, this part explores the physical and systemic components that make up the modern electric vehicle landscape. The review will examine the advancements in charging technology, the crucial integration with smart grids, and the in-vehicle systems that are key to the functionality and security of EVs.

3.2.1. Vehicle Charging

The article discusses fast-charging techniques for electric vehicle batteries, emphasizing that the use of dynamic control algorithms can reduce cell degradation [58]. The paper focuses on wireless charging methods, with studies confirming that inductive charging systems can achieve high efficiency when properly calibrated [59]. The authors analyze the problem of cell overheating during fast charging and propose a cooling system based on dielectric fluids [60].

The article investigates the impact of different charging methods on EV battery lifespan, showing that algorithmically optimized charging increases the number of operational cycles [61]. The paper describes a charging management system based on artificial intelligence, highlighting its potential to balance grid loads effectively [62]. The study focuses on the interoperability of different charging protocols, underlining the importance of standards such as ISO 15118 for the advancement of electromobility [63].

ISO 15118 is a key standard supporting the development of electromobility [64]. In practice, ISO 15118-2 has been the baseline implementation, providing message structures for AC and DC charging sessions, including support for Plug & Charge and Smart Charging. The newer ISO 15118-20 refines this further by improving session renegotiation, supporting bidirectional power flow (V2G), and extending functionalities such as price communication and advanced load control [65]. One of the most critical aspects is the cryptographically secured handshake, which relies on a Public Key Infrastructure (PKI) with contract certificates and station certificates. This mechanism enables Plug & Charge, where the electric vehicle authenticates seamlessly without requiring RFID cards or mobile apps. Known implementation pitfalls include the complexity of certificate lifecycle management, backward compatibility with legacy standards (DIN SPEC, IEC 61851), and ensuring resilience in case of expired or corrupted certificates. ISO 15118-20 addresses several of these challenges by strengthening the diagnostic and security mechanisms, ensuring interoperability in heterogeneous charging environments.

Complementary to ISO 15118, the Open Charge Point Protocol (OCPP) 2.0.1 has become the de-facto standard for communication between charging stations and central management systems (CSMS). The 2.0.1 release introduces the Device Model for consistent station inventories, the new TransactionEvent structure for flexible session handling, and native support for ISO 15118 certificate management workflows. Importantly, OCPP 2.0.1 defines three distinct security profiles that prescribe TLS (Transport Layer Security) configurations and mutual authentication methods suitable for different deployment contexts. It also introduces secure firmware updates (signed firmware), a Security Event Log, and explicit certificate management messages (e.g., SignCertificate, CertificateSigned, Get15118EVCertificate, InstallCertificate), which allow CSMS operators to manage the entire lifecycle of certificates used in ISO 15118 Plug & Charge [64]. These enhancements significantly reduce the risk of unauthorized access, tampered firmware, and data integrity attacks.

While OCPP remains the dominant implementation today, ongoing international standardization efforts in IEC 63110 and IEC 63119 are expected to shape future deployments. IEC 63110 addresses charging management, including V2G operations, roaming, metering, and maintenance, while IEC 63119 standardizes roaming interfaces between e-mobility service providers (eMSPs) and charge point operators (CPOs). Together, they aim to harmonize EVSE–CSMS and CSMS–market communication at a global scale. Importantly, telemetry and monitoring requirements defined in IEC 63110 and IEC 63119 can be aligned with the monitoring and information exchange mechanisms of IEC 62351-7, which is already referenced in the conclusions of this work [25]. This linkage highlights how secure interoperability can be extended beyond station-level communication to encompass market and grid services.

To provide a concise overview, Table 1 maps the key communication interfaces in the EV charging ecosystem to their respective protocols, trust anchors, common threats, and recommended security controls. The article examines the design of high-power charging stations, stressing the necessity of advanced cooling and energy management systems [66]. The paper describes the architecture of charging stations powered by renewable energy sources, noting that integration with photovoltaic panels reduces grid load [67]. The authors present a model of a charging station equipped with an energy storage system, with research confirming that buffer batteries enhance station stability [68].

The article analyzes the economic aspects of building charging stations, pointing out that profitability depends on location, energy tariffs, and infrastructure utilization rates [69]. The paper discusses dynamic power allocation systems for simultaneous multi-vehicle charging, showing that this solution minimizes the risk of grid overload [70]. The article investigates the impact of EV charging on CO_2 emissions depending on the energy mix, with results indicating that renewable energy significantly reduces emissions [71]. The study examines the availability of charging stations in urban areas, highlighting that infrastructure density is a key factor for EV user adoption [72]. The article focuses on the cybersecurity of charging stations, with findings showing that the lack of encrypted communication protocols exposes systems to manipulation [73]. The paper discusses the development of national-level charging infrastructure, emphasizing the need for strategic planning and collaboration between the public and private sectors [74].

3.2.2. Grid Integration

The article discusses the concept of V2G as a tool for stabilizing the power grid, emphasizing that vehicles can act as decentralized energy storage units [75]. The paper analyzes the impact of V2G on battery lifespan, with results showing that appropriate control algorithms minimize battery degradation [76]. The authors present a mathematical model assessing V2G efficiency in urban conditions, with simulations demonstrating its ability to reduce peak load demand [77].

The article focuses on the impact of EV charging on transmission networks, highlighting that uncontrolled mass charging can lead to overloads [78]. The paper presents methods for forecasting loads in transmission networks within the context of electromobility, showing that predictive algorithms improve system stability. The study analyzes EV integration with smart grids, proving that bidirectional communication enables effective energy management [79]. The article discusses the role of smart grid technologies in reducing transmission losses, noting that the deployment of smart meters improves energy balance [80]. The paper presents a smart grid architecture supporting EV charging, proposing decentralized control algorithms [81]. The article examines the development of smart energy networks in Europe, underlining the importance of standardization and interoperability [82].

The study analyzes the resilience of smart grids to cyber attacks, stressing the need for IDS systems and secure communication [83]. The article presents control methods for power systems under fluctuating demand, emphasizing the role of adaptive algorithms in stabilizing the grid [84]. The paper focuses on energy flow management in hybrid RES–EV (Renewable energy sources EV) systems, showing that dynamic load management increases efficiency [85].

The study discusses the impact of electromobility on power grid reliability, noting the risks of local overloads [86]. The article analyzes the necessity of modernizing transmission networks for the development of electromobility, highlighting investments in automation and digitalization [87]. The paper describes the potential of energy sharing between vehicles and the grid within V2G, proving that this approach supports renewable energy integration [12].

The article presents a business model for V2G services, pointing to new revenue opportunities for both grid operators and EV owners [88]. The study analyzes the implementation of V2G in Japan and Europe, emphasizing the key role of regulatory frameworks in its development [89]. The paper focuses on communication in V2G systems compliant with ISO 15118, underlining the importance of secure vehicle and station authentication [90]. The article discusses methods of peer-to-peer energy transfer between vehicles, highlighting that this approach supports decentralized energy models [91]. The study analyzes real-time energy transfer technologies in V2G networks, with findings showing high transmission efficiency when using dynamic communication protocols [92].

3.2.3. Vehicle Systems

The article discusses the role of embedded systems in electric vehicles, highlighting that their miniaturization and energy efficiency are crucial for the development of electromobility [93]. The study analyzes embedded system architectures in the context of security, emphasizing that hardware module isolation enhances resilience against cyber attacks [94].

The article presents methods for battery health diagnostics in Battery Management System (BMS), showing that artificial intelligence algorithms improve prediction accuracy [95]. The paper describes the development of BMS with cell-balancing functions, demonstrating that active balancing extends battery lifespan [96]. The study focuses on BMS safety, underlining the importance of protection mechanisms against overcharging and overheating [97].

The article analyzes vehicle-to-vehicle (V2V) communication as a tool for improving traffic safety, showing that information exchange between vehicles helps prevent collisions [98]. The paper focuses on communication standards for V2V systems, stressing the importance of protocols designed for low-latency transmission [99]. The article discusses the integration of electric vehicles with public transportation systems, showing that such solutions improve urban mobility efficiency [100]. The study analyzes the architecture of autonomous vehicle systems, highlighting the importance of redundancy in control and sensor systems [101]. The article presents the use of vehicular networks in traffic management, showing that dynamic data exchange reduces congestion [102]. The paper analyzes the security of communication in vehicular networks, warning that the lack of encryption in vehicle-to-everything (V2X) protocols creates risks of transmission interception [103].

The article discusses fleet management of electric vehicles using predictive systems, showing that machine learning algorithms enhance operational efficiency [104]. The study focuses on route optimization for EV fleets, demonstrating that intelligent planning algorithms reduce energy consumption [105]. The article analyzes the domain architecture of electric vehicles, showing that function integration within domains improves system management efficiency [106].

The paper describes challenges related to consolidating control units in vehicles, emphasizing the importance of standardizing communication interfaces [107]. The article discusses the impact of electromobility on transportation systems, showing that EV integration with public transport requires expansion of charging infrastructure [108]. The study focuses on the sustainable development of transportation systems, highlighting the role of EVs in reducing CO_2 emissions [109].

The article analyzes the integration of IoT with electric vehicles, showing that real-time connectivity enables continuous vehicle monitoring [65]. The paper focuses on IoT security in vehicle environments, stressing the need for encryption and device authentication [110]. The study discusses IoT applications in EV fleet management, showing that IoT systems optimize logistics processes and monitor energy consumption [111].

3.3. Summary

Overall, the research demonstrates the convergence of security and infrastructure, emphasizing the necessity of strong authentication and cryptographic mechanisms. These mechanisms, such as Plug & Charge and ISO 15118, must balance low latency with privacy protection. Based on the conducted analysis, the following findings can be drawn:

1. The threat landscape encompasses DoS/DDoS, MITM/replay, data manipulation, and backend protocol vulnerabilities (e.g., OCPP). Coordinated load attacks may escalate from charging stations to the entire power system.
2. Intrusion and anomaly detection systems increasingly rely on ML/DL approaches (LSTM, autoencoders), hybrid designs, and edge/federated solutions. These are used to monitor vehicle buses, EVSE, and 5G-EVCS with low latency and fewer false positives.
3. Blockchain and smart contracts support settlements, access control, and energy traceability (including in V2G and dynamic charging). However, they require lightweight consensus mechanisms and careful balancing of scalability and costs.
4. On the technological side, rapid and wireless charging, modular EVCS, and integration with RES and storage are advancing; integration with the grid (V2G, Smart Grid, advanced control) improves stability but expands the attack surface.

In vehicle systems, AI enhances BMS and diagnostics, V2V/V2X and IoT improve fleet and transport management. Overall progress depends on standardization, formal verification, segmentation/zero-trust, continuous monitoring, and response readiness to ensure secure, resilient, and privacy-preserving electromobility.

Table 2 provides a comprehensive review of the most frequently applied machine learning algorithms in the field of electromobility, highlighting representative use cases and corresponding references. Regression-based methods are primarily employed for forecasting energy consumption and optimizing charging schedules, whereas neural networks and hybrid approaches facilitate the modeling of highly dynamic processes and enable adaptive power management. Tree-based algorithms and support vector machines (SVMs) are widely utilized for pattern classification, peak demand prediction, and grid overload risk assessment. Unsupervised learning techniques contribute to user segmentation and dimensionality reduction, while federated learning enables the construction of distributed models with strict preservation of data privacy. Furthermore, generative adversarial networks (GANs) have emerged as powerful tools for generating synthetic datasets, detecting rare anomalies, and evaluating model robustness under adverse conditions. Collectively, the literature indicates that the judicious selection and integration of machine learning methods significantly enhance energy management, improve system security, and optimize the development of electromobility infrastructures.

Table 3 presents the diverse applications of machine learning algorithms in electric vehicle energy management and grid integration. In the area of charging management, algorithms such as neural networks, SVM, and k-NN support demand forecasting, optimization of charging schedules, and dynamic pricing strategies. In vehicle energy management, decision trees, boosting, and recurrent neural networks are applied to optimize battery usage, predict loads, and control BMS. In the area of user and data security, autoencoders, Isolation Forest, and federated learning are employed for anomaly detection, access control, and privacy protection. Grid integration relies on LSTM algorithms, ensemble learning, and deep reinforcement learning, enabling power flow control, availability forecasting, and load balancing. Finally, V2X communication systems leverage Bayesian networks, CNNs, and Random Forest to optimize protocols, enhance transmission efficiency, and detect interference.

Table 3. Application of machine learning algorithms in electric vehicle energy management and grid integration.

Application Area	Objective/Function	Applied ML Algorithms	Example Publications
Electric vehicle charging management	Charging optimization, demand forecasting, dynamic pricing	Artificial Neural Networks, Support Vector Machines, K-Nearest Neighbors	[3,8,112]
In-vehicle energy management	Battery usage optimization, load prediction, BMS control	Decision Trees, Gradient Boosting, Recurrent Neural Networks	[3,6,9]
User and data security	Anomaly detection, access control, privacy protection	Autoencoders, Isolation Forest, Federated Learning	[4,8,113]
Grid integration	Power flow control, availability prediction, grid balancing	Long Short-Term Memory, Ensemble Learning, Deep Reinforcement Learning	[5,7,10]
Vehicle communication systems	V2X protocol optimization, transmission efficiency, interference detection	Bayesian Networks, Convolutional Neural Networks, Random Forest	[1,11,22]

4. Statistical Overview

This article summarizes 102 scientific publications, divided into two time periods: 2017–2020 (26 publications) and 2021–2024 (76 publications). A significant intensification of research activity has been observed in recent years, which may indicate the growing importance of cybersecurity and digital technologies in the context of electromobility. The percentage of publications from 2021–2024 was over 74%.

Figure 5 presents document types, with the dominance of two main publication forms: scientific articles (Articles) and conference papers (Conference Papers). A total of 26 papers were published between 2017 and 2020—including 15 as conference papers and 11 as articles. In the 2021–2024 period, the number of these documents increased significantly—as many as 37 publications were assigned to each of these categories. Additionally, there are two publications classified as "Other," which may include reports, review, or other non-standard forms. This means that conference proceedings and peer-reviewed articles accounted for over 98% of the total publication output. It can therefore be concluded that research was presented primarily in formats that have the greatest impact on the development and dissemination of knowledge—at scientific conferences and in professional journals. The increase in the number of scientific articles from 11 to 37 may indicate the maturation of topics and the transition from the exploratory phase to established knowledge.

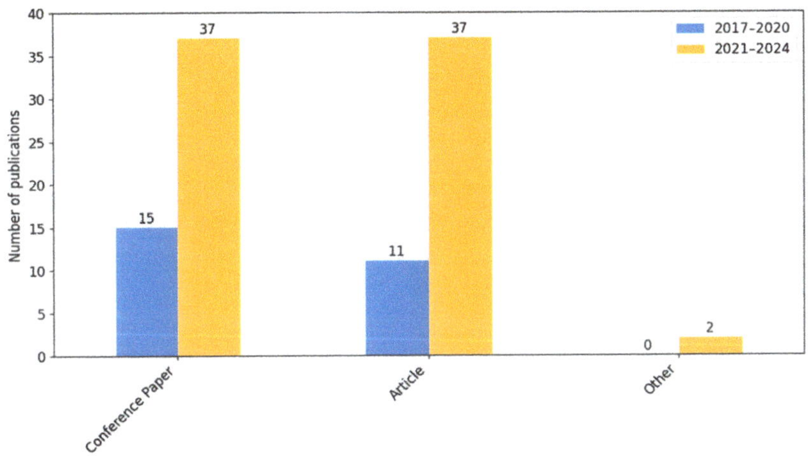

Figure 5. Number of publications by document type.

4.1. Research Trends and Dynamics

Figure 6 focuses on digital cybersecurity and is structured into four primary thematic categories, each exhibiting distinct growth dynamics:

- Authentication and Cryptography: The number of publications increased from 21 to 27, indicating a stable yet moderate development in this area. It is noteworthy that this category was dominant during the earlier time period and maintained a high level of academic interest in subsequent years;
- Cyber Attacks: This category experienced a sharp increase in publication volume—from 6 to 35 articles—making it the fastest-growing research area among the four. This surge likely reflects the rising cybersecurity threats associated with the proliferation of networked charging systems and the integration of electric vehicles (EVs) into digital infrastructures.
- Intrusion and Anomaly Detection: The number of publications grew from 3 to 19, suggesting a growing relevance of behavioral detection techniques and increased interest in machine learning (ML)-based solutions for cybersecurity challenges in EV ecosystems.
- Blockchain: Publications in this category rose from 3 to 11, which may indicate expanding research opportunities in the application of blockchain technologies for transaction security, identity verification, and access management in electric vehicles.

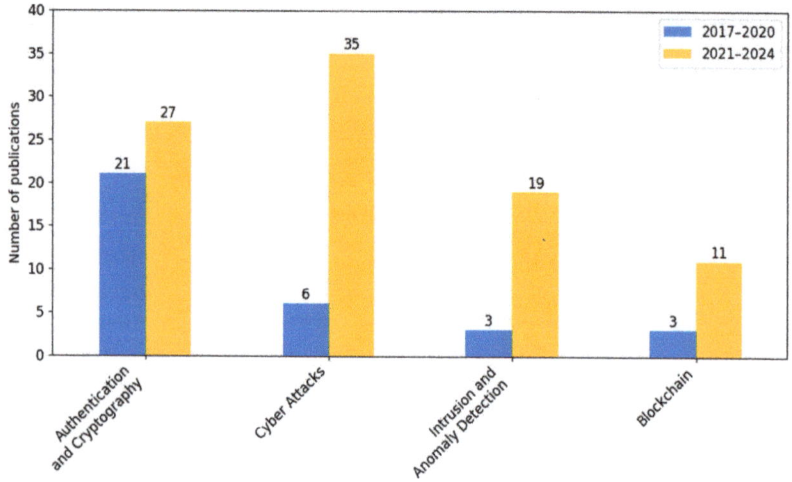

Figure 6. Digital cybersecurity.

The particularly dynamic growth in the areas of Cyber Attacks and Intrusion and Anomaly Detection suggests a trend in which traditional cryptographic approaches are being increasingly complemented—or even supplanted—by predictive and adaptive security systems. The analysis of research types revealed a clear shift in the nature of studies conducted over the analyzed period. Specifically:

- Experimental Studies: The number of publications increased from 15 to 46, representing a 207% growth. This trend indicates a transition from theoretical investigations toward practical implementation and hypothesis testing in real-world or simulated environments.
- Literature Analyses: A rise from 6 to 26 publications was observed, reflecting a growing interest among researchers in synthesizing existing knowledge and identifying prevailing trends within the field.
- Case Studies: No publications of this type were recorded between 2017 and 2020. However, since 2021, seven case studies have been published, suggesting an increasing demand for application-oriented research, particularly in commercial deployments or pilot implementations of charging infrastructure.
- Conceptual Research: Conceptual studies rose from 20 to 48 publications, highlighting continued efforts in design and development aimed at introducing new models for energy management and cybersecurity alongside empirical validation.

This methodological distribution suggests that the field is reaching a higher level of maturity, supporting both theoretical exploration and the practical deployment of machine learning (ML) models. Figure 7 shows the number of scientific publications by research methodology in two time periods: 2017–2020 and 2021–2024. The second half of the decade (2021–2024) saw a significant increase in publications across all categories, particularly in experimental research (from 15 to 46) and conceptual research (from 20 to 48). It is also worth noting the emergence of case studies after 2020, which may indicate a growing interest in practical implementation.

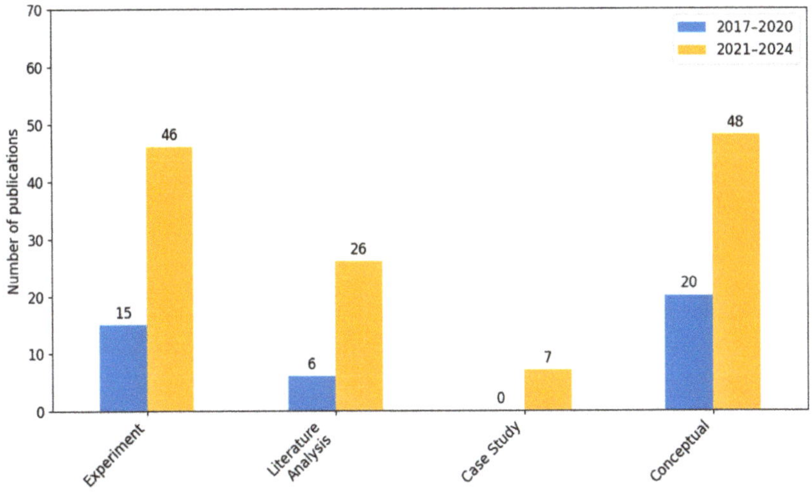

Figure 7. Research methodology.

The classification of studies by application domains (Figure 8) revealed three dominant research areas:

- Vehicle Charging: This was the most frequently investigated topic across both analyzed periods, with the number of publications increasing from 21 to 63. This trend likely reflects the growing need for optimization of charging strategies and flexible energy cycle management in electric vehicle (EV) systems.
- Grid Integration: The number of studies increased from 16 to 35, indicating a rising focus on Vehicle-to-Grid (V2G) interactions, dynamic pricing models, and energy balancing within power systems involving EVs.
- Vehicle Systems: Publications in this category grew from 7 to 29, suggesting increased research interest in onboard energy and safety management systems. This is particularly relevant to the deployment of embedded machine learning (ML) algorithms and edge computing within EV architectures.

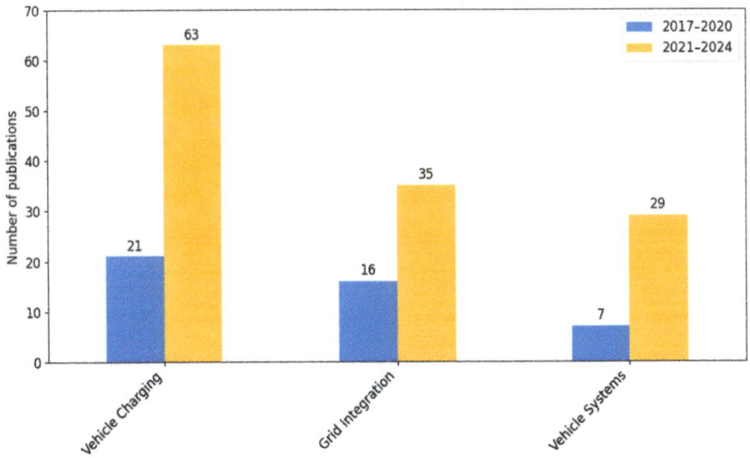

Figure 8. Division of research by application areas.

These findings confirm that the most significant advancements in research have occurred in areas directly related to EV operation and their integration with external infrastructure systems.

Table 4 presents a detailed analysis of the distribution of scientific publications by cybersecurity topic and their relationship to technologies, infrastructure, and research methodology. Four main categories of security issues are considered: Authentication and Cryptography, Cyber Attacks, Intrusion and Anomaly Detection, and Blockchain. Additionally, a statistical analysis was performed using the chi-square test in two groups: (1) technologies and infrastructure, and (2) research methodology.

Table 4. Publications by year in all categories.

Name	2017–2020	2021–2024	All Years	Share [%]	Chi–Square
Total	26	76	102	100.0	
Document Type					
Conference Paper	15	37	52	50.98	$\chi^2 = 1.16$ ($df = 2, p = 0.56$)
Article	11	37	48	47.06	
Other	0	2	2	1.96	
Cybersecurity					
Authentication and Cryptography	21	27	48	47.06	$\chi^2 = 12.38$ ($df = 3, p = 0.01$)
Cyber Attacks	6	35	41	40.2	
Intrusion and Anomaly Detection	3	19	22	21.57	
Blockchain	3	11	14	13.73	
Technologies and Infrastructure					
Vehicle Charging	21	63	84	82.35	$\chi^2 = 1.62$ ($df = 2, p = 0.45$)
Grid Integration	16	35	51	50.0	
Vehicle Systems	7	29	36	35.29	
Research Methodology					
Experiment	15	46	61	59.8	$\chi^2 = 3.74$ ($df = 3, p = 0.29$)
Literature Analysis	6	26	32	31.37	
Case Study	0	7	7	6.86	
Conceptual	20	48	68	66.67	

A total of 102 publications were included in the analysis, distributed as follows:

- Authentication and Cryptography: 48 studies;
- Cyber Attacks: 41 studies;
- Intrusion and Anomaly Detection: 22 studies;
- Blockchain: 14 studies.

Authentication and Cryptography emerged as the most represented category (47% of all publications), confirming the critical role of traditional security mechanisms in electromobility systems. At the same time, the significant share of publications addressing Cyber Attacks highlights the increasing vulnerability of EV systems and energy infrastructure in open network environments.

The distribution of publications by technological and infrastructural domains revealed the following patterns. This was the most frequently studied domain, with 84 total publications. With in this area:

- Authentication and Cryptography appeared in 41 studies.
- Cyber Attacks in 32 studies.
- Intrusion and Anomaly Detection in 18 studies.
- Blockchain in 9 studies.

This distribution indicates a high level of complexity and associated risk in charging interfaces, which serve as a critical point of interaction between users and the network. These systems require robust authentication, anomaly detection, and transaction-level security mechanisms.

A total of 51 publications addressed this domain:

- Authentication and Cryptography: 25 occurrences.
- Cyber Attacks: 23 occurrences.
- Intrusion and Anomaly Detection: 9 occurrences.
- Blockchain: 6 occurrences.

Integration of EVs with power grids requires secure bidirectional exchange of data and energy, as well as resilience against external attacks, particularly within dynamic grid environments.

This domain included 36 publications, with a more balanced distribution:

- Authentication and Cryptography: 15 studies;
- Cyber Attacks: 16 studies;
- Intrusion and Anomaly Detection: 8 studies;
- Blockchain: 8 studies.

This even distribution suggests that all four cybersecurity aspects are considered equally critical in onboard systems, with particular emphasis placed on protection against direct attacks and internal threats.

The analysis of methodological approaches revealed a strong presence of conceptual and experimental studies, alongside limited use of case studies:

- Experiments dominated in each security category, especially in Authentication and Cryptography (30 studies) and Cyber Attacks (27 studies). In the case of Blockchain, 9 experimental studies were identified, indicating growing efforts to implement this technology in practical settings.
- Literature Analyses were primarily conducted for Authentication and Cryptography (14 studies) and Cyber Attacks (14 studies), suggesting a well-established theoretical foundation in these areas.
- Case Studies were limited to only 7 publications overall, with 4 in Cyber Attacks and 3 in Intrusion and Anomaly Detection, indicating a relatively low degree of real-world deployment or documented implementations.
- Conceptual Studies were dominant in Authentication and Cryptography (37 studies) and Cyber Attacks (23 studies), and notably present in Blockchain (12 studies), reflecting the ongoing development of theoretical models and architectures in these areas.

Table 4 provides a comprehensive statistical overview of the 102 publications included in the review, categorized according to document type, cybersecurity domain, technological and infrastructural area, and research methodology.

In terms of document type, conference papers (52; 50.98%) and journal articles (48; 47.06%) dominate the dataset, while other formats account for only a marginal share (2; 1.96%). Within the cybersecurity categories, Authentication and Cryptography emerges as the most frequently studied area, represented in 48 publications (47.06%). This is followed by Cyber Attacks (41; 40.20%), Intrusion and Anomaly Detection (22; 21.57%), and Blockchain (14; 13.73%).

From the perspective of technologies and infrastructure, the majority of studies focus on Vehicle Charging (84; 82.35%), while Grid Integration (51; 50.00%) and Vehicle Systems (36; 35.29%) are also well represented, confirming the strong research emphasis on charging processes as the primary interface between users and the network.

Methodologically, Conceptual works (68; 66.67%) and Experimental studies (61; 59.80%) dominate, supported by Literature analyses (32; 31.37%) and a limited number of Case studies (7; 6.86%). This distribution reflects a field that combines strong theoretical development with increasing tendencies toward practical validation, while still showing relatively few real-world implementation studies.

The results of the chi-square tests, presented in parentheses in Table 4, indicate whether the observed distributions differ significantly across categories. The analysis shows no statistically significant differences for document type ($\chi^2 = 1.16$, $df = 2$, $p = 0.56$), technologies and infrastructure ($\chi^2 = 1.62$, $df = 2$, $p = 0.45$), or research methodology ($\chi^2 = 3.74$, $df = 3$, $p = 0.29$). However, a significant difference is observed for cybersecurity categories ($\chi^2 = 12.38$, $df = 3$, $p = 0.01$), suggesting that research efforts are unevenly distributed across security domains, with a clear predominance of authentication- and cryptography-related studies. Similarly, this result suggests no statistically significant differences in research methodologies applied across the various security categories. This implies that the choice of research approach was relatively balanced regardless of the topic area.

The statistical analyses based on χ^2 tests reported were performed using the categorical distributions presented in Table 4. This table provides the raw counts of publications assigned to specific categories (e.g., Authentication and Cryptography, Cyber Attacks, Intrusion and Anomaly Detection, Blockchain, etc.), which allows for full replication of the test setup. The reported χ^2 values, degrees of freedom, and p-values were calculated directly from these category counts. Since the underlying frequencies are included in the tables, the results can be independently verified and reproduced without additional preprocessing.

Based on the conducted analysis, the following conclusions can be drawn:

5. Authentication and Cryptography remains the most frequently studied security domain. However, there is a noticeable increase in research activity related to Cyber Attacks and Anomaly Detection, indicating an expanding focus on behavioral and adaptive security mechanisms.
6. Charging infrastructure has been identified as the most critical area in terms of cybersecurity concerns. This domain accounts for the highest number of studies and experimental implementations, underscoring its importance as a user-network interface vulnerable to various threats.
7. Blockchain, although currently less represented in the dataset, shows increasing potential and is beginning to be investigated not only in vehicle-level applications but also in infrastructure-level contexts. This suggests an emerging trend of exploring decentralized and transparent security frameworks.
8. The absence of statistically significant differences across categories may reflect the maturity of the field, where research is distributed evenly with respect to both methodology and technological context. This balanced distribution may indicate a well-established foundation for interdisciplinary and multi-domain research efforts in the area of electromobility cybersecurity.

The collected data revealed that a total of 26 studies were published between 2017 and 2020, while 76 publications appeared during the period from 2021 to 2024, representing an increase in research activity of nearly 192%. This growth can be interpreted as a direct reflection of the rising global importance of electromobility and cybersecurity as interrelated areas of scientific and technological interest.

4.2. Geographic Distribution and Statistical Analysis

Figure 9 and Table 5 present a detailed analysis of the geographical distribution of publications. The highest number of publications was attributed to:

- The United States, which increased its output from 11 publications in the initial period to 22 in the subsequent one. This accounted for 32.35% of the total body of work, thereby maintaining its dominant position as the leading research hub in the field.
- China, which recorded a substantial increase—from 2 to 17 publications—resulting in a total share of 18.63%, positioning it as the second-largest contributor. This growth may be interpreted as a consequence of national policies promoting research in advanced transportation and digitalization.
- Canada, whose contribution also grew significantly, from 5 to 13 publications, representing 17.65% of the total.
- Germany, the only European country with relatively stable output across both periods—5 publications prior to 2020 and 6 afterward—contributing 10.78% overall.
- India, Brazil, South Korea which was unrepresented in the first period, produced 10 publications between 2021 and 2024, thereby securing the fifth position globally.
- Australia, Italy, the United Kingdom, and Turkey all exhibited activity exclusively during the 2021–2024 period, each publishing between 4 and 5 studies.
- Brazil was the only country whose contributions were limited solely to the 2017–2020 period (4 publications), potentially indicating a decline in research initiatives or a shift in national priorities.
- South Korea expanded its research activity moderately, with 1 publication before 2020 and 2 after this time.
- Turkey expanded its research activity moderately, with 1 publication.
- The "Other" category, representing additional countries, included a total of 11 publications—3 from the first period and 8 from the second—indicating the internationalization of the topic and the growing involvement of emerging academic and industrial centers.

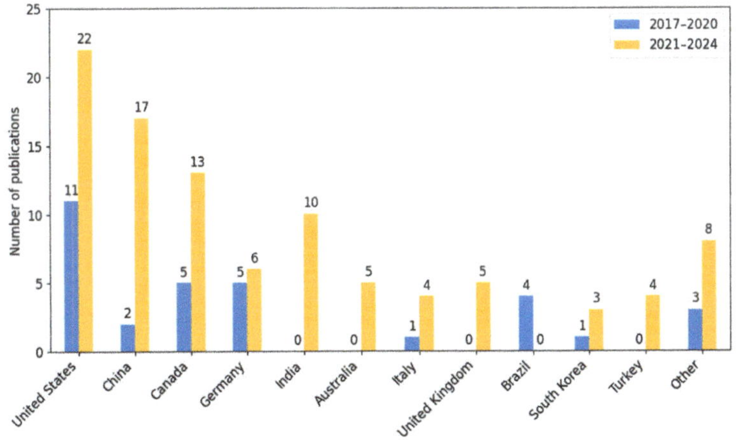

Figure 9. Number of publications by country and period.

Table 5. Publications by year in Countries.

Country	2017–2020	2021–2024	All Years	Share [%]	Chi–Square
All countries	26	76	102	100.0	26.1 ($df = 11,\ p = 0.01$)
United States	11	22	33	32.35	
China	2	17	19	18.63	
Canada	5	13	18	17.65	
Germany	5	6	11	10.78	
India	0	10	10	9.8	
Australia	0	5	5	4.9	
Italy	1	4	5	4.9	
United Kingdom	0	5	5	4.9	
Brazil	4	0	4	3.92	
South Korea	1	3	4	3.92	
Turkey	0	4	4	3.92	
Other	3	8	11	10.78	

To determine whether the changes in the number of publications across the analyzed years and countries were statistically significant, a chi-square test was performed (Table 5):

- Chi-square value (χ^2): 26.1
- Degrees of freedom: 11
- p-value: 0.01

Since the p-value was less than 0.05, a statistically significant relationship was observed between the year of publication and the country of origin. In other words, the variation in publication distribution was not random and may be attributed to political, financial, or infrastructural conditions that supported research activities in specific countries.

In summary, the United States maintained its leading position in the field of electromobility and digital security research. China and Canada demonstrated a marked increase in research activity after 2020, which may be attributed to the implementation of aggressive national technology strategies. A significant rise in publication volume was also observed in India, as well as in countries such as the United Kingdom, Australia, Italy, and Turkey, which had previously not been represented in this research domain. Germany exhibited a stable publication output, reinforcing its long-standing commitment to engineering and energy-related research. Conversely, Brazil was the only country that did not continue its publishing activity beyond 2020, potentially indicating a shift in research priorities or funding. The chi-square statistical analysis confirmed that the observed geographic redistribution of publications was statistically significant and likely reflects deliberate research policies and national initiatives aimed at advancing innovation in the domains of electric vehicles and cybersecurity.

4.3. Methodological and Thematic Relationships

Figure 10 presents a detailed analysis of the distribution of research methods used in scientific publications in the field of cybersecurity in the context of electromobility.

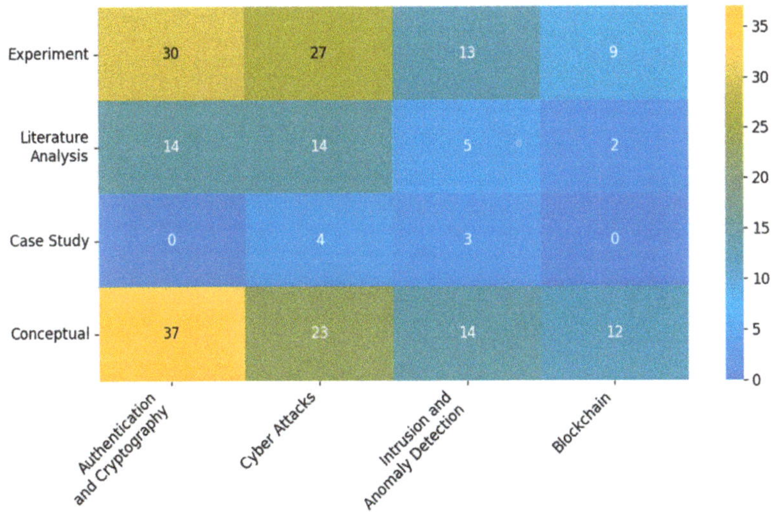

Figure 10. Heatmap of the scope of cybersecurity in the context of electromobility.

The data were structured along two dimensions: the type of research methodology applied (horizontal axis) and the security-related thematic category (vertical axis). The analysis incorporated four primary methodological approaches:

- Experiment;
- Literature Analysis;
- Case Study;
- Conceptual;

As well as four thematic categories:

- Authentication and Cryptography;
- Cyber Attacks;
- Intrusion and Anomaly Detection;
- Blockchain.

The Authentication and Cryptography thematic area accounted for the largest number of publications, with 81 studies divided into:

- 37 conceptual works;
- 30 experimental studies;
- 14 literature reviews;
- 0 case studies.

This distribution indicates a strong theoretical foundation in the domain of authentication and cryptographic mechanisms, accompanied by a substantial volume of experimental research. The relatively high number of experimental publications suggests that authentication and encryption algorithms are actively being tested and deployed within real-world EV system environments. The absence of case studies may reflect a limited number of institution-specific implementations documented through in-depth field analysis.

The category of Cyber Attacks encompassed a total of 68 publications, distributed as follows:

- 27 experimental studies;
- 23 conceptual papers;
- 14 literature analyses;
- 4 case studies.

This area demonstrates a strong focus on applied research, particularly through experimental approaches, which is justified by the need to validate security mechanisms against real-world attack scenarios. The presence of case studies, although limited in number, suggests that specific incidents or implementations have been analyzed in detail, providing valuable insights into actual system vulnerabilities and defense strategies.

In the thematic category of Intrusion and Anomaly Detection, 35 publications were identified, derived in a manner:

- 14 conceptual studies
- 13 experimental studies
- 5 literature analyses
- 3 case studies

Compared to the other categories, Intrusion and Anomaly Detection exhibits a moderate methodological distribution. The relatively high number of conceptual and experimental studies indicates an ongoing phase of active development and testing of anomaly detection algorithms. In contrast, the limited number of case studies suggests that the implementation of these technologies remains at an early stage in real-world systems.

The Blockchain category was the least represented topic category, containing 25 publications, including:

- 12 conceptual studies
- 9 experimental studies
- 2 literature reviews
- 0 case studies

The field of Blockchain is characterized by a strong dominance of theoretical research, suggesting that its application to electromobility security is still in the preliminary development stage, with limited practical implementation (as evidenced by the absence of case studies). Nevertheless, the presence of experimental studies indicates initial attempts at integrating blockchain with transaction systems or charging infrastructure.

In summary, the Authentication and Cryptography and Cyber Attacks categories are characterized by the largest number of experimental studies, which highlights their technological maturity and advanced level of practical application. In contrast, Blockchain and Intrusion and Anomaly Detection remain dominated by conceptual research, indicating that these areas are still being formulated and tested and are less frequently deployed in actual EV systems.

The case study methodology was minimally represented across all categories, suggesting a need for further documentation of real-world implementations and comparative analyses between different deployment environments. Literature reviews were most frequently applied in traditional security domains (e.g., Authentication and Cyber Attacks), which reflects a well-developed knowledge base and the existence of a large body of prior research in these areas. Overall, the methodological matrix points toward a shift from conceptual research to experimental validation, which may indicate the increasing maturity of machine learning and cybersecurity technologies within the electromobility domain.

Figure 11 presents a heat map matrix illustrating the relationship between cybersecurity categories and technological/infrastructural domains in the context of electromobility. For each intersection between a technological area and a security category, the matrix indicates the number of publications that addressed both topics simultaneously. This visualization allows for a clearer assessment of the technical contexts in which issues such as authentication, cyber attacks, anomaly detection, and blockchain have been most frequently studied.

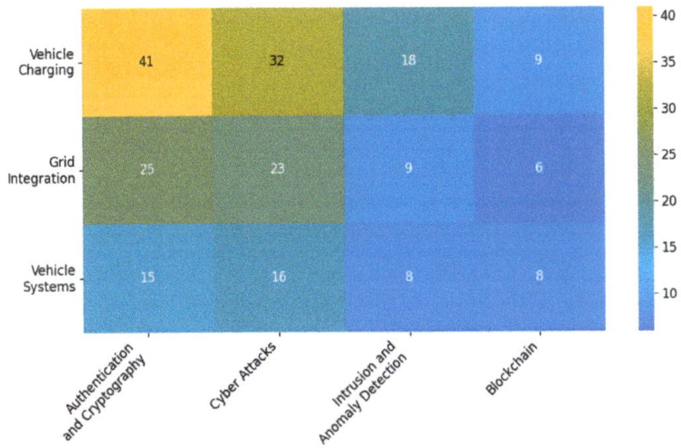

Figure 11. Heat map illustrating the relationship between cybersecurity categories and technologies/infrastructure in the field of electromobility.

The Vehicle Charging category emerged as the most represented in the literature, underscoring its significance as a critical node within the electromobility infrastructure. Within this group, the following distribution of publications was observed:

- 41 publications focused on Authentication and Cryptography;
- 32 related to Cyber Attacks;
- 18 addressed Intrusion and Anomaly Detection;
- 9 covered Blockchain.

These high values indicate that charging stations represent a highly sensitive element of the EV system from a cybersecurity perspective. This context requires robust mechanisms for both user and vehicle authentication, as well as protection against attacks such as spoofing, denial-of-service (DoS), and malware. The moderate representation of Blockchain suggests that this technology is being explored for identity verification, energy transaction settlement, and traceability applications.

The second most frequently studied area involved the integration of EVs with the power grid (V2G—vehicle-to-grid). This area featured the following number of publications:

- 25 focused on Authentication and Cryptography;
- 23 related to Cyber Attacks;
- 9 addressing Intrusion and Anomaly Detection;
- 6 concerning Blockchain.

This distribution highlights the critical need to secure the communication between electric vehicles and the grid. Ensuring confidentiality and integrity of real-time data exchange is essential, along with protection against manipulation in energy accounting and scheduling. The relatively lower presence of anomaly detection algorithms and blockchain applications suggests that these are still emerging areas of research rather than fully implemented technologies.

In onboard systems, i.e., those located directly within electric vehicles (EVs), the total number of publications was lower, yet the distribution was more balanced:

- 15 publications addressed Authentication and Cryptography;
- 16 focused on Cyber Attacks;
- 8 were related to Intrusion and Anomaly Detection;
- 8 discussed Blockchain.

This relatively even distribution suggests a comprehensive approach to securing the internal layers of the vehicle, where multiple layers of protection must be implemented simultaneously. Onboard systems—such as the Battery Management System (BMS), motor control units, and user interfaces—are vulnerable to both physical tampering and remote Cyber attacks. Blockchain is gaining traction as a potential platform for permission management and logging of diagnostic and service operations.

Table 6 highlights the methodological distribution of studies across four cybersecurity domains. Authentication and Cryptography (81 publications) and Cyber Attacks (68 publications) dominate, with conceptual and experimental works prevailing in both categories. In contrast, Intrusion and Anomaly Detection (35 publications) and Blockchain (25 publications) remain less developed, showing more balanced but still emerging research profiles with limited case study evidence. Across all three categories, Authentication and Cryptography remained the most prevalent topic, clearly confirming that identity, data, and access protection are fundamental requirements in all electromobility architectures. The use of Blockchain remains moderate but steadily increasing, indicating that its full potential has yet to be fully exploited. Future research may further explore this technology as a component for data transparency and non-repudiation.

Table 6. Publications by cybersecurity in other categories.

Name	Authentication and Cryptography	Cyber Attacks	Intrusion and Anomaly Detection	Blockchain	Total
Total	48	41	22	14	102
Technologies and Infrastructure					
Vehicle Charging	41	32	18	9	84
Grid Integration	25	23	9	6	51
Vehicle Systems	15	16	8	8	36
Research Methodology					
Experiment	30	27	13	9	61
Literature Analysis	14	14	5	2	32
Case Study	0	4	3	0	7
Conceptual	37	23	14	12	68

In Section 4, a detailed quantitative analysis of the scientific literature on cybersecurity in electromobility was conducted. This analysis was based on the data presented in tabular format and corresponding visualizations, categorized by two time periods: 2017–2020 and 2021–2024. Additionally, chi-square tests were applied to assess the statistical significance of changes in data distributions.

The first step involved examining the overall trends in the number of scientific publications. Both the table and the corresponding bar chart revealed that the number of publications increased from 26 in 2017–2020 to 76 in 2021–2024. This represents an almost threefold increase in research activity, confirming growing interest in digital security in the context of EV development. This rise was deemed significant, and its statistical relevance was confirmed in subsequent analyses.

Further classification was performed according to document types. The chart indicates that journal articles and conference proceedings were the most frequently published formats. A total of 11 journal articles were recorded before 2020, compared to 37 after 2020. Likewise, conference papers increased from 15 to 37 over the same periods. Although other document types remained marginal, their appearance in the more recent period may reflect a diversification of knowledge sources within this emerging research domain.

Another classification focused on the thematic scope of the analyzed publications. Four main categories were identified: Authentication and Cryptography, Cyber Attacks, Intrusion and Anomaly Detection, and Blockchain. The most significant increase in the number of publications was observed in the area of Cyber Attacks, which grew from 6 to 35 publications. While Authentication and Cryptography experienced a less dynamic rise (from 21 to 27), it remained the most frequently represented category. This distribution is considered logical given the increasing risk of attacks in intelligent transportation systems and the ongoing development of defensive mechanisms.

The subsequent methodological analysis addressed the type of research approach used. The largest increase was observed in experimental studies, which rose from 15 to 46 publications, and in conceptual papers, which increased from 20 to 48. Literature reviews also grew, from 6 to 26, while case studies appeared only after 2020, totaling 7 publications. This trend suggests growing research maturity in the field, alongside a stronger focus on practical implementation and validation efforts.

A significant portion of the analysis also addressed the technological and infrastructural context. The highest number of publications was related to vehicle charging, increasing from 21 to 63, highlighting the central role of this area. Grid integration grew from 16 to 35 publications, and vehicle systems from 7 to 29. These trends indicate that cybersecurity concerns must be considered both in localized contexts (e.g., the EV itself) and at the infrastructure level (i.e., energy networks and charging stations).

The geographic analysis revealed that the highest number of publications originated from the United States (33 in total), followed by China (19 in total) and Canada (18 in total). An increase in publications was also noted in countries previously unrepresented in the literature, including India, Australia, Italy, and Turkey. This increase was confirmed to be statistically significant, as evidenced by a chi-square test result of $\chi^2 = 26.1$ with 11 degrees of freedom and a p-value of 0.01. This indicates that the changes in publication distribution over time and across countries were not random, but likely influenced by national research policies and strategic investments.

Further analyses provided insight into the relationships between research methodologies and thematic categories, as well as between technological areas and cybersecurity domains. In the context of Authentication and Cryptography, conceptual and experimental approaches predominated. In contrast, studies on Cyber Attacks employed a diverse array of methods, including case studies, which highlights the applied and practical nature of this topic.

In terms of infrastructure, cybersecurity was most frequently addressed in the context of vehicle charging, as confirmed by a total of 100 publications in this area. While less research attention was dedicated to grid integration and vehicle systems, both areas also showed notable increases in scholarly interest.

In conclusion, based on the analyzed tables and visualizations, it can be stated that recent years have witnessed a marked increase in research related to cybersecurity in electromobility. These studies have grown in complexity, demonstrated broader engagement from emerging economies, and revealed strong connections to technical infrastructure. The application of chi-square testing confirmed that these changes were statistically significant, thus reflecting genuine developmental trends in the scientific literature.

5. Discussion

This review set out to synthesize where cybersecurity risks concentrate across the EV ecosystem and which countermeasures the literature treats as credible at today's deployment scale. Across the corpus, the charging interface emerges as the primary exposure point: it couples users, vehicles, backend services and the distribution grid, and it concentrates authentication, metering and payment flows. Quantitatively, publication activity after 2020 accelerated and shifted from foundations (crypto, access control) toward attack modelling and intrusion/anomaly detection, which is consistent with the sector's move from pilots to operation at scale. The dataset shows a near-threefold rise in publications from 26 (2017–2020) to 76 (2021–2024), with charging, grid integration and onboard systems all trending upward in security coverage.

5.1. What the Evidence Says About the Threat Surface

Three layers concentrate risk:

- Protocol and session layer (EV–EVSE–backend). Work dissecting ISO 15118 highlights weaknesses around certificate lifecycle, session take-over and availability of Plug & Charge when misconfigured; wireless access proposals (WAS) add new benefits but also new abuse paths if proximity and signaling are not robustly verified [11,13]. Schemes hardening mutual authentication with PUF and lightweight crypto reduce cost/latency and raise spoofing resistance in dynamic and wireless charging [36], while zero-knowledge proof (ZKP) approaches target privacy of use-patterns without involving a central authority [25,48].
- Operations and market layer. Work on tariff and billing integrity shows how manipulation of metering/price signals or free-riding during wireless charging can distort settlements or incentivize abusive behaviors; blockchain-backed contracts (and proof-of-behaviour (PoB)-type consensus) are explored to ensure auditability and deter misuse while preserving user confidentiality [16]. Architectural reviews of Electric vehicle supply equipment (EVSE) platforms point to gaps in security monitoring and incident response that keep recurring across brands and operators [26].
- Grid-coupling layer. Several studies model coordinated attacks that exploit the aggregate flexibility of EV fleets. Examples include distributed denial-of-service (DDoS) on EVSE control backbones or mobile apps, false data injection attacks (FDIA)on state estimation, and modal/oscillatory forcing against coordinated charging that propagates into frequency/voltage control; the literature reports material stability and latency impacts in simulation and co-simulation settings [14,33,34,41,43,45]. In vehicle-to-grid (V2G) scenarios, adversarial price manipulation can steer charging/discharging to harm feeders in IEEE test networks [19].

5.2. Defenses That Scale with Deployment

Defensive themes coalesce around four families:

1. Hardened authentication and crypto. Beyond ISO 15118's Public Key Infrastructure (PKI), the trend is toward hardware-rooted identities, short-lived credentials and context-bound sessions (location, proximity, Radio Frequency (RF) traits). Lightweight, post-quantum-ready designs are called for at the edge where compute and energy are tight [11,15,36,42].
2. Detection and response. Intrusion/anomaly detection blends physics-aware features with learned behaviors. Results favor deep models (autoencoders, LSTM/GRU) and edge/fog deployment to push latency down and resilience up; hybrid schemes combining signatures with behavior improve coverage against novel patterns [28,37,43,46].
3. Privacy-preserving settlement. Blockchain and smart contracts appear whenever transparency, non-repudiation and privacy of mobility traces must be balanced; zero-

knowledge proofs (ZKP) and pseudonymous protocols reduce traceability while keeping audits feasible [16,25,42,48].

4. Architecture and zero-trust. The literature and our quantitative split both point to architectures that assume every interface can be hostile: verify each request end-to-end, segment control from metering and payments, and instrument runtime monitoring compliant with IEC 62351-7-style telemetry [25].

5.3. Methods, Data and Where the Evidence Is Thin

Methodologically, the field leans on experimental and conceptual work, with case studies still scarce. That imbalance matters: simulated datasets and lab prototypes can overstate detector performance and understate integration/operational friction. The geographic split (rise of US/China/Canada; new entrants like India, Australia, Italy, Turkey) suggests targeted public investment is steering output rather than random drift (χ^2 analysis reported in Section 4). It should be noted that the χ^2 analyses reported earlier can be replicated directly from the category counts shown in Tables 3 and 4, as these tables provide the complete underlying distributions used in the tests.

Throughout the article, and particularly in the sections presenting analysis and conclusions, the text indicates whether the evidence discussed originates from simulation-based studies or from real-world implementations. This distinction helps clarify the strength and applicability of the findings, as simulation results may not fully capture the practical challenges of deployment, while field evidence provides direct insights into operational contexts. By noting the type of underlying data, the synthesis ensures that readers can interpret the recommendations with appropriate consideration of their evidentiary basis.

5.4. Practical Implications

The analysis highlights that electric vehicle (EV) charging infrastructures constitute a critical vulnerability point within the broader smart grid ecosystem. Attacks targeting these infrastructures not only threaten service continuity but also risk cascading effects on distribution networks and consumer trust. To strengthen resilience, practical measures must be adopted by different stakeholder groups. For DSOs (Distribution System Operators) and TSOs (Transmission System Operators), ensuring the integrity of control signals to EVSE fleets is essential. Public Key Infrastructure (PKI)-based authentication should be implemented to prevent malicious command injection [81,114]. Monitoring load curves through feeder-level anomaly detection allows for early identification of synchronized charging events that may stem from coordinated botnet attacks [115,116]. Stress-testing tariff broadcast and demand-response systems using penetration testing techniques ensures robustness against spoofing and replay attacks [14,117]. Further, blockchain-enabled auditing frameworks can support transparent verification of load balancing commands and pricing adjustments [118]. Simulation studies demonstrate how improper communication security may destabilize the grid [119], while field evidence from smart grid pilots confirms that authenticated, encrypted control channels reduce operational risks [120]. DSOs should also leverage intrusion detection tailored to cyber–physical energy systems, which combine anomaly recognition with real-time telemetry [121]. Finally, cross-domain lessons from industrial control systems and healthcare infrastructures highlight the importance of harmonized standards and continuous monitoring in preventing cascading failures [19,122]. CPOs (Charge Point Operators) and eMSPs (e-Mobility Service Providers) must automate certificate lifecycle management, using OCPP 2.0.1 certificate provisioning to reduce manual errors and enable rapid revocation of compromised devices [123,124]. Rate-limiting and API protection mechanisms should be systematically applied to mitigate denial-of-service (DoS) attacks [125]. Telemetry auditing—including charging duration, energy flow, and billing logs—can detect mismatches indicative of fraud or data tampering [11]. Proximity-based authentication, such as Wi-Fi Access Systems (WAS) or Ultra-Wideband (UWB), must be reinforced with secondary cryptographic verification via ISO 15118 Plug & Charge to ensure robust identity validation [126]. Field studies have shown that insecure APIs in charging management systems can be exploited for fraudulent session initiation, underlining the need for layered defenses. Research into blockchain-based settlement demonstrates the feasibility of tamper-resistant billing and roaming authentication mechanisms that directly support eMSPs in managing contractual obligations [47]. Experience from campus networks and banking systems further illustrates the risk of overload strategies and fraudulent access patterns, both of which are transferable to EVSE contexts [127]. Thus, operators should incorporate anomaly detection modules trained on domain-specific datasets to strengthen fraud prevention and ensure continuity of mobility services [128].

Original Equipment Manufacturers (OEMs) and Tier Suppliers must embed secure hardware roots of trust, such as Trusted Platform Modules or Secure Elements, to guarantee that only authenticated firmware can be executed on EVSE or vehicle components [129]. Secure boot processes reduce exposure to unauthorized firmware injection, while explainable intrusion detection systems enhance operator situational awareness by issuing interpretable alerts (e.g., "unauthorized firmware flash detected") instead of generic fault codes. Minimalist ISO 15118 stack implementations, combined with regular third-party code audits, are crucial to avoid vulnerabilities arising from unnecessary complexity [123]. Blockchain can also be adopted at the OEM level to maintain transparent firmware update histories and to manage distributed trust across multiple suppliers [118]. Lessons learned from anomaly detection in IoT and autonomous vehicles suggest that adaptive, machine learning-based IDS should be incorporated into EV architectures to capture evolving threats [47,121]. Finally, cross-referencing security practices from adjacent sectors—such as financial services and critical infrastructure—demonstrates the importance of standardized certification schemes and coordinated disclosure processes [122,127].

Together, these measures emphasize that cybersecurity in EV charging infrastructure cannot be addressed in isolation. Practical safeguards must combine strong cryptographic foundations, robust anomaly detection, and transparent auditing mechanisms, applied consistently across DSOs/TSOs, CPOs/eMSPs, and OEMs/Tier suppliers.

6. Conclusions

Cybersecurity is no longer a side constraint in electromobility; it defines whether charging networks can scale without eroding grid stability, user trust or market integrity. The literature mapped in this review converges on the following consistent points:

1. The attack surface grows with interoperability. As EVSE density rises and V2G matures, coordinated abuse of scheduling, pricing and control channels becomes plausible at system scale [14,19,33,43].
2. The most credible defense posture combines robust, hardware-anchored authentication with continuous, physics-aware anomaly detection, and with privacy-preserving settlement. This looks like a zero-trust stance applied to a cyber–physical grid: verify identities and intents at each hop; offer the least privilege; monitor continuously; and fail safely.
3. Quantitatively, research attention has shifted from basic crypto to attacks and IDS, mirroring deployment realities, while case-based evidence remains limited and deserves priority in the next wave of studies.

The practical value of these findings spans operators, manufacturers and service providers: enforcing identity and session hardening at the edge, instrumenting EVSEs for runtime monitoring, and making settlement auditable without leaking mobility patterns are all tractable near-term steps. The barriers are equally concrete: computational cost at the edge, DDoS susceptibility of backend and field networks, inconsistent standards across stacks, and the immaturity of explainable AI for anomaly response.

The following avenues remain valid: Lightweight cryptographic and authentication algorithms (including post-quantum solutions) to secure communication in resource-constrained environments [8–11,15,22]; advanced anomaly and attack detection leveraging deep neural networks, federated learning and edge/fog computing [16,19,20,32–48]; blockchain and smart contracts for transparent settlements, and to access control and privacy [17,49–57]; integration with smart grids and expansion of V2G services [12,18,75–92]; secure V2X/IoT communication with low-latency, zero-trust standards [65,98–111]. These avenues should be undertaken with an emphasis on operational pilots and interoperability so that results transfer across vendors and regions.

Given the evidence synthesized in the articles, we recommend prioritizing the following: hardening ISO 15118 stacks and certificate workflows; evaluating WAS/UWB proximity only with multi-factor checks [11,13,47]; privacy-preserving authentication and billing using ZKP/PUF and selective-disclosure contracts to curb traceability and free-riding [16,36,42,48,81]; and grid-aware IDS tuned for EV-induced oscillations, FDIA and price-manipulation in V2G [14,19,33,34,43,45]. The findings presented reflect only those aspects of cybersecurity directly related to attacks, anomaly detection, and authentication in EV charging systems, without extending to the broader cybersecurity landscape.

Finally, the evidence base needs more field-grade datasets and open, standardized telemetry—preferably aligned with IEC 62351-7—to allow operators to benchmark detectors, share indicators and automate response without vendor lock-in.

Supplementary Materials: The following supporting information can be downloaded at: https://www.mdpi.com/article/10.3390/en18184847/s1, scopus.csv (raw Table S1. Scopus query results), Table S2. table4.csv and Table S3. table4_chi.csv (data and statistical analysis for Table 4), Table S4. table5.csv and Table S5. table5_chi.csv (data and statistical analysis for Table 5).

Author Contributions: Conceptualization, G.W.-J.; methodology, G.W.-J.; software, L.P.; validation, L.P.; formal analysis, J.L.W.-J.; investigation, P.T.G. and L.P.; resources, L.P.; data curation, L.P.; writing—original draft preparation, L.P., J.L.W.-J. and G.W.-J.; final writing—review and editing, P.T.G.; visualization, L.P.; supervision, J.L.W.-J., G.W.-J. and L.P.; project administration, J.L.W.-J. and L.P.; funding acquisition, J.L.W.-J. All authors have read and agreed to the published version of the manuscript.

Funding: This research received no external funding. This is the waiver paper. No APC.

Data Availability Statement: The original contributions presented in this study are included in the article. Further inquiries can be directed to the corresponding author.

Conflicts of Interest: The authors declare no conflicts of interest.

References

1. Christoforidis, C.; Christakis, I.; Kotzasavva, F.; Fousekis, G.; Rimpas, D. Electric Vehicles Charging Stations and Their Influence on the Electricity Utility Grid. *Arch. Automot. Eng. Arch. Motoryz.* **2025**, *107*, 102–126. [CrossRef] [PubMed]
2. Dhandapani, S.; Raja, T.; Murugan, V.; Selvaraj, J.; Rathinasamy, V. Comprehensive Review on Evolution, Progression, Design, and Exploration of Electric Bicycle. *Arch. Automot. Eng. Arch. Motoryz.* **2024**, *105*, 5–43. [CrossRef]
3. Lučić, M.; Lukić, J.; Grujić, I. Statistical Analysis of Trends in Battery Electric Vehicles: Special Reference to Vehicle Weight Reduction, Electric Motor, Battery, and Interior Space Dimensions. *Arch. Automot. Eng. Arch. Motoryz.* **2024**, *104*, 63–96. [CrossRef]
4. Małek, A.; Marciniak, A.; Bartnik, G. The Selection of an Electric Vehicle for the Existing Photovoltaic System—Case Study in Polish Climatic Conditions. *Arch. Automot. Eng. Arch. Motoryz.* **2024**, *103*, 38–56. [CrossRef]
5. Małek, A.; Marciniak, A. Selection of the Photovoltaic System Power for the Electric Vehicle. *Arch. Automot. Eng. Arch. Motoryz.* **2023**, *100*, 44–61. [CrossRef]

6. Szumska, E.; Skuza, A.; Jurecki, R. The Analysis of Energy Recovered by an Electric Vehicle during Selected Braking Manoeuvres. *Arch. Automot. Eng. Arch. Motoryz.* **2023**, *99*, 18–29. [CrossRef]
7. Kieracińska, A.; Piórkowski, P. Analysis of the Reasonableness of Using Electric Drives in Motor Vehicles. *Arch. Automot. Eng. Arch. Motoryz.* **2016**, *74*, 47–58. [CrossRef]
8. Lin, Q.; Li, T.; Zhao, Y.; Guan, J.; Zhang, W.; Wang, X. Research on Charging Infrastructure Related Detection Technology Based on GAN. In Proceedings of the 2023 2nd International Conference on Clean Energy Storage and Power Engineering (CESPE), Xi'an, China, 15–16 December 2023; Institute of Electrical and Electronics Engineers Inc.: Interlaken, Switzerland, 2023; pp. 57–62.
9. Mousavian, S.; Erol-Kantarci, M.; Wu, L.; Ortmeyer, T. A Risk-Based Optimization Model for Electric Vehicle Infrastructure Response to Cyber Attacks. *IEEE Trans. Smart Grid* **2018**, *9*, 6160–6169. [CrossRef]
10. Hamied, M.H.A. Authentication of the Electric Vehicle with the Charging Station Using AAA Framework. In Proceedings of the 2023 3rd International Conference on Computing and Information Technology (ICCIT), Tabuk, Saudi Arabia, 13–14 September 2023; Institute of Electrical and Electronics Engineers Inc.: Interlaken, Switzerland, 2023; pp. 186–191.
11. Jakó, Z.; Knapp, Á.; Sayed, N.E. Wireless Authentication Solution and TTCN-3 Based Test Framework for ISO-15118 Wireless V2G Communication. *Infocommunications J.* **2019**, *11*, 39–47. [CrossRef]
12. Basnet, M.; Ali, M.H. Deep Learning-Based Intrusion Detection System for Electric Vehicle Charging Station. In Proceedings of the 2020 2nd International Conference on Smart Power & Internet Energy Systems (SPIES), Bangkok, Thailand, 15–18 September 2020; Institute of Electrical and Electronics Engineers Inc.: Interlaken, Switzerland, 2020; pp. 408–413.
13. Bao, K.; Valev, H.; Wagner, M.; Schmeck, H. A Threat Analysis of the Vehicle-to-Grid Charging Protocol ISO 15118. *Comput. Sci. Res. Dev.* **2018**, *33*, 3–12. [CrossRef]
14. Aljohani, T.; Almutairi, A. Modeling Time-Varying Wide-Scale Distributed Denial of Service Attacks on Electric Vehicle Charging Stations. *Ain Shams Eng. J.* **2024**, *15*, 102860. [CrossRef]
15. Li, Z.; Sun, Z. A Cross-Domain Authentication Scheme for Electric Vehicle Intelligent Charging for Multiple Power Service Providers. In Proceedings of the 2023 13th International Conference on Communication and Network Security, Fuzhou, China, 6–8 December 2023; Association for Computing Machinery: New York, NY, USA, 2023; pp. 26–31.
16. Zhang, X.; Liu, C.; Chai, K.K.; Poslad, S. A Privacy-Preserving Consensus Mechanism for an Electric Vehicle Charging Scheme. *J. Netw. Comput. Appl.* **2021**, *174*, 102908. [CrossRef]
17. Sharma, G.; Joshi, A.M.; Mohanty, S.P. sTrade: Blockchain Based Secure Energy Trading Using Vehicle-to-Grid Mutual Authentication in Smart Transportation. *Sustain. Energy Technol. Assess.* **2023**, *57*, 103296. [CrossRef]
18. Cirimele, V.; Diana, M.; Bellotti, F.; Berta, R.; El Sayed, N.; Kobeissi, A.; Guglielmi, P.; Ruffo, R.; Khalilian, M.; La Ganga, A.; et al. The Fabric ICT Platform for Managing Wireless Dynamic Charging Road Lanes. *IEEE Trans. Veh. Technol.* **2020**, *69*, 2501–2512. [CrossRef]
19. Yin, H.; Liu, D.; Weng, J. Risk Analysis of Cyber Physical Distribution System Considering Cyber Attacks on V2g System. In Proceedings of the 10th Renewable Power Generation Conference (RPG 2021), Online, 14–15 October 2021; Institution of Engineering and Technology: Stevenage, UK, 2021; Volume 2021, pp. 841–846.
20. Basnet, M.; Ali, M.H. Exploring Cybersecurity Issues in 5G Enabled Electric Vehicle Charging Station with Deep Learning. *IET Gener. Transm. Distrib.* **2021**, *15*, 3435–3449. [CrossRef]
21. van Eck, N.J.; Waltman, L. Software Survey: VOSviewer, a Computer Program for Bibliometric Mapping. *Scientometrics* **2010**, *84*, 523–538. [CrossRef]
22. Kanz Abdillah, H.; Wildan Rizkia, N.A.H.; Sidharta, S. Exploring the Role of Artificial Intelligence in Enhancing Battery Performance and Mitigating Cybersecurity Threats in Electric Vehicles: A Systematic Literature Review. *Procedia Comput. Sci.* **2024**, *245*, 155–165. [CrossRef]
23. Pazos-Revilla, M.; Alsharif, A.; Gunukula, S.; Guo, T.N.; Mahmoud, M.; Shen, X. Secure and Privacy-Preserving Physical-Layer-Assisted Scheme for EV Dynamic Charging System. *IEEE Trans. Veh. Technol.* **2018**, *67*, 3304–3318. [CrossRef]
24. Gabay, D.; Akkaya, K.; Cebe, M. Privacy-Preserving Authentication Scheme for Connected Electric Vehicles Using Blockchain and Zero Knowledge Proofs. *IEEE Trans. Veh. Technol.* **2020**, *69*, 5760–5772. [CrossRef]
25. Reghunath, R.; Sayed, M.A.; Sarieddine, K.; Atallah, R.; Jafarigiv, D.; Kassouf, M.; Assi, C.; Ghafouri, M. A Real-Time Monitoring Architecture for Enhanced Cybersecurity in the EV Ecosystem. In Proceedings of the IECON 2024—50th Annual Conference of the IEEE Industrial Electronics Society, Chicago, IL, USA, 3–6 November 2024; IEEE Computer Society: Washington, DC, USA, 2024.
26. Ronanki, D.; Karneddi, H. Electric Vehicle Charging Infrastructure: Review, Cyber Security Considerations, Potential Impacts, Countermeasures, and Future Trends. *IEEE J. Emerg. Sel. Top. Power Electron.* **2024**, *12*, 242–256. [CrossRef]
27. Sepehrzad, R.; Faraji, M.J.; Al-Durra, A.; Sadabadi, M.S. Enhancing Cyber-Resilience in Electric Vehicle Charging Stations: A Multi-Agent Deep Reinforcement Learning Approach. *IEEE Trans. Intell. Transp. Syst.* **2024**, *25*, 18049–18062. [CrossRef]
28. Basnet, M.; Ali, M.H. Deep-Learning-Powered Cyber-Attacks Mitigation Strategy in the EV Charging Infrastructure. In Proceedings of the 2023 IEEE Power & Energy Society General Meeting (PESGM), Orlando, FL, USA, 16–20 July 2023; IEEE Computer Society: Washington, DC, USA, 2023.

29. Fard, M.F.; Huo, X.; Liu, M. Exploration of For-Purpose Decentralized Algorithmic Cyber Attacks in EV Charging Control. In Proceedings of the 2023 IEEE 32nd International Symposium on Industrial Electronics (ISIE), Helsinki/Espoo, Finland, 19–21 June 2023; Institute of Electrical and Electronics Engineers Inc.: Interlaken, Switzerland, 2023.
30. Sarkar, A.; Ray, S.K. Smart Contract-Based Electric Vehicle Charging: A Practice-Based Economy of Things Application. In Proceedings of the Pacific Asia Conference on Information Systems (PACIS 2020), Dubai, United Arab Emirates, 22–24 June 2020; Association for Information Systems: Atlanta, GA, USA, 2020.
31. Yaqub, R. Architecture and Protocols for Toll-Free Electric Vehicle Charging. *World Electr. Veh. J.* **2019**, *10*, 10. [CrossRef]
32. Guo, S.; Chen, H.; Rahman, M.; Qian, X. DCA: Delayed Charging Attack on the Electric Shared Mobility System. *IEEE Trans. Intell. Transp. Syst.* **2023**, *24*, 12793–12805. [CrossRef]
33. Gumrukcu, E.; Arsalan, A.; Muriithi, G.; Joglekar, C.; Aboulebdeh, A.; Zehir, M.A.; Papari, B.; Monti, A. Impact of Cyber-Attacks on EV Charging Coordination: The Case of Single Point of Failure. In Proceedings of the 2022 4th Global Power, Energy and Communication Conference (GPECOM), Cappadocia, Turkey, 14–17 June 2022; Institute of Electrical and Electronics Engineers Inc.: Interlaken, Switzerland, 2022; pp. 506–511.
34. Zhou, Y.; Xu, H.; Liu, W.; Li, Y. Malicious Mode Attack on Electric Vehicle Coordinated Charging and Its Defense Strategy. *Sustain. Energy Grids Netw.* **2024**, *39*, 101440. [CrossRef]
35. Chen, C.-M.; Miao, Q.; Khan, F.; Srivastava, G.; Kumari, S. Sustainable Secure Communication in Consumer-Centric Electric Vehicle Charging in Industry 5.0 Environments. *IEEE Trans. Consum. Electron.* **2024**, *70*, 1544–1555. [CrossRef]
36. Alshaeri, A.; Younis, M. Lightweight Authentication and Authorization Protocol for Dynamic Charging of Electric Vehicles. In Proceedings of the 2022 IEEE 19th Annual Consumer Communications & Networking Conference (CCNC), Las Vegas, NV, USA, 8–11 January 2022; Institute of Electrical and Electronics Engineers Inc.: Interlaken, Switzerland, 2022; pp. 550–556.
37. Feng, Q.; Li, H.; Zhou, Y.; Feng, D.; Wang, Y.; Su, Y. Review of Electric Vehicles' Charging Data Anomaly Detection Based on Deep Learning. In Proceedings of the 2022 Power System and Green Energy Conference (PSGEC), Shanghai, China, 25–27 August 2022; Li, G., Li, Z., Eds.; Institute of Electrical and Electronics Engineers Inc.: Interlaken, Switzerland, 2022; pp. 337–341.
38. Wang, Y.; Zhao, Z.; Qi, B.; Cheng, Y.; Tang, K.; Li, B. Vulnerability Analysis of an Electric Vehicle Fleet for Car-Sharing Service under Cyber Attacks. *Sustain. Energy Grids Netw.* **2024**, *37*, 101207. [CrossRef]
39. Fahim, S.R.; Atat, R.; Kececi, C.; Takiddin, A.; Ismail, M.; Davis, K.R.; Serpedin, E. Generalized FDIA Detection in Power Dependent Electrified Transportation Systems. In Proceedings of the 2024 32nd European Signal Processing Conference (EUSIPCO), Lyon, France, 26–30 August 2024; IEEE: Washington, DC, USA, 2024; pp. 1851–1855.
40. Wadi, M.; Shobole, A.; Elmasry, W.; Kucuk, I. Load Frequency Control in Smart Grids: A Review of Recent Developments. *Renew. Sustain. Energy Rev.* **2024**, *189*, 114013. [CrossRef]
41. Attia, M.; Senouci, S.M.; Aglzim, E.-H. New Optimization and Security Approaches to Enhance the Smart Grid Performance and Reliability. In Proceedings of the 2016 7th International Conference on the Network of the Future (NOF), Búzios, Brazil, 16–18 November 2016; Duarte, O.C.M.B., Secci, S., Rubinstein, M.G., de Amorim, M.D., Eds.; Institute of Electrical and Electronics Engineers Inc.: Interlaken, Switzerland, 2017.
42. Li, H.; Dan, G.; Nahrstedt, K. Portunes+: Privacy-Preserving Fast Authentication for Dynamic Electric Vehicle Charging. *IEEE Trans. Smart Grid* **2017**, *8*, 2305–2313. [CrossRef]
43. Sarieddine, K.; Sayed, M.A.; Torabi, S.; Atallah, R.; Assi, C. Edge-Based Detection and Localization of Adversarial Oscillatory Load Attacks Orchestrated by Compromised EV Charging Stations. *Int. J. Electr. Power Energy Syst.* **2024**, *156*, 109735. [CrossRef]
44. Peng, X.; Duan, S.; Sankavaram, C.; Jin, X. Unsupervised Adaptive Fleet Battery Pack Fault Detection with Concept Drift under Evolving Environment. *IEEE Trans. Autom. Sci. Eng.* **2024**, *21*, 2276–2288. [CrossRef]
45. Kern, D.; Krauß, C. Analysis of E-Mobility-Based Threats to Power Grid Resilience. In Proceedings of the 5th ACM Computer Science in Cars Symposium, Ingolstadt, Germany, 30 November 2021; Spencer, S.N., Ed.; Association for Computing Machinery Inc.: New York, NY, USA, 2021.
46. Lin, Z.; Li, J. FedEVCP: Federated Learning-Based Anomalies Detection for Electric Vehicle Charging Pile. *Comput. J.* **2024**, *67*, 1521–1530. [CrossRef]
47. Rinaldi, S.; Depari, A.; Ferrari, P.; Flammini, A.; Mondini, E.; Pasetti, M.; Sisinni, E. An Evaluation of UWB for Location-Based Hands-Free Authentication Charging of Electric Vehicles. In Proceedings of the 2022 IEEE 31st International Symposium on Industrial Electronics (ISIE), Anchorage, AK, USA, 1–3 June 2022; Institute of Electrical and Electronics Engineers Inc.: Interlaken, Switzerland, 2022; pp. 696–701.
48. Gabay, D.; Cebe, M.; Akkaya, K. Poster: On the Overhead of Using Zero-Knowledge Proofs for Electric Vehicle Authentication. In Proceedings of the WiSec'19: 12th ACM Conference on Security and Privacy in Wireless and Mobile Networks, Miami, FL, USA, 15–17 May 2019; Association for Computing Machinery, Inc.: New York, NY, USA, 2019; pp. 347–348.
49. Wei, F.; Lin, X. Cyber-Physical Attack Launched from EVSE Botnet. *IEEE Trans. Power Syst.* **2024**, *39*, 3603–3614. [CrossRef]

50. Sureshkumar, V.; Chinnaraj, P.; Saravanan, P.; Amin, R.; Rodrigues, J.J.P.C. Authenticated Key Agreement Protocol for Secure Communication Establishment in Vehicle-to-Grid Environment with FPGA Implementation. *IEEE Trans. Veh. Technol.* **2022**, *71*, 3470–3479. [CrossRef]
51. Zhang, L.; Zhu, Y.; Ren, W.; Wang, Y.; Choo, K.-K.R.; Xiong, N.N. An Energy-Efficient Authentication Scheme Based on Chebyshev Chaotic Map for Smart Grid Environments. *IEEE Internet Things J.* **2021**, *8*, 17120–17130. [CrossRef]
52. Mavikumbure, H.S.; Cobilean, V.; Wickramasinghe, C.S.; Phillips, T.; Varghese, B.J.; Carlson, B.; Rieger, C.; Pennington, T.; Manic, M. Physical Anomaly Detection in EV Charging Stations: Physics-Based vs. ResNet AE. In Proceedings of the 2023 IEEE 32nd International Symposium on Industrial Electronics (ISIE), Helsinki, Finland, 19–21 June 2023; Institute of Electrical and Electronics Engineers Inc.: Interlaken, Switzerland, 2023.
53. Linnartz, P.; Winkens, A.; Ulbig, A. Analysis of Malicious Operation of Electric Vehicle CHARGING stations to Disrupt Stable Operation of Distribution Grids. In Proceedings of the CIRED Porto Workshop 2022: E-Mobility and Power Distribution Systems, Porto, Portugal, 2–3 June 2022; Institution of Engineering and Technology: Stevenage, UK, 2022; Volume 2022, pp. 990–994.
54. Babu, P.R.; Amin, R.; Reddy, A.G.; Das, A.K.; Susilo, W.; Park, Y. Robust Authentication Protocol for Dynamic Charging System of Electric Vehicles. *IEEE Trans. Veh. Technol.* **2021**, *70*, 11338–11351. [CrossRef]
55. Soares, A.A.Z.; Mattos, D.M.F.; Lopes, Y.; Medeiros, D.S.V.; Fernandes, N.C.; Muchaluat-Saade, D.C. An Efficient Authentication Mechanism Based on Software-Defined Networks for Electric Vehicles. In Proceedings of the 2019 IEEE 28th International Symposium on Industrial Electronics (ISIE), Vancouver, BC, Canada, 12–14 June 2019; Institute of Electrical and Electronics Engineers Inc.: Interlaken, Switzerland, 2019; pp. 2471–2476.
56. Kern, D.; Krauß, C.; Hollick, M. Detection of Anomalies in Electric Vehicle Charging Sessions. In Proceedings of the 39th Annual Computer Security Applications Conference, Austin, TX, USA, 4–8 December 2023; Association for Computing Machinery: New York, NY, USA, 2023; pp. 298–309.
57. Kilic, A. TLS-Handshake for Plug and Charge in Vehicular Communications. *Comput. Networks* **2024**, *243*, 110281. [CrossRef]
58. Kilari, V.T.; Yu, R.; Misra, S.; Xue, G. Robust Revocable Anonymous Authentication for Vehicle to Grid Communications. *IEEE Trans. Intell. Transp. Syst.* **2020**, *21*, 4845–4857. [CrossRef]
59. Khan, O.G.M.; Elghitani, F.; Youssef, A.; Salama, M.; El-Saadany, E.F. Real-Time Congestion-Aware Charging Station Assignment Model for EVs. *IEEE Internet Things J.* **2024**, *11*, 11723–11736. [CrossRef]
60. Akbarian, A.; Bahrami, M.; Vakilian, M.; Lehtonen, M. Vulnerability of EV Charging Stations to Cyber Attacks Manipulating Prices. In Proceedings of the 2023 International Conference on Future Energy Solutions (FES), Vaasa, Finland, 12–14 June 2023; Institute of Electrical and Electronics Engineers Inc.: Interlaken, Switzerland, 2023.
61. Nguyen, T.-V.; Sun, H.; Wang, H.; Hu, R.Q. Authentication and PHY-Security Schemes for Electric Vehicle Dynamic Wireless Charging. *IEEE Trans. Veh. Technol.* **2024**, *73*, 1698–1712. [CrossRef]
62. Afrin, S.; Kwasinski, A. A Privacy-Preserving Method with Flexible Charging Schedules for Electric Vehicles in the Smart Grid. In Proceedings of the 2017 IEEE International Conference on Advanced Networks and Telecommunications Systems (ANTS), Bhubaneswar, India, 17–20 December 2017; Institute of Electrical and Electronics Engineers Inc.: Interlaken, Switzerland, 2018; pp. 1–6.
63. Yazdanipour, S.; Arani, M.F.M.; Jahromi, A.A. Investigating Cyberattacks Against Off-Grid Solar-Powered Electric Vehicle Charging Stations. In Proceedings of the 2024 IEEE/PES Transmission and Distribution Conference and Exposition (T&D), Anaheim, CA, USA, 6–9 May 2024; Institute of Electrical and Electronics Engineers Inc.: Interlaken, Switzerland, 2024.
64. *ISO 15118-20:2022(En)*; Road Vehicles—Vehicle to Grid Communication Interface—Part 20: 2nd Generation Network Layer and Application Layer Requirements. International Organization for Standardization. International Electrotechnical Commission: Geneva, Switzerland, 2022. Available online: https://www.iso.org/obp/ui/en/#iso:std:iso:15118:-20:ed-1:v1:en (accessed on 1 September 2025).
65. Plappert, C.; Jäger, L.; Irrgang, A.; Potluri, C. Secure Multi-User Contract Certificate Management for ISO 15118-20 Using Hardware Identities. In Proceedings of the 18th International Conference on Availability, Reliability and Security, Benevento, Italy, 29 August–1 September 2023; Association for Computing Machinery: New York, NY, USA, 2023.
66. Yang, Y.; Li, J. Electric Vehicle Charging Anomaly Detection Method Based on Multivariate Gaussian Distribution Model. In Proceedings of the 2022 Asia Conference on Electrical, Power and Computer Engineering, Shanghai, China, 22–24 April 2022; Association for Computing Machinery: New York, NY, USA, 2022.
67. Fattahi, J. A Federated Byzantine Agreement Model to Operate Offline Electric Vehicle Supply Equipment. *IEEE Trans. Smart Grid* **2024**, *15*, 2004–2016. [CrossRef]
68. Jahangir, H.; Lakshminarayana, S.; Poor, H.V. Charge Manipulation Attacks Against Smart Electric Vehicle Charging Stations and Deep Learning-Based Detection Mechanisms. *IEEE Trans. Smart Grid* **2024**, *15*, 5182–5194. [CrossRef]
69. Kapassa, E.; Themistocleous, M.; Christodoulou, K.; Iosif, E. Blockchain Application in Internet of Vehicles: Challenges, Contributions and Current Limitations. *Future Internet* **2021**, *13*, 313. [CrossRef]
70. Chen, C.-M.; Hao, Y.; Wu, T.-Y. Discussion of "Ultra Super Fast Authentication Protocol for Electric Vehicle Charging Using Extended Chaotic Maps" . *IEEE Trans. Ind. Appl.* **2023**, *59*, 2091–2092. [CrossRef]

71. Eghtessad, S.; Hasanli, T.; Huynh, R.; Workneh, K.; Ibrahem, M.I.; Fouda, M.M. A Survey on Privacy Preservation Methods in Future Vehicular Networks. In Proceedings of the 2023 11th International Conference on Information and Communication Technology (ICoICT), Melaka, Malaysia, 23–24 August 2023; IEEE: Washington, DC, USA, 2023; pp. 319–325.
72. Van Den Brink, H. The Need for Cybersecurity within the Electric Vehicle Infrastructure. In Proceedings of the EVS—International Electric Vehicle Symposium and Exhibition, Stuttgart, Germany, 9–11 October 2017; Landesmesse Stuttgart GmbH: Stuttgart, Germany, 2017.
73. Albanese, E.; Terruggia, R. Secure and Resilient Iot and Cloud-Based Infrastructure for Electric Vehicles Recharge Systems. *IET Conf. Proc.* **2023**, *2023*, 2059–2063. [CrossRef]
74. Bourass, A.; Cherkaoui, S.; Khoukhi, L. Secure Communication Scheme for Electric Vehicles in the Smart Grid. In Proceedings of the GLOBECOM 2017—2017 IEEE Global Communications Conference, Singapore, 4–8 December 2017; IEEE: Washington, DC, USA, 2017; pp. 1–5.
75. Kern, D.; Krauß, C.; Hollick, M. Attack Analysis and Detection for the Combined Electric Vehicle Charging and Power Grid Domains. In Proceedings of the 19th International Conference on Availability, Reliability and Security, Vienna, Austria, 30 July–2 August 2024; Association for Computing Machinery: New York, NY, USA, 2024.
76. Girdhar, M.; Hong, J.; Lee, H.; Song, T.-J. Hidden Markov Models-Based Anomaly Correlations for the Cyber-Physical Security of EV Charging Stations. *IEEE Trans. Smart Grid* **2022**, *13*, 3903–3914. [CrossRef]
77. Patel, T.; Jhaveri, R.H.; Sidana, S.; Benedetto, F. Empowering Energy Transition: Detecting Cyber Threats in EV Charging Infrastructure through ML and XAI Analysis. In Proceedings of the 2024 International Conference on Sustainable Energy: Energy Transition and Net-Zero Climate Future (ICUE), Pattaya City, Thailand, 21–23 October 2024; Institute of Electrical and Electronics Engineers Inc.: Interlaken, Switzerland, 2024.
78. Garg, S.; Kaur, K.; Kaddoum, G.; Gagnon, F.; Rodrigues, J.J.P.C. An Efficient Blockchain-Based Hierarchical Authentication Mechanism for Energy Trading in V2G Environment. In Proceedings of the 2019 IEEE International Conference on Communications Workshops (ICC Workshops), Shanghai, China, 20–24 May 2019; Institute of Electrical and Electronics Engineers Inc.: Interlaken, Switzerland, 2019.
79. Streubel, T.; Kattmann, C.; Eisenmann, A.; Rudion, K. Detection and Monitoring of Supraharmonic Anomalies of an Electric Vehicle Charging Station. In Proceedings of the 2019 IEEE Milan PowerTech, Milan, Italy, 23–27 June 2019; Institute of Electrical and Electronics Engineers Inc.: Interlaken, Switzerland, 2019.
80. Li, M.; Ni, M.; Tong, H.; Ru, Y.; Xu, J.; Wu, Y. Risk Analysis of Electric Vehicles Connected to the Cyber-Physical Power System. In Proceedings of the 2021 IEEE Sustainable Power and Energy Conference (iSPEC), Nanjing, China, 23–25 December 2021; Institute of Electrical and Electronics Engineers Inc.: Interlaken, Switzerland, 2021; pp. 4203–4208.
81. Sarieddine, K.; Sayed, M.A.; Torabi, S.; Attallah, R.; Jafarigiv, D.; Assi, C.; Debbabi, M. Uncovering Covert Attacks on EV Charging Infrastructure: How OCPP Backend Vulnerabilities Could Compromise Your System. In Proceedings of the 19th ACM Asia Conference on Computer and Communications Security, Singapore, 1–5 July 2024; Association for Computing Machinery, Inc.: New York, NY, USA, 2024; pp. 987–999.
82. Li, P.; Ou, W.; Liang, H.; Han, W.; Zhang, Q.; Zeng, G. A Zero Trust and Blockchain-Based Defense Model for Smart Electric Vehicle Chargers. *J. Netw. Comput. Appl.* **2023**, *213*, 103599. [CrossRef]
83. Sayed, M.A.; Atallah, R.; Assi, C.; Debbabi, M. Electric Vehicle Attack Impact on Power Grid Operation. *Int. J. Electr. Power Energy Syst.* **2022**, *137*, 107784. [CrossRef]
84. Yang, F.; Yang, Y.; Wang, X.; Ouyang, X.; Shuai, C. Electric Bikes Charging Anomaly Detection from Alternating Current Side Based on Big Data. *Eng. Appl. Artif. Intell.* **2024**, *136*, 109042. [CrossRef]
85. Nisar, F.; Ramachandran, G.; Vilathgamuva, M.; Jurdak, R. Manipulation of Actual Demand in Electric Vehicles (MaD EV): A Cyber-Security Perspective. In Proceedings of the 2022 IEEE 7th Southern Power Electronics Conference (SPEC), Nadi, Fiji, 5–8 December 2022; Institute of Electrical and Electronics Engineers Inc.: Interlaken, Switzerland, 2022.
86. ElGhanam, E.; Ahmed, I.; Hassan, M.; Osman, A. Authentication and Billing for Dynamic Wireless EV Charging in an Internet of Electric Vehicles. *Futur. Internet* **2021**, *13*, 257. [CrossRef]
87. Rahman, S.; Aburub, H.; Mekonnen, Y.; Sarwat, A.I. A Study of EV BMS Cyber Security Based on Neural Network SOC Prediction. In Proceedings of the 2018 IEEE/PES Transmission and Distribution Conference and Exposition (T&D), Denver, CO, USA, 16–19 April 2018; Institute of Electrical and Electronics Engineers Inc.: Interlaken, Switzerland, 2018.
88. Bourass, A.; Cherkaoui, S.; Khoukhi, L. Secure Optimal Itinerary Planning for Electric Vehicles in the Smart Grid. *IEEE Trans. Ind. Inf.* **2017**, *13*, 3236–3245. [CrossRef]
89. Girdhar, M.; Hong, J.; Yoo, Y.; Song, T.-J. Machine Learning-Enabled Cyber Attack Prediction and Mitigation for EV Charging Stations. In Proceedings of the 2022 IEEE Power & Energy Society General Meeting (PESGM), Denver, CO, USA, 17–21 July 2022; IEEE Computer Society: Washington, DC, USA, 2022.
90. Kilic, A. Plug and Charge Solutions with Vehicle-to-Grid Communication. *Electr. Power Comp. Syst.* **2023**, *51*, 1786–1814. [CrossRef]

91. Mookherji, S.; Odelu, V.; Prasath, R. Secure Ultra Fast Authentication Protocol for Electric Vehicle Charging. *Comput. Electr Eng.* **2024**, *119*, 109512. [CrossRef]
92. Xu, J.; Yu, X.; Tian, L.; Wang, J.; Zhou, L.; Liu, C. A Lightweight Security Authentication Method for the Charging Protocol of Smart Charging Pile. In Proceedings of the 2020 International Conference on Aviation Safety and Information Technology, Weihai, China, 14–16 October 2020; Association for Computing Machinery: New York, NY, USA, 2020; pp. 376–382.
93. Shao, H.; Ma, Y.; Shao, B. A Lightweight Privacy-Preserving Authentication Scheme for Vehicle-to-Grid. In Proceedings of the 2024 6th Asia Energy and Electrical Engineering Symposium (AEEES), Chengdu, China, 28–31 March 2024; Institute of Electrical and Electronics Engineers Inc.: Interlaken, Switzerland, 2024; pp. 278–282.
94. Singh, N.; Agarwal, R. Intrusion Detection System for Smart Vehicles Using Machine Learning Algorithms. In Proceedings of the 2023 International Conference on Communication, Security and Artificial Intelligence (ICCSAI), Greater Noida, India, 23–25 November 2023; Institute of Electrical and Electronics Engineers Inc.: Interlaken, Switzerland, 2023; pp. 749–753.
95. Kaur, K.; Kaddoum, G.; Zeadally, S. Blockchain-Based Cyber-Physical Security for Electrical Vehicle Aided Smart Grid Ecosystem. *IEEE Trans. Intell. Transp. Syst.* **2021**, *22*, 5178–5189. [CrossRef]
96. Tiwari, P.; Ronanki, D. Cyber-Resilient Grid-Interactive Renewable Powered Wireless Charging of Electric Vehicles. In Proceedings of the 2022 IEEE International Conference on Power Electronics, Drives and Energy Systems (PEDES), Jaipur, India, 14–17 December 2022; Institute of Electrical and Electronics Engineers Inc.: Interlaken, Switzerland, 2022.
97. Liu, Y.; Ardakanian, O.; Nikolaidis, I.; Liang, H. False Data Injection Attacks on Smart Grid Voltage Regulation with Stochastic Communication Model. *IEEE Trans. Ind. Inf.* **2023**, *19*, 7122–7132. [CrossRef]
98. Gandhi, K.; Morsi, W.G. Impact of the Open Charge Point Protocol Between the Electric Vehicle and the Fast Charging Station on the Cybersecurity of the Smart Grid. In Proceedings of the 2022 IEEE Canadian Conference on Electrical and Computer Engineering (CCECE), Halifax, NS, Canada, 18–20 September 2022; Institute of Electrical and Electronics Engineers Inc.: Interlaken, Switzerland, 2022; pp. 235–240.
99. Sayed, N.E. A Prototypical Implementation of an ISO-15118-Based Wireless Vehicle to Grid Communication for Authentication over Decoupled Technologies. In Proceedings of the 2019 AEIT International Conference of Electrical and Electronic Technologies for Automotive (AEIT AUTOMOTIVE), Turin, Italy, 2–4 July 2019; Institute of Electrical and Electronics Engineers Inc.: Interlaken, Switzerland, 2019.
100. Wang, W.; Han, Z.; Alazab, M.; Gadekallu, T.R.; Zhou, X.; Su, C. Ultra Super Fast Authentication Protocol for Electric Vehicle Charging Using Extended Chaotic Maps. *IEEE Trans. Ind. Appl.* **2022**, *58*, 5616–5623. [CrossRef]
101. Alshaeri, A.; Younis, M. A Blockchain-Based Energy Trading Scheme for Dynamic Charging of Electric Vehicles. In Proceedings of the 2021 IEEE Global Communications Conference (GLOBECOM), Madrid, Spain, 7–11 December 2021.
102. Roman, L.F.A.; Gondim, P.R.L.; Lloret, J. Pairing-Based Authentication Protocol for V2G Networks in Smart Grid. *Ad Hoc Netw.* **2019**, *90*, 101745. [CrossRef]
103. Mohammadi, M.; Mohasefi, J.B. Availability-Based and Risk-Less Optimization Model for Electric Vehicles Optimal Itinerary Planning in Smart Grid. *Sustain. Energy Grids Netw.* **2022**, *30*, 100642. [CrossRef]
104. Grgic, D.; Podobnik, V.; Capuder, T. EV Charging Station Monitoring Framework—A Case Study of Croatia: Performance Analysis and Anomaly Detection under Pandemic Situation. *Int. J. Electr. Power Energy Syst.* **2022**, *141*, 108095. [CrossRef]
105. Yang, B.; Guo, L.; Ye, J. Physics-Based Attack Detection for Traction Motor Drives in Electric Vehicles Using Random Forest. In Proceedings of the 2021 IEEE Applied Power Electronics Conference and Exposition (APEC), Phoenix, AZ, USA, 14–17 June 2021; Institute of Electrical and Electronics Engineers Inc.: Interlaken, Switzerland, 2021; pp. 849–854.
106. Sayed, M.A.; Ghafouri, M.; Debbabi, M.; Assi, C. Dynamic Load Altering EV Attacks Against Power Grid Frequency Control. In Proceedings of the 2022 IEEE Power & Energy Society General Meeting (PESGM), Denver, CO, USA, 17–21 July 2022; IEEE Computer Society: Washington, DC, USA, 2022.
107. Yazdanipour, S.; Oruganti, V.S.R.V.; Son, J.; Arani, M.F.M.; Williamson, S.S. State-of-the-Art in Cyber-Physical Security for Dynamic Wireless Charging of Electric Vehicles. In Proceedings of the IECON 2024—50th Annual Conference of the IEEE Industrial Electronics Society, Chicago, IL, USA, 3–6 November 2024; IEEE Computer Society: Washington, DC, USA, 2024.
108. Parameswarath, R.P.; Gope, P.; Sikdar, B. A Privacy-Preserving Authenticated Key Exchange Protocol for V2G Communications Using SSI. *IEEE Trans. Veh. Technol.* **2023**, *72*, 14771–14786. [CrossRef]
109. Babu, P.R.; Reddy, A.G.; Palaniswamy, B.; Das, A.K. EV-PUF: Lightweight Security Protocol for Dynamic Charging System of Electric Vehicles Using Physical Unclonable Functions. *IEEE Trans. Netw. Sci. Eng.* **2022**, *9*, 3791–3807. [CrossRef]
110. Zhao, X.; Lin, J.; Li, H. Privacy-Preserving Billing Scheme against Free-Riders for Wireless Charging Electric Vehicles. *Mob. Inf. Syst.* **2017**, *2017*, 1–9. [CrossRef]
111. Hussain, R.; Son, J.; Kim, D.; Nogueira, M.; Oh, H.; Tokuta, A.O.; Seo, J. PBF: A New Privacy-Aware Billing Framework for Online Electric Vehicles with Bidirectional Auditability. *Wirel. Commun. Mob. Comput.* **2017**, *2017*. [CrossRef]
112. Małek, A.; Kośko, M.; Łusiak, T. Urban Logistics of Small Electric Vehicle Charged from a Photovoltaic Carport. *Arch. Automot. Eng. Arch. Motoryz.* **2018**, *82*, 63–75. [CrossRef]

113. Małek, A.; Taccani, R. Long-Term Test of an Electric Vehicle Charged from a Photovoltaic Carport. *Arch. Automot. Eng. Arch. Motoryz.* **2019**, *86*, 55–63. [CrossRef]
114. Li, M.; Xu, H.; Xu, J.; Ding, Z.; Liu, Q.; Yin, Z. Risk Assessment of Cyber Physical Power System Considering Attack Model. In Proceedings of the 2022 IEEE 5th International Electrical and Energy Conference (CIEEC), Nanjing, China, 27–29 May 2022.
115. Team, P.M. EV Cybersecurity & ISO 15118 for Secure Charging Infrastructure. PlaxidityX. 29 May 2025. Available online: https://plaxidityx.com/blog/blog-post/iso-15118-ev-cybersecurity-guide/ (accessed on 1 September 2025).
116. Driivz. Glossary: ISO 15118. Available online: https://driivz.com/glossary/iso-15118/ (accessed on 1 September 2025).
117. Zhou, M. Anomaly Detection in Smart Distribution Grids with Deep Neural Network. Master's Thesis, University of Alberta, Edmonton, AB, Canada, 2022.
118. Explainable Anomaly Detection for Electric Vehicles Charging Stations. Available online: https://arxiv.org/html/2507.15718v1 (accessed on 1 September 2025).
119. *Stress Testing by Large Financial Institutions: Current Practice and Aggregation Issues*; BIS: Basel, Switzerland, 2000; ISBN 978-92-9131-601-4.
120. European Central Bank. *Stress Testing with Multiple Scenarios: A Tale on Tails and Reverse Stress Scenarios*; Publications Office: Luxembourg, 2024.
121. Grid Operators: TSO and DSO Explained—gridX. Available online: https://www.gridx.ai/knowledge/what-is-a-grid-operator (accessed on 1 September 2025).
122. CORPORATIVA, I. DSO, ¿cómo Transformar La Gestión de Redes Hacia Un Modelo Más Inteligente? Available online: https://www.iberdrola.com/about-us/our-innovation-model/distribution-system-operation (accessed on 1 September 2025).
123. Complete Guide to EV Charging Protocols (OCPP, OICP, OHCP). Available online: https://solidstudio.io/blog/ev-charging-protocols (accessed on 1 September 2025).
124. Coppoletta, G.; Gjomemo, R.; Kaur, A.; Valizadeh, N.; Rana, O.; Venkatakrishnan, V. OCPPStorm: A Comprehensive Fuzzing Tool for OCPP Implementations. In Proceedings of the Symposium on Vehicle Security & Privacy, San Diego, CA, USA, 26 February 2024; Internet Society: San Diego, CA, USA, 2024.
125. How to Increase Charging Session Money Collection and EV Driver Satisfaction with AMPECO Data Validation Feature. Available online: https://www.ampeco.com/blog/ampeco-session-data-validation-feature/ (accessed on 1 September 2025).
126. Unlocking the Future: How UWB Is Reshaping Secure Access in the Connected World. Available online: https://www.qorvo.com/design-hub/blog/unlocking-the-future-how-uwb-is-reshaping-secure-access-in-the-connected-world (accessed on 1 September 2025).
127. Stevens, B. *Implementing Secure Boot with a Root of Trust*; Semiconductor Engineering: Silicon Valley, CA, USA, 2021.
128. Secure Boot—Silicon Labs. Available online: https://www.silabs.com/security/secure-boot (accessed on 1 September 2025).
129. Position Papers & Regulation—CharIN. Available online: https://www.charin.global/technology/knowledge-base/ (accessed on 1 September 2025).

energies

MDPI

Review

Advances in Energy Storage, AI Optimisation, and Cybersecurity for Electric Vehicle Grid Integration

Muhammed Cavus [1,2,3,*], Huseyin Ayan [4,5], Margaret Bell [4] and Dilum Dissanayake [6,*]

1 The Department of Engineering, Durham University, Durham DH1 3LE, UK
2 Department of Mathematics, Physics and Electrical Engineering, Northumbria University, Newcastle Upon Tyne NE1 8SA, UK
3 Department of Engineering, Iskenderun Technical University, Hatay 31200, Turkey
4 School of Engineering, Newcastle University, Newcastle Upon Tyne NE1 7RU, UK; h.ayan2@newcastle.ac.uk (H.A.); margaret.bell@ncl.ac.uk (M.B.)
5 School of Engineering, Istanbul University-Cerrahpasa, Istanbul 34320, Turkey
6 School of Geography, Earth and Environmental Sciences, University of Birmingham, Birmingham B15 2TT, UK
* Correspondence: muhammed.cavus@northumbria.ac.uk (M.C.); d.dissanayake@bham.ac.uk (D.D.)

Abstract

The integration of electric vehicles (EVs) into smart grids (SGs) is reshaping both energy systems and mobility infrastructures. This review presents a comprehensive and cross-disciplinary synthesis of current technologies, methodologies, and challenges associated with EV–SG interaction. Unlike prior reviews that address these aspects in isolation, this work uniquely connects three critical pillars: (i) the evolution of energy storage technologies, including lithium-ion, second-life, and hybrid systems; (ii) optimisation and predictive control techniques using artificial intelligence (AI) for real-time energy management and vehicle-to-grid (V2G) coordination; and (iii) cybersecurity risks and post-quantum solutions required to safeguard increasingly decentralised and data-intensive grid environments. The novelty of this review lies in its integrated perspective, highlighting how emerging innovations, such as federated AI models, blockchain-secured V2G transactions, digital twin simulations, and quantum-safe cryptography, are converging to overcome existing limitations in scalability, resilience, and interoperability. Furthermore, we identify underexplored research gaps, such as standardisation of bidirectional communication protocols, regulatory inertia in V2G market participation, and the lack of unified privacy-preserving data architectures. By mapping current advancements and outlining a strategic research roadmap, this article provides a forward-looking foundation for the development of secure, flexible, and grid-responsive EV ecosystems. The findings support policymakers, engineers, and researchers in advancing the technical and regulatory landscape necessary to scale EV–SG integration within sustainable smart cities.

Keywords: electric vehicles (EVs); smart grids; vehicle-to-grid (V2G); energy storage systems (ESSs); artificial intelligence (AI); cybersecurity; digital twins (DTs); wireless charging; renewable energy integration; post-quantum cryptography

1. Introduction

1.1. Background and Motivation

The rapid urbanisation of the 21st century has driven global efforts to develop smart cities, leveraging digital innovation, environmental sustainability, and integrated infrastructure to improve the quality of urban life. This evolution comes with a dual imperative:

to decarbonise transportation and to modernise power systems. Smart cities epitomise the convergence of these goals, embedding Electric Vehicles (EVs) within intelligent and resilient energy networks [1,2].

EVs have transitioned from niche alternatives to mainstream mobility solutions, largely fuelled by the urgent quest to achieve net-zero emissions and diminish reliance on fossil fuels. Recent years have witnessed exponential growth in EV adoption, with projections forecasting a significant increase in global EV stock by 2030 [3]. Concurrently, smart grids (SGs) enhanced by sensors, real-time data, and intelligent control offer dynamic frameworks capable of absorbing fluctuating energy demand, facilitating the integration of renewable energy, and improving resilience through features such as self-healing and fault management [4].

A profound shift occurs when EVs transcend their traditional role as mobile transportation units to become mobile energy storage devices. Through bidirectional charging technologies along with Grid-to-Vehicle (G2V) and Vehicle-to-Grid (V2G), EVs can absorb surplus energy during low-demand periods and discharge power back into the grid when needed [5]. This capability supports peak-load levelling, frequency regulation, and even participation in energy markets via demand response mechanisms [5,6].

Nevertheless, integrating EVs into SGs introduces complex challenges:

1. **Energy storage efficiency and longevity**: Battery degradation rates, charge/discharge cycles, and second-life applications present significant constraints on system sustainability and economic feasibility [7].
2. **Management and optimisation**: Advanced algorithms are required for tasks such as decentralised control, real-time scheduling, and predictive energy forecasting, with many approaches increasingly harnessing artificial intelligence (AI) and open-source energy models [8].
3. **Cybersecurity vulnerabilities**: As EV-grid architectures gain connectivity, they become susceptible to cyber–physical threats, including communication interception, malware injection, and denial-of-service attacks. Research indicates that even smart-charging stations and V2G systems may harbour critical security flaws [9–12].

The Biden Administration's cybersecurity roadmap underscores the urgency of securing EVs, charging infrastructure, and grid-support systems, highlighting a pressing need for "secure-by-design" mindsets [13].

1.2. Objectives of the Review

The primary aim of this review is to present a comprehensive and interdisciplinary evaluation of the integration of EVs into SGs, with a particular focus on technical advancements in energy storage, system-level optimisation, and cybersecurity. While existing studies often examine these dimensions in isolation, focusing either on EV infrastructure, grid operations, or data systems, this review uniquely synthesises them into a unified framework. Such integration is vital for understanding the operational dynamics, technological enablers, and systemic challenges of EV-grid interaction, especially in the context of evolving smart cities.

Specifically, this review sets out to:

- Critically examine the technological landscape of energy storage systems (ESSs) within EV–SG ecosystems, evaluating the advantages, limitations, and application contexts of lithium-ion, solid-state, and second-life batteries, as well as hybrid configurations involving supercapacitors.
- Analyse the role of AI, including machine learning (ML) and predictive analytics, in enhancing demand forecasting, charging optimisation, V2G scheduling, and real-time grid interaction.

- Assess emerging cybersecurity challenges, exploring how blockchain, cryptographic protocols, and quantum-safe algorithms can safeguard EV–grid communications, protect user data, and maintain system resilience.
- Investigate the integration of EVs with renewable energy sources (RESs), digital twins (DTs), wireless charging systems (WCSs), and other enabling technologies to improve operational efficiency, decentralisation, and infrastructure scalability.
- Identify and articulate a forward-looking research roadmap, highlighting unresolved technical and regulatory challenges, and calling for cross-disciplinary collaboration among engineers, data scientists, urban planners, and policymakers.

In addressing these objectives, this review aims to serve as a strategic reference for academic researchers, industry developers, and policy stakeholders seeking to advance intelligent and sustainable EV-grid integration.

To guide this effort, the following research questions are proposed:

1. *How can EVs be optimally integrated into SG infrastructures to enable bi-directional energy flows, dynamic load management, and decentralised energy exchange, while maximising interoperability and system resilience?*
2. *What are the comparative performance characteristics, deployment challenges, and lifecycle considerations of emerging energy storage technologies, including second-life and hybrid systems. within V2G frameworks?*
3. *Which AI and cybersecurity approaches offer the most robust, scalable, and secure solutions for EV–grid coordination, and how can these be deployed with minimal latency, strong privacy protection, and resilience against evolving threats?*
4. *How can enabling technologies, such as DTs, renewable energy integration, and wireless charging, along with regulatory and socio-technical strategies, be combined to accelerate large-scale, sustainable EV–SG adoption?*

By exploring these questions, the review not only surveys the state of the art but also maps a critical path for future research and policy development in EV–SG convergence.

1.3. Structure of the Paper

The structure of this paper has been carefully designed to provide a comprehensive and logical progression through the interdisciplinary dimensions of EV integration into smart electric grids. Following the introduction, Section 2 establishes the conceptual and technological foundations of EV–grid synergy by examining bidirectional energy flows, with a focus on G2V and V2G mechanisms. Section 3 delves into ESSs, exploring advanced battery technologies, supercapacitor integration, and the reuse of second-life EV batteries in stationary applications, while addressing issues such as energy density, degradation, lifecycle cost, and scalability. Section 4 presents energy management and optimisation strategies, covering both centralised and decentralised control frameworks, ML-based scheduling techniques designed to improve system responsiveness and reduce peak loads. Section 5 turns to cybersecurity, identifying major threats, including data breaches and communication hijacking, and evaluating countermeasures such as cryptographic protocols, secure communication standards, and blockchain applications. Section 6 surveys emerging technologies that are shaping the future of EV–SG systems, including AI-powered data transmission, renewable energy integration, wireless charging infrastructure, and quantum-resilient security techniques. Section 7 synthesises current limitations and highlights unresolved technical, regulatory, and economic challenges, while identifying key research gaps related to interoperability, standardisation, and policy development. Finally, Section 8 concludes the paper by summarising major insights and proposing a forward-looking research agenda that calls for interdisciplinary collaboration to realise the full potential of EV–grid integration within sustainable smart cities.

1.4. Methodology

To ensure the comprehensiveness and reproducibility of this review, a structured literature search was conducted across multiple scientific databases, including IEEE Xplore, ScienceDirect, Web of Science, Scopus, and MDPI. The search was carried out between *[August 2019]* and *[August 2025]*, covering publications from 2019 to 2025 to capture both foundational concepts and the latest advancements.

Search queries combined keywords and Boolean operators such as:

"electric vehicle" AND "smart grid",
"vehicle-to-grid" OR "V2G",
"energy storage" AND "second-life batteries",
"blockchain" AND "EV charging",
"cybersecurity" AND "smart grid",
"federated learning" AND "energy management".

Inclusion criteria:

- Peer-reviewed journal articles, conference papers, and authoritative reports.
- Works presenting original research, substantial reviews, or validated field demonstrations relevant to EV–SG integration.
- Studies in English with sufficient methodological detail.

Exclusion criteria:

- Non-peer-reviewed sources unless providing essential technical standards or policy documentation.
- Publications lacking clear methodology or relevance to at least one of the review's core themes (energy storage, optimisation, cybersecurity, enabling technologies).

The initial search yielded 1000 articles. Titles and abstracts were screened to remove duplicates and irrelevant works. Full-text reviews were then conducted to assess methodological quality and thematic relevance. Reference lists of key articles were also examined to identify additional pertinent studies through backward snowballing.

This structured approach ensures that the analysis presented is grounded in a balanced and representative body of literature, allowing replication and verification of the review's scope and depth.

2. Electric Vehicles and Smart Grid Synergy

The integration of EVs within SG systems represents a paradigm shift in energy management and distribution. The characteristics of EVs, their interactions with the grid, and the technological frameworks supporting these interactions are crucial for the successful deployment of future energy systems. This section outlines the various classifications of EVs, explains the fundamentals of SGs, and examines the bidirectional energy flows enabled by V2G technologies, facilitating a dynamic interaction that enhances grid stability and maximises energy utilisation.

Figure 1 illustrates the multi-layered architecture of EV integration within a SG ecosystem, highlighting the key technologies and interactions that underpin dynamic and secure energy management. At the user level, EV owners engage with intelligent charging services that are supported by AI-based optimisation and cybersecurity mechanisms, ensuring both operational efficiency and data protection. The communication and control layer coordinates grid interaction through an energy management system (EMS), cloud-based services, and blockchain platforms, enabling real-time data exchange and secure transaction management. This layer facilitates bidirectional power flow, such as G2V and V2G, allowing EVs to function not only as energy consumers but also as distributed storage assets contributing to grid stability. At the infrastructure level, RESs, ESSs, and the utility grid are integrated to enhance sustainability and flexibility. The layered synergy depicted in this figure exemplifies how advanced digital tools, decentralised architectures, and AI-driven optimisation converge to realise an intelligent, resilient, and user-centric EV–SG ecosystem.

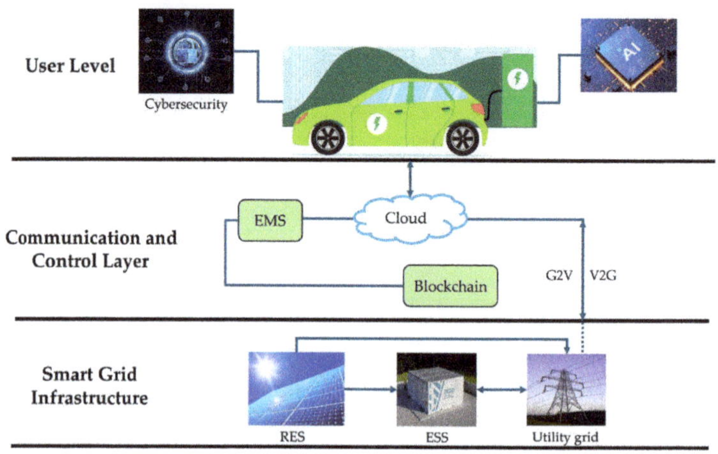

Figure 1. Multi-layered architecture for EV integration within AI-enabled and blockchain-secured SG ecosystems.

2.1. Classification of EVs

EVs can be classified into several distinct categories based on their power source and charging methodologies, including Battery Electric Vehicles (BEVs), Plug-in Hybrid Electric Vehicles (PHEVs), Hybrid Electric Vehicles (HEVs), and Fuel Cell Electric Vehicles (FCEVs). BEVs operate solely on electric power stored in rechargeable batteries, thus contributing to reduced greenhouse gas emissions compared to traditional combustion engine vehicles [14]. PHEVs combine conventional internal combustion engines with electric propulsion, offering flexibility in energy usage as they can switch between gasoline and electric power, allowing for extended driving ranges while meeting some emission reduction requirements [15].

HEVs use a combination of an internal combustion engine and an electric motor to optimise fuel efficiency, where the electric motor assists the engine, improving overall vehicle performance without the need for an external power source [16]. FCEVs utilise hydrogen as a fuel source, converting it into electricity through chemical reactions in fuel cells, thereby emitting only water vapour, showcasing an alternative avenue for zero-emission vehicles [17]. FCEVs harness hydrogen fuel cells, most commonly proton-exchange membrane fuel cells (PEMFCs), to generate electricity onboard, enabling rapid refuelling and extended driving range compared to battery-only vehicles. Recent developments have significantly advanced this technology. Notably, a 2025 study introduces an energy-management strategy that improves fuel cell efficiency by approximately 5.4% over existing approaches, highlighting gains in both operational performance and system durability. Nevertheless, FCEVs retain potential, particularly in niche use cases such as long-haul transport or heavy-duty applications, where higher energy density and fast refuelling may outweigh efficiency trade-offs. These developments suggest that FCEVs could serve as a complementary technology rather than a direct substitute for BEVs, contributing to diversified decarbonisation pathways in smart mobility systems [18,19]. This classification requires an understanding of the varying implications for electricity demand when integrated with SGs and the subsequent management of these loads.

The increasing penetration of EVs, particularly BEVs and PHEVs, influences the energy landscape; thus, policymakers are encouraged to analyse their potential impacts on energy supply, demand profiles, and infrastructure requirements. The synergy between various vehicle types significantly impacts how SGs should be developed to handle increased loads while optimising renewable energy integration [20].

2.2. Smart Grid Fundamentals

SGs comprise intelligent network architectures and functional layers that enable efficient electricity management between suppliers and consumers, particularly under fluctuating demand conditions. Essentially, a SG enhances the reliability of electricity distribution through advanced communication technologies and automation [21]. A typical SG architecture comprises several layers, including the physical layer, which consists of power infrastructure, the data layer for communication and information flow, and the application layer, where control and management systems reside [22].

The integration of physical and virtual components is paramount for enabling real-time monitoring and control of the electrical grid. Such capabilities enable improved efficiency in addressing dynamic loads generated by EVs and other renewable energy systems. For instance, vehicle-grid interaction relies heavily on efficient communication technologies that can transmit data regarding grid conditions and demand across all layers of operation, enabling the optimisation of resources [23]. The functional layers of SGs provide vital support for demand response initiatives, where EVs can be utilised as flexible resources to assist with load levelling during peak demands [24].

In particular, the architecture must adequately address the dual functionality of charging EVs and providing energy back to the grid. This requires advanced power management systems capable of ensuring energy quality and reliability while accommodating the high variability associated with RESs [25]. Such challenges necessitate integrating various technological approaches, including distributed energy generation, storage systems, and EV management strategies, to achieve operational effectiveness in SGs [26].

2.3. Vehicle-to-Grid and Grid-to-Vehicle Interactions

The paradigms of V2G and G2V interactions significantly transform the dynamics of energy management. V2G refers to the bidirectional flow of energy, where EVs cannot only draw power from the grid but also discharge energy back into it, effectively acting as distributed energy resources (DERs) [27]. This interaction enhances grid stability and peak load management. During periods of high electricity demand, idle vehicle batteries can provide additional power to balance supply and demand, facilitating energy reliability [28].

V2G systems enable utilities to manage energy loads and provide ancillary services, such as frequency regulation and voltage control, through leveraging EV batteries. It entails intelligent scheduling protocols that determine when and how much energy can be drawn or sent to the grid while considering vehicle usage patterns and local grid conditions [26]. These interactions help in achieving peak shaving, reducing electricity consumption during peak hours and valley filling, using excess energy during off-peak hours, thus optimising overall distribution network performance [29].

In contrast, G2V interactions involve charging infrastructure facilitating the process of energy flow to vehicles. In this scenario, smart charging methods adapt the vehicle charging process based on real-time network conditions and electricity pricing signals, ensuring that EV owners benefit financially while contributing to grid efficiency [30]. The implementation of time-of-use pricing can serve as a deterrent against charging during peak hours while encouraging users to recharge their vehicles when energy costs are lower, thereby aligning individual consumption with grid requirements [31]. V2G technologies, therefore, offer not just a technical solution for energy distribution but also an opportunity for societal engagement, where consumers actively participate in EMSs. By incentivising individuals to engage in these practices, EVs can serve as an integral component of a sustainable energy ecosystem, fostering the integration of renewable energy and the reliability of conventional power generation [32].

The financial implications of V2G systems are notable, as EV owners could receive compensation for their services to the grid. Additionally, the deployment of such technology can reduce levels of carbon emissions, thereby contributing to climate goals [33]. Effective communication protocols, data management systems, and regulatory frameworks will be essential for the widespread adoption of these technologies, ensuring that all stakeholders can participate optimally in the evolving energy landscape [30]. Considering the increasing electrification of transportation, it is crucial for energy systems to grow through innovative frameworks, such as SGs. The classification of EVs highlights the various interactions and demands placed upon the grid, necessitating robustness in structural design and operational strategies. Moreover, the bidirectional energy flows facilitated by V2G and G2V paradigms illustrate how consumer participation can significantly enhance grid stability, optimise energy utilisation, and advance sustainability initiatives. Future research must address the economic, environmental, and technical challenges of deploying these systems on a scale, ensuring a seamless integration into our energy systems.

3. Energy Storage Technologies in EV–SG Integration

As EV adoption accelerates, the significance of resilient and adaptable energy storage technologies within SGs becomes ever more critical. Storage systems not only buffer energy between variable RESs and grid demand but also enable bi-directional interaction with EVs themselves [20]. The following subsections examine the principal storage technologies, covering in-vehicle batteries, second-life reuse, hybridisation with supercapacitors, strategic deployment, and the challenges that persist.

3.1. Battery Technologies

Lithium-ion (Li-ion) batteries are currently the cornerstone of both EV propulsion systems and grid-connected energy storage due to their high gravimetric energy density, relatively efficient charge/discharge cycles, and improving cost trajectories. Within this domain, lithium iron phosphate chemistries have gained prominence in grid applications, offering enhanced safety at a lower cost, which is crucial for large-scale adoption [34,35].

Despite these advantages, Li-ion batteries face challenges, including thermal runaway risks, dependency on critical raw materials (such as cobalt), and inherent capacity fade. The pursuit of next-generation solid-state batteries, which replace organic electrolytes with solid ones, promises enhancements in safety and energy density; however, manufacturing complexity and cost remain significant barriers to widespread deployment [36,37].

An alternative is the deployment of flow batteries, such as vanadium redox systems, which offer decoupled power and energy scaling, extended cycle life, and minimal capacity fade. Their comparatively low energy density renders them better suited to stationary applications rather than vehicular use. Nevertheless, when aggregated with EV charging infrastructure, flow batteries can help balance grid load, reduce reliance on peak-power plants, and support long-duration storage needs [38,39].

Figure 2 presents a comparative assessment of three prominent battery technologies along with li-ion, solid-state, and flow batteries across four key performance parameters: energy density, cycle life, cost, and response time. Solid-state batteries exhibit superior energy density and extended cycle life, making them attractive for future EV applications, though they remain costly and technically challenging to commercialise. Li-ion batteries continue to offer a well-balanced profile, combining high response times and moderate cost with proven scalability and market readiness. In contrast, flow batteries, while offering promising cycle life and safety advantages, generally lag in energy density and are better suited to stationary grid applications where size and weight are less critical. This comparison highlights the importance of selecting battery chemistries based on specific operational contexts within the EV–SG ecosystem.

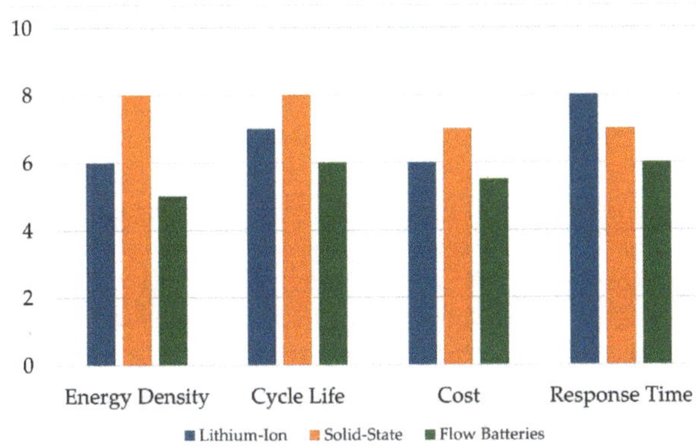

Figure 2. Comparative performance metrics of Li-ion, Solid-state, and Flow Batteries for EV–SG applications.

3.2. *Second-Life EV Batteries in Grids*

Once EV batteries degrade to approximately 70–80% of their original capacity, typically after 8–10 years, they may no longer serve as primary propulsion systems, but they retain substantial utility for stationary energy storage. This reuse, often termed *second-life*, is gaining traction due to its potential to reduce lifecycle costs and underpin circular economy principles [40]. Estimates suggest that by 2030, second-life EV batteries could contribute over 200 GWh to global stationary storage—an amount comparable to projected total utility-scale BESS demand [41]. Pilot projects, such as the reuse of retired Nissan Leaf batteries for street-lighting or community energy storage, illustrate both environmental benefits and extended lifespans of up to 12 years in non-vehicular settings [42].

However, technical challenges persist. Standardising performance metrics is challenging due to the diverse chemistries and usage histories [43]. Economic viability is also sensitive to testing protocols, warranty offerings, and degradation-aware management strategies. Moreover, ensuring reliability and safety in second-life applications requires sophisticated battery management systems (BMSs) to monitor health and manage cell-to-cell variability [44,45].

Beyond technical performance, the economic viability of second-life EV batteries warrants deeper scrutiny. Heterogeneous degradation profiles, influenced by varied driving behaviours, climate conditions, and charging practices, complicate the standardisation of testing protocols and the prediction of residual value. Remanufacturing costs, including disassembly, refurbishment, module reconfiguration, and quality assurance, can significantly reduce profit margins, particularly in smaller-scale deployments. In addition, implementing robust safety concepts, such as fire suppression systems, enhanced thermal management, and fault isolation, is essential to meet regulatory compliance and mitigate liability risks. A full lifecycle cost–benefit analysis, incorporating logistics, warranty structures, and end-of-life recycling, is therefore critical to establish credible business cases for large-scale second-life battery adoption in stationary storage markets.

3.3. Supercapacitors and Hybrid Storage Systems

Supercapacitors, also known as ultracapacitors, excel in power density and cycle life, offering rapid charge and discharge capabilities, as well as robust performance across a wide range of temperatures. Their strengths are particularly suitable for applications involving frequent load fluctuations or short bursts of energy, such as regenerative braking, voltage smoothing, and frequency regulation [46,47].

Figure 3 illustrates a typical hybrid energy storage system (HESS) architecture, designed to enhance the performance and flexibility of EV–grid interactions. In this configuration, the EV connects to a charging station that interfaces with both a battery and a supercapacitor through a bidirectional converter. The battery provides sustained energy storage, managing long-duration charging and discharging cycles, while the supercapacitors address high-power, short-duration demands such as regenerative braking or fast grid stabilisation events. A bidirectional inverter facilitates power exchange between the combined storage system and the utility grid, supporting both G2V and V2G operations. This hybrid arrangement reduces stress on the primary battery, extends its operational lifespan, and improves system efficiency by allowing the supercapacitors to handle rapid power fluctuations. The architecture shown here exemplifies how the integration of diverse storage technologies can optimise both grid services and EV performance.

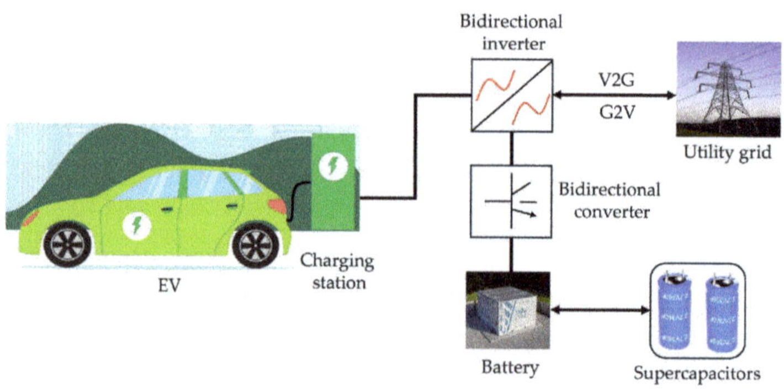

Figure 3. HESS architecture for EV–SG integration.

HESSs, which integrate both batteries and supercapacitors, are emerging as promising configurations for charging infrastructure and V2G deployment. Supercapacitors can absorb transient power surges, thereby reducing mechanical and thermal stress on batteries. Meanwhile, batteries handle sustained energy exchanges [48]. A recent design tool (BaSiS model) calibrates supercapacitor/battery combinations to optimise cost, ageing, and performance metrics. Demonstrations of fuzzy-logic controllers and bidirectional converter designs further emphasise the feasibility of hybrid architectures in real-world systems [49,50].

3.4. Energy Storage Sizing and Placement in Grids

Effective sizing and strategic placement of storage assets are key to operational efficiency and economic maximisation. Storage capacity and location must align with local demand profiles, renewable energy generation patterns, grid topology, and levels of EV penetration. Under-sized systems may fail to deliver necessary load-shifting benefits, while over-sizing results in underutilised investments [51,52].

To address these issues, optimisation models incorporate real-time data feeds, predictive analytics, and financial constraints. With the rise of distributed architectures, such as rooftop solar, local EV fleets, and microgrids, modelling must increasingly account for decentralised decision-making, dynamic pricing, and resilience against grid perturbations [38,53,54].

3.5. Challenges in Storage Deployment

Figure 4 presents a structured overview of the key barriers impacting the large-scale deployment of ESSs within EV–SG integration. These challenges are categorised into three principal domains: economic barriers, technical challenges, and environmental and sustainability concerns. Economic barriers include high upfront capital costs, expenses associated with power electronics and installation, and limited access to favourable financing or incentive schemes. Technical challenges encompass battery degradation over time, reduced capacity and safety, the need for accurate degradation modelling, and a lack of integration with advanced BMSs. Environmental and sustainability concerns focus on the dependence on critical raw materials, ethical issues related to resource extraction, limited reuse at end-of-life, and the current absence of closed-loop recycling systems. Together, these factors underscore the need for holistic approaches that address financial, technological, and ecological dimensions in order to scale storage deployment effectively.

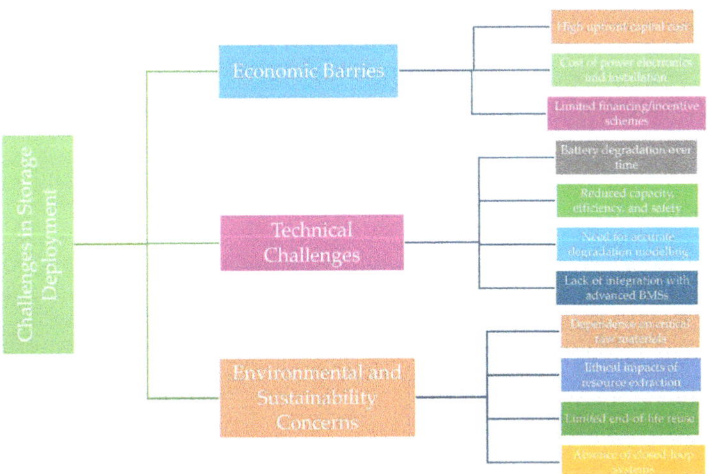

Figure 4. Hierarchical representation of key challenges in energy storage deployment for EV–SG integration.

Despite technological and economic progress, several barriers continue to hinder the full-scale deployment of energy storage. Upfront capital costs for batteries and associated infrastructure remain significant, although Li-ion prices have fallen by approximately 90% since 2010. Continued cost reduction is essential for equitable adoption, especially in emerging economies [55]. Another critical challenge is storage degradation, which compromises capacity, efficiency, and safety over time. Reliable modelling of ageing under variable charging patterns and integrating these models into BMS is key to predicting lifecycle performance and economic return [56]. Finally, issues relating to environmental sustainability, such as recyclability, critical mineral sourcing, and end-of-life protocols, must be addressed to align storage deployment with climate and resource policy objectives. The continued development of closed-loop recycling and certification frameworks will be indispensable in mitigating these concerns [57].

4. Management and Optimisation Techniques

The management and optimisation techniques relevant to smart charging strategies for EVs are critical in ensuring the efficiency and reliability of power systems, especially with the increasing penetration of renewable energy [58–60]. This exploration delves into smart charging strategies, scheduling and load forecasting utilising ML and DL, EMSs, multi-objective optimisation approaches, and comparisons between decentralised and centralised control structures, particularly focusing on edge computing and the Internet of Things (IoT), as shown in Table 1.

Table 1. Management and optimisation techniques for EV–SG Integration.

Category	Function	Key Technologies/ Methods	Challenges	Refs.
Smart Charging Strategies	Schedule charging to minimise grid impact and costs	Time-of-Use (ToU) pricing, dynamic tariffs, demand response	User participation, charger availability, regulatory consistency	[61]
Energy Forecasting	Predict EV demand, renewable availability, and grid load	ML, DL (LSTM, RNN), statistical models	Forecast accuracy, data sparsity, privacy concerns	[62]
Load Management & Peak Shaving	Predict EV demand, renewable availability, and grid load	Coordinated charging, peak clipping algorithms, and aggregation platforms	Real-time coordination, network latency, uneven EV distribution	[63,64]
V2G Scheduling	Optimise bidirectional energy exchange for grid support and user benefit	Reinforcement Learning (RL), Genetic Algorithms (GAs), Game Theory (GT)	Battery degradation, incentive design, energy market integration	[65,66]
EMS	Monitor and control power flows across distributed energy nodes	Model Predictive Control (MPC), rule-based logic, agent-based systems	Complexity of real-time control, scalability across multiple assets	[67]
Centralised Control	Top-down optimisation based on aggregated system data	Cloud-based servers, supervisory control algorithms	Communication delays, single point of failure, limited local adaptability	[62]
Decentralised/Distributed Control	Localised decision-making with minimal central intervention	Multi-agent systems (MASs), edge computing, and federated learning (FL)	Coordination consistency, protocol interoperability, and data privacy	[68]
Charging Station Optimisation	Maximise utilisation, minimise queue times, and integrate with grid signals	Queuing theory, priority-based scheduling, and real-time data analytics	Variability in arrival times, limited grid feedback, and infrastructure limitations	[69]

4.1. Smart Charging Strategies

Figure 5 illustrates the architecture of a smart charging system designed to optimise EV charging behaviour in response to multiple dynamic inputs. The smart charger serves as the central control element, processing real-time electricity pricing, user preferences, and prevailing utility grid conditions to generate an adaptive charging schedule. This system enables demand-responsive charging, allowing EVs to draw power at times that are both economically advantageous for users and operationally beneficial for the grid. By integrating wireless communication, the smart charger continuously updates its control logic, responding to fluctuations in energy markets and grid load. The adaptive approach depicted in this figure supports grid stability, enhances the utilisation of renewable energy, and delivers cost savings to end-users while improving the overall efficiency of EV–grid interactions.

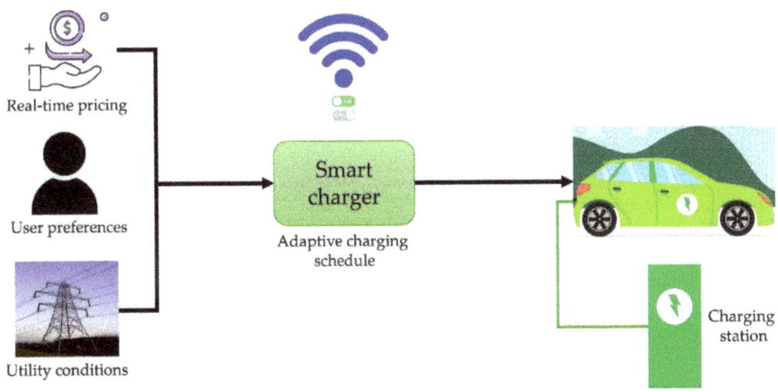

Figure 5. Architecture of a smart charging system with adaptive scheduling for EV-SG interaction.

Smart charging strategies aim to optimise EV charging by adapting to real-time electricity prices and grid conditions. ToU pricing is one approach where electricity prices vary at different times to encourage users to charge during off-peak hours, thereby flattening the load curve [70]. Dynamic pricing strategies can further optimise charging by adjusting prices based on real-time demand and supply conditions. For instance, algorithms that incorporate ML techniques, such as RL, can develop dynamic demand-aware pricing models that effectively allocate charging loads according to fluctuating energy prices and user demand [71,72]. These strategies promote efficient energy use while facilitating the integration of RESs [73].

Additionally, real-time charging algorithms propose methods that enable EVs to be charged at different rates depending on price signals from the grid. This not only incentivises users to alter their charging behaviours but also aligns with energy suppliers' strategies in managing peak demand periods [70]. Various models demonstrate that real-time pricing can lead to increased customer savings, improved grid performance, and enhanced utilisation of energy resources [74].

4.2. Scheduling and Load Forecasting

Effective scheduling and load forecasting are pivotal in maintaining grid stability and efficiency. ML and DL models have emerged as powerful tools, with techniques such as regression models, time-series analysis, and reinforcement learning being employed to predict EV charging loads and grid demand [75,76]. Advanced models leverage historical data, coupled with algorithmic learning processes, to understand and accurately anticipate power requirements. For instance, employing LSTM networks has shown proficiency in forecasting short-term demand fluctuations, enabling pre-emptive adjustments to energy supply strategies [76].

Hybrid approaches that integrate time-series analysis and ML provide robust forecasts that account for seasonal patterns and sudden demand shifts, which are increasingly common due to the proliferation of EVs and intermittent RESs [75]. These predictive models contribute significantly to optimising EMS, allowing for better resource allocation and operational efficiencies.

Figure 6 illustrates the role of advanced data-driven techniques in predicting EV charging loads to support intelligent grid scheduling and resource management. Various modelling approaches, including ML, regression models, DL architectures such as LSTM and CNN, and hybrid techniques, contribute to accurately forecasting EV charging demand. These predictions enable grid operators to anticipate power requirements, optimise resource allocation, and enhance operational efficiencies. By leveraging historical data and real-time inputs, these forecasting tools enable EMSs to maintain grid stability, support DR programmes, and better integrate RESs into EV charging strategies.

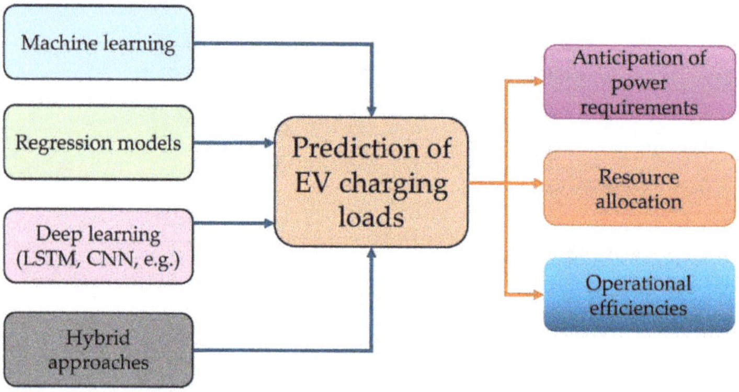

Figure 6. ML-based prediction of EV charging loads for enhanced SG resource optimisation.

4.3. Power Flow and Energy Management Systems

EMSs form the backbone of modern power systems, incorporating intelligent control mechanisms to facilitate the integration of renewable energy, energy storage, and demand-side management (DSM) programs. Optimal power flow (OPF) analysis is crucial in EMS, allowing for the real-time assessment of energy distribution and ensuring system reliability while minimising operational costs [3,6]. Algorithms utilising GAs and particle swarm optimisation (PSO) have proven effective in optimising scheduling in SGs, enhancing both responsiveness and efficiency in energy distribution [8]. By deploying real-time data from IoT devices, EMS can more effectively balance supply and demand while responding to variable costs associated with energy procurement. This intelligent management also incorporates predictive analytics to simulate and analyse power flow scenarios, enabling proactive decision-making during peak loads or emergencies [3,6].

Figure 7 presents a control system block diagram that illustrates how an EMS coordinates various functional components within the EV–SG ecosystem. At the core of the system, the EMS integrates data from IoT devices, DSM programmes, RESs, and ESSs. IoT devices, such as sensors, controllers, and gateways, provide real-time system monitoring and feed dynamic data back into the EMS. Predictive analytics and optimisation algorithms enable the EMS to perform OPF analysis, ensuring efficient distribution of energy resources. DSM strategies, as illustrated through various load shaping techniques, further enhance grid flexibility by modulating demand patterns in response to operational requirements. The continuous data exchange between these subsystems enables adaptive, intelligent control of power flows, thereby improving grid stability, enhancing the integration of RESs, and facilitating cost-effective EV charging strategies.

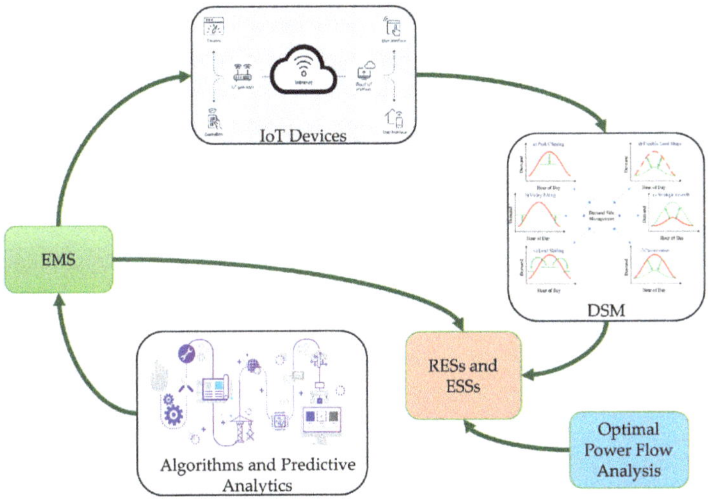

Figure 7. Integrated control system architecture for optimised power flow in EV–SG ecosystems.

4.4. Multi-Objective Optimisation Approaches

Multi-objective optimisation in energy management considers various competing objectives such as cost minimisation, emission reduction, and grid reliability enhancement. Optimisation models are developed to balance these goals while ensuring efficient resource utilisation. Approaches integrating environmental impacts along with economic objectives have been widely researched [77]. The utilisation of algorithms such as NSGA-II (Non-dominated Sorting Genetic Algorithm II) has been highlighted for multi-objective problems, offering robust solutions that respect all objectives simultaneously [78].

Additionally, real-time data analytics paired with advanced optimisation techniques can dynamically adjust operations to align with fluctuating market environments and regulatory standards, thereby reinforcing grid stability and sustainability [78]. Such interdisciplinary approaches promote a deeper understanding of the various factors that impact system performance, extending beyond mere economic considerations, and lead to more holistic energy strategies.

4.5. Decentralised vs. Centralised Control

Figure 8 compares the characteristics of decentralised and centralised control models within EV–SG systems. In decentralised control architectures, edge computing enables localised decision-making at distributed nodes, such as individual EV chargers, microgrids, or home EMSs. This enhances system responsiveness, resilience, and flexibility, particularly in environments with high penetration of DERs. In contrast, centralised control architectures rely on unified oversight, typically managed by a central authority or control centre, which optimises resource allocation across the entire grid. While this approach can ensure coordinated management and system-wide optimisation under stable conditions, it may face limitations in terms of adaptability and scalability in dynamic or highly decentralised networks. The comparative visualisation in this figure highlights the complementary advantages of each paradigm, suggesting that future SGs may benefit from hybrid architectures that leverage both centralised coordination and decentralised intelligence.

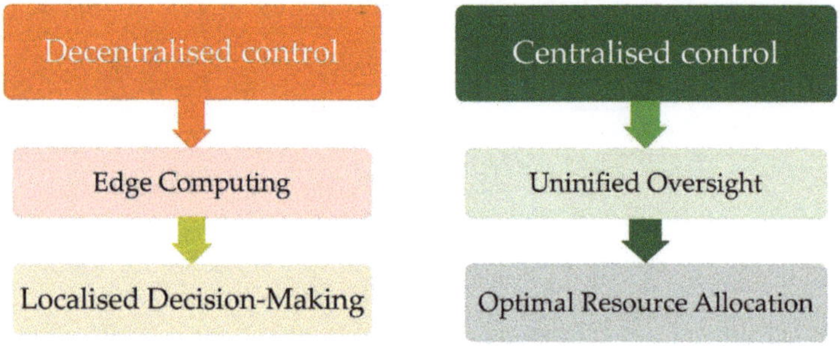

Figure 8. Comparative framework of centralised and decentralised control models in EV–SG systems.

Control mechanisms in energy systems can be broadly categorised into decentralised and centralised models. Decentralised control, enhanced by edge computing, employs localised decision-making at resource nodes, facilitating real-time responses to grid conditions and user demands [79,80]. This model supports the integration of DERs and enhances grid resilience by ensuring that local generation can autonomously meet local demand during disturbances or outages.

In contrast, centralised control systems rely on a central authority to manage power flow and enforce policies across the grid. While providing benefits such as unified oversight and potentially optimal resource allocation under stable conditions, centralised systems may struggle with dynamic adaptability in rapidly changing environments, particularly when RESs are interconnected [79]. The advent of IoT technologies introduces significant advantages for decentralised systems by enabling efficient data collection and processing at the edge, which has a substantial impact on charging strategies and DR initiatives [78,81].

Through analysing these dynamics, it becomes evident that both control paradigms have distinct advantages depending on specific situations. Future developments in control architectures should aim to leverage hybrid solutions that combine the strengths of both models, ensuring performance reliability and flexibility in the energy management landscape. In conclusion, optimising charging strategies, leveraging ML for scheduling and forecasting, employing advanced EMSs, balancing multi-objective requirements, and understanding the implications of control paradigms all play crucial roles in modern energy systems. Continuous research and technological advancements in these areas are crucial for developing more sustainable, resilient, and economically viable systems as we transition to a low-carbon future.

5. Cybersecurity Challenges and Solutions

The integration of EVs with SGs creates a sophisticated cyber–physical system that combines mobility, energy, and digital connectivity. While this synergy enables intelligent energy management and improved efficiency, it also significantly enlarges the system's attack surface [82–86]. The digital interconnectivity of EVs, charging infrastructure, and backend control systems introduces complex cybersecurity challenges that must be addressed to safeguard grid stability, user privacy, and operational continuity, as illustrated in Table 2. This section examines the primary threat vectors, outlines the foundational security principles, and explores technological and regulatory solutions to address these threats.

Table 2. Cybersecurity challenges and solutions in EV–SG integration.

Cybersecurity Domain	Threat/ Vulnerability	Impact/Risk	Mitigation Strategies	Enabling Technologies	Ref.
Communication Security	Data interception, man-in-the-middle (MITM) attacks	Tampering with control signals, unauthorised access to the charging infrastructure	End-to-end encryption, secure communication protocols (TLS, VPN)	Public Key Infrastructure (PKI), TLS, VPN	[87]
Authentication and Access Control	Weak identity verification, unauthorised device access	Impersonation of users or charging points, false energy transactions	Multi-factor authentication, digital certificates, and role-based access	Blockchain, Digital Signatures, ISO/IEC 15118	[88]
Firmware and Software Integrity	Malware injection, unauthorised updates	Malicious firmware leading to control override or device hijacking	Secure boot, firmware signing, and over-the-air update validation	Cryptographic hash functions, secure hardware modules	[89]
Data Privacy	Exposure of personal user data and vehicle behaviour	Violation of user privacy, profiling, and data misuse	Data anonymisation, differential privacy, access control	SMPC, Differential Privacy, Trusted Execution Environments (TEEs)	[90]
Energy Fraud and Tampering	Fake energy transactions, manipulated billing	Financial losses, imbalance in energy accounting	Smart contracts, immutable logging, anomaly detection	Blockchain, AI-based anomaly detection	[91]
System Availability and Resilience	Denial-of-Service (DoS) attacks on charging or control systems	Service disruption, operational instability	Redundancy planning, intrusion detection systems (IDS), rate limiting	IDS, AI/ML intrusion detection, edge security gateways	[92]
Post-Quantum Threats	Future attacks using quantum decryption	Breakage of conventional encryption (RSA, ECC), long-term data exposure	Migration to quantum-safe cryptographic algorithms	Lattice-based encryption, Quantum Key Distribution (QKD)	[93]

5.1. Threat Vectors in EV–Grid Ecosystem

The EV–SG ecosystem is vulnerable to a broad spectrum of cyber threats, encompassing physical, software-based, and human-centric attacks. Key risks include data breaches, malware infiltration, DoS attacks, and communication hijacking. Notably, public charging infrastructure has already been exploited. Instances of firmware tampering have resulted in the display of offensive content and temporary system failures, revealing weaknesses in physical and software security [92,93]. Moreover, power-line communication (PLC) channels used for real-time control in fast-charging networks have been hijacked via "man-in-the-middle" attacks, allowing unauthorised modification of authentication credentials or charge commands. Sophisticated false-data injection attacks also threaten grid stability by manipulating sensing data, disrupting load forecasting, and interfering with frequency control mechanisms. These threats underscore the urgent need for a security-first design across the entire stack, from onboard EV units and charging points to cloud-based energy management platforms [89,90].

To systematically assess cybersecurity risks, an EV–SG threat model can be employed to classify attacks according to likelihood, impact, and attack surface. Common attack surfaces include:

- EV firmware updates, which may be exploited to inject malicious code or override control logic.
- Communication protocols, susceptible to interception, spoofing, or replay attacks.
- User data repositories, which could be targeted for location tracking, charging behaviour profiling, or identity theft.

High-likelihood, high-impact threats, such as compromised firmware or protocol-level manipulation, demand stringent, multi-layered defences, including cryptographic signing, secure boot mechanisms, and intrusion detection. Lower-likelihood but high-impact attacks, such as large-scale data breaches, necessitate strong privacy safeguards and regulatory compliance. Using such a threat model enables prioritisation of mitigation measures according to both risk severity and technical feasibility.

Figure 9 provides an overview of the diverse cyber threat landscape facing EV–SG ecosystems. As illustrated, multiple attack vectors target key components across the communication network that links EVs, charging stations, cloud-based EMSs, and utility grids. Notable threats include firmware injection at both the EV and charger level, man-in-the-middle attacks during data transmission, communication hijacking, malware infiltration, and false data injection, which can disrupt load forecasts and grid control signals. DoS attacks may impact cloud EMS and utility grid interfaces, while data breaches remain a significant risk across user-facing services. The figure highlights the critical importance of designing cybersecurity frameworks that address vulnerabilities spanning the entire cyber–physical architecture of EV–SG systems.

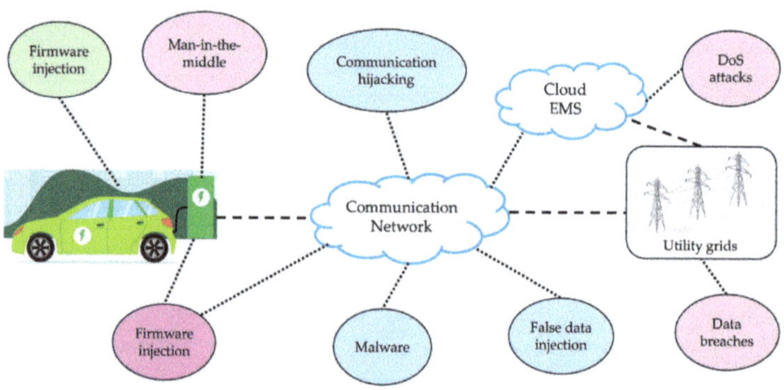

Figure 9. EV–SG Cyber Threat Landscape across Communication and Control Networks.

5.2. Security Requirements

Addressing the cybersecurity vulnerabilities of EV–SG systems requires strict adherence to the confidentiality, integrity, and availability (CIA) triad—core principles in information security. Confidentiality ensures that sensitive user and operational data remain protected from unauthorised access. Integrity safeguards against manipulation of control signals and stored data, while availability ensures that charging services and grid operations remain reliable even during active cyber threats. To implement these principles, cybersecurity frameworks such as NIST IR 7628 recommend a "defence-in-depth" strategy. This includes network segmentation, encrypted communication protocols, access control policies, IDSs, and secure firmware management [84]. Real-world assessments by institutions such as Argonne National Laboratory have revealed common vulnerabilities in commercial EV chargers, including unsecured interfaces and the absence of encrypted data channels. These findings underscore the importance of integrating secure software update mechanisms and compliance verification into the system's lifecycle [10,90,94].

5.3. Role of Blockchain and Cryptographic Techniques

Blockchain provides a decentralised and tamper-resistant platform for managing transactions and interactions within the EV–SG environment. By using a distributed ledger, blockchain ensures non-repudiation, immutability, and data provenance, which are vital for applications such as peer-to-peer energy trading, decentralised billing, and energy sharing among EV fleets [85]. Smart contracts built on blockchain automate energy transactions based on predefined rules—e.g., disbursing payments when an EV supplies power to the grid under V2G protocols. Cryptographic tools such as elliptic curve cryptography (ECC) and zero-knowledge proofs enhance transaction security and user anonymity. These are well-suited for constrained devices, along with EV controllers and smart meters, due to their lightweight computational requirements [92,93].

However, practical adoption of blockchain faces barriers, including scalability limitations, transaction delays, and interoperability issues with existing grid management systems [95]. Furthermore, concerns over data privacy compliance (e.g., GDPR) and the environmental impact of certain consensus algorithms (such as proof-of-work) have led to increased interest in energy-efficient alternatives, including proof-of-authority or DAG-based ledgers. Despite these challenges, blockchain remains a promising tool for building trusted and autonomous EV–SG platforms [91].

Figure 10 depicts a blockchain-based framework for securing V2G energy transactions. In this architecture, EVs interact with utility grids through energy exchanges governed by cryptographic techniques and smart contracts. Blockchain technology ensures that all energy transactions are recorded on a distributed ledger, providing transparency, immutability, and non-repudiation. Smart contracts automate the terms of energy exchanges, such as dynamic pricing or grid balancing, enabling seamless V2G integration without reliance on centralised intermediaries. Cryptographic safeguards, applied throughout the process, maintain data integrity and user privacy. By embedding blockchain into V2G operations, this model enhances trust and security within EV-SG ecosystems, supporting scalable and decentralised energy markets.

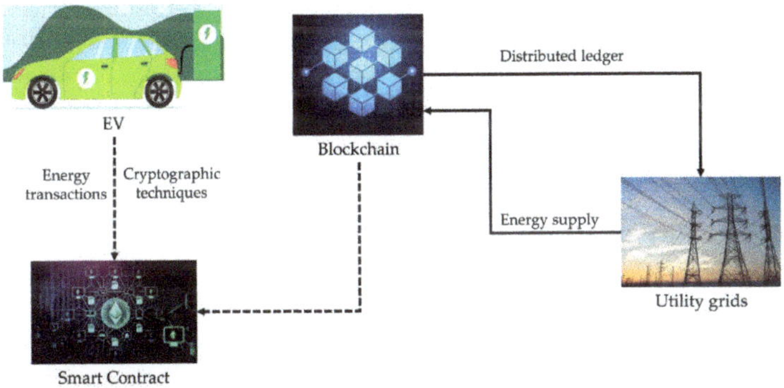

Figure 10. Blockchain-Enabled Secure Transactions between EVs and Utility Grids.

While blockchain offers significant advantages in transparency, immutability, and decentralised trust, its deployment in V2G ecosystems must be critically evaluated against practical constraints. Certain consensus mechanisms, particularly proof-of-work, can incur substantial energy overhead, which may undermine the environmental gains associated with EV–grid integration. Moreover, transaction latency, the time taken to validate and record energy exchanges, can limit the responsiveness of real-time grid balancing, particularly in high-frequency ancillary service applications. From a governance perspective, the regulatory acceptance of blockchain-based energy trading remains uneven across jurisdictions, with questions over legal enforceability of smart contracts, cross-border transaction compliance, and data protection requirements. Differentiating between conceptual innovations, which demonstrate strong theoretical potential, and solutions with demonstrated field viability is therefore essential for stakeholders. Pilot projects and real-world deployments should be prioritised to validate scalability, interoperability, and compliance under operational conditions.

Blockchain offers clear benefits for V2G applications, including tamper-resistant transaction logging, decentralised trust management, and automated energy exchange via smart contracts. However, these advantages must be weighed against practical limitations. Depending on the consensus algorithm, blockchain networks can introduce energy overhead and transaction latency, which may be incompatible with the sub-second response times needed for real-time grid balancing. Regulatory acceptance remains inconsistent, with uncertainties over legal enforceability of blockchain-mediated transactions and cross-border compliance. Furthermore, scalability and interoperability challenges persist, especially when integrating blockchain with existing grid management systems. A balanced approach involves differentiating between conceptual designs with strong theoretical promise and those with demonstrated field viability, ensuring that adoption decisions are informed by operational evidence rather than purely by novelty.

5.4. *Privacy-Preserving Data Analytics*

As EVs become increasingly connected, they generate rich datasets containing personal and behavioural information, including location histories, charge times, and payment details. This creates potential for misuse if data privacy is not rigorously protected. Thus, privacy-preserving data analytics has emerged as a priority in EV–SG research and implementation [93]. Trusted Execution Environments (TEEs), such as Intel SGX or ARM TrustZone, offer hardware-based isolation for sensitive computations, safeguarding data even in the presence of compromised host systems [90]. Techniques such as Secure Multi-Party Computation (SMPC) enable collaborative analytics, such as optimising fleet charging or grid balancing, without revealing private input data among participating entities [92].

Differential privacy, by injecting controlled noise into datasets, further protects user identities while allowing accurate system-level insights. It is increasingly being used in demand response analytics, usage pattern modelling, and decentralised load forecasting [93]. Nevertheless, these methods must be optimised to balance analytical accuracy, processing latency, and regulatory compliance, particularly under data protection laws including the GDPR in Europe or the CCPA in the United States [91].

5.5. *Regulatory Frameworks and Standards*

Effective cybersecurity in the EV–grid ecosystem depends not only on technology but also on robust governance and regulatory compliance. Frameworks such as NIST IR 7628, the North American Electric Reliability Corporation Critical Infrastructure Protection (NERC CIP) standards, and the IEC 62443 series provide comprehensive guidelines for protecting SG infrastructure [84]. The Open Charge Point Protocol (OCPP) 2.0, widely adopted in Europe, includes key provisions for secure communication, certificate management, and over-the-air firmware updates. However, many older systems still operate on outdated versions lacking these features, exposing them to well-documented vulnerabilities [94].

National security policies are also evolving. In the UK, government advisories have cautioned against the use of certain imported EV brands, citing risks associated with embedded surveillance tools [10]. Meanwhile, in the USA, the Biden Administration's 2023 Clean Energy Cybersecurity Strategy mandates secure-by-design requirements for EVSE (Electric Vehicle Supply Equipment), including encryption, access control, and system monitoring. Still, global harmonisation of cybersecurity standards remains incomplete, and real-world enforcement varies widely by region and manufacturer [96].

6. Emerging Technologies and Trends

The EV–SG ecosystem is advancing rapidly, shaped by a host of emerging technologies that promise to improve system intelligence, interoperability, resilience, and sustainability. These innovations are not simply incremental upgrades but signal a paradigm shift in how mobility and energy networks co-evolve. From AI to quantum-safe cybersecurity, the following trends represent key enablers of the next-generation SG infrastructure, as demonstrated in Table 3.

Table 3. Emerging technologies and trends in EV–SG integration.

Technology/Trend	Primary Application	Key Advantages	Associated Challenges	Refs.
AI	Predictive analytics, load forecasting, V2G optimisation, anomaly detection	Real-time decision-making, adaptive control, predictive maintenance	Model explainability, data privacy, and training data quality	[97]
Renewable Energy Integration	Coordinated EV charging with solar/wind generation	Reduced curtailment, enhanced grid flexibility, local energy self-use	Intermittency of supply, synchronisation complexity, and forecast accuracy	[98]
DTs	Real-time simulation of EV-SG components and control validation	Predictive diagnostics, scenario testing, cybersecurity simulation	High data requirements, model complexity, and integration with physical systems	[99]
WPT	Static and dynamic inductive EV charging	Contactless convenience, reduced infrastructure wear, and autonomy-friendly	High cost, electromagnetic safety, and lack of standardisation	[100]
Dynamic Inductive Roads	On-the-move EV charging via embedded coils in road infrastructure	Smaller battery requirements, continuous energy access, and range extension	Infrastructure costs, regulatory approvals, and urban planning constraints	[101]
Quantum-Safe Cryptography	Securing EV–SG communications against quantum-era attacks	Long-term security, futureproofing, and regulatory resilience	Computational overhead, standardisation in progress, limited hardware support	[102]
Blockchain Applications	Secure transaction logging, V2G payments, and access control	Transparency, tamper-resistance, and decentralised trust models	Scalability, energy usage (depending on consensus), and GDPR compliance	[103]
Edge Computing	Localised decision-making at charging points and EV control units	Low latency, reduced cloud dependency, better bandwidth use	Security at the edge, processing limitations, integration with centralised EMS	[104]
Federated Learning	Distributed AI model training across charging networks	Privacy-preserving intelligence, no raw data transfer	Model drift, synchronisation issues, computational load on local devices	[104,105]

6.1. Artificial Intelligence and Predictive Analytics

AI is playing an increasingly precise role in enhancing the efficiency, reliability, and autonomy of EV–SG integration. Rather than acting as a broad management layer, AI is being tailored to specific tasks that require high responsiveness and predictive accuracy. One of the most impactful applications is short-term load forecasting, where ML algorithms, such as LSTM networks, predict the timing, location, and volume of EV charging demand. These forecasts enable dynamic grid reconfiguration and pre-emptive energy dispatch to avoid overloads, and voltage drops in distribution networks [106,107].

In V2G operations, deep RLL models are being trained to determine optimal charge–discharge schedules for fleets of EVs, balancing grid needs against economic incentives. These models adapt to real-time electricity prices, demand signals, and battery health constraints, making them particularly suitable for frequency regulation and ancillary services markets. AI is also driving innovation in EV infrastructure diagnostics. Predictive maintenance models, built on anomaly detection algorithms and temporal pattern recognition, can identify subtle degradations in charger performance or battery thermal stability, minimising downtime and extending asset lifespans [107,108].

Moreover, AI-enhanced cyber threat detection systems are being embedded into EVSE and energy management software. These systems utilise unsupervised learning to flag irregular communication patterns or sensor readings, providing early warnings against malware, spoofing, or false-data injection. To ensure safe deployment, there is a growing emphasis on explainable AI (XAI) frameworks. These tools provide interpretable decision pathways, which are crucial in regulated environments where model accountability and traceability are required [108].

While federated learning (FL) offers privacy-preserving model training by keeping raw data local, its practical deployment in EV–SG contexts face notable limitations. Communication overhead, resulting from frequent parameter exchanges between participating nodes, can strain network resources, particularly when bandwidth is constrained. FL is also vulnerable to poisoning attacks, where compromised nodes inject malicious updates, degrading global model accuracy or biasing system behaviour. Furthermore, heterogeneity across nodes, in terms of data distribution, processing capability, and connectivity, complicates model aggregation, potentially slowing convergence and reducing performance consistency. These challenges necessitate the integration of secure aggregation protocols, robust anomaly detection in parameter updates, and adaptive synchronisation schemes to ensure FL viability in large-scale, decentralised EV–SG networks.

6.2. Integration with Renewable Energy Sources

The convergence of EVs with renewable energy systems is not only environmentally strategic but also operationally synergistic. EVs offer dynamic storage capabilities that can be co-optimised with intermittent energy sources, particularly solar photovoltaics (PV) and wind power, to enhance grid flexibility and decarbonisation. One of the most promising applications lies in co-located solar-EV systems, where smart chargers are synchronised with rooftop or community solar output. These systems utilise real-time solar irradiance forecasts and inverter telemetry to modulate charging rates, ensuring that EVs draw power when solar generation is highest. This PV-following charging strategy reduces both grid dependency and energy export losses [109,110]. EVs also support renewable load balancing through vehicle-to-home (V2H) and vehicle-to-building (V2B) schemes. During periods of low electricity generation or grid outages, EVs can supply stored energy to critical building loads. This is being actively trialled in residential settings using bidirectional inverters with islanding capabilities, enabling energy continuity without the need for diesel generators [20].

In wind-integrated systems, EV fleets provide ramping support by absorbing sudden generation spikes, particularly during off-peak hours. In some deployments, day-ahead wind generation forecasts are fed into fleet scheduling algorithms to align charging sessions with availability profiles, thus reducing the need for curtailment. Furthermore, in remote microgrids or islanded systems, EVs are being piloted as stabilisation assets. By working in conjunction with BESSs and diesel offset controllers, EVs help maintain frequency and voltage levels within permissible bounds. At the systems level, integration is increasingly coordinated through multi-agent EMSs, which use decision-theoretic algorithms to balance renewable output, EV charging flexibility, and user driving requirements. These agents operate within predefined operational boundaries to ensure grid stability while prioritising renewable energy consumption [98,111].

6.3. Digital Twins and Simulation Models

DT technology provides a real-time, data-driven virtual representation of physical assets and systems, enabling enhanced observability, control, and predictive insight across EV–SG infrastructures. By continuously syncing sensor data from physical components, such as EV batteries, inverters, charging stations, and distribution transformers, DTs enable stakeholders to assess performance degradation, predict component failure, and dynamically optimise energy scheduling. The integration of DTs with simulation tools, including OpenDSS for distribution networks, MATLAB/Simulink for system dynamics, and GridLAB-D for SG applications, enables multi-domain co-simulation. These hybrid models simulate the interactions of electrical, thermal, mechanical, and even economic systems across distributed energy systems, supporting the validation of AI control algorithms, the coordination of DERs, and stress testing under rare grid events or cyberattack scenarios [112].

A particularly emerging use case is the development of cyber-physical security twins, which simulate the effects of cyber intrusions such as data spoofing or control hijacking on grid operations. This enables operators to rehearse mitigation strategies, verify cyber defence protocols, and maintain operational continuity. As SGs become more decentralised and reliant on IoT, the role of DTs is expanding into edge environments, where real-time data ingestion, processing, and feedback are performed locally to reduce latency and improve system responsiveness. When combined with predictive maintenance, demand forecasting, and V2G coordination, DTs serve as essential enablers of next-generation, self-aware energy systems [113].

A notable example is the *Flemish Living Lab Electric Vehicles* project in Belgium, where DT models were integrated with real-time charging data from over 200 public charging points. This allowed operators to forecast load peaks with a mean absolute percentage error (MAPE) below 5% and proactively adjust V2G schedules to smooth distribution network loads. Similarly, a DT implementation in the *Auckland City microgrid pilot* enabled predictive maintenance of 50 EV chargers, reducing unplanned outages by 27% compared to baseline operation. These results demonstrate the potential of DTs not only for scenario analysis but for continuous optimisation of operational assets in EV–SG ecosystems.

6.4. Wireless Charging and Inductive Roads

Wireless power transfer (WPT) represents a significant advancement in EV charging technology, addressing key barriers associated with physical connectors, human intervention, and urban congestion. Using inductive coupling, energy is transmitted from embedded coils in the ground to a receiver coil in the vehicle, enabling a fully contactless and automated charging experience. Static wireless charging is already being deployed in urban taxi hubs, airport shuttles, and fleet depots, where vehicles stop periodically and can benefit from consistent top-up charging. This approach reduces maintenance costs and enhances the feasibility of fully autonomous EV fleets, where plug-in options are impractical [114].

The more revolutionary concept, dynamic inductive charging, allows vehicles to charge while driving on electrified lanes. This could significantly downsize battery capacity requirements, reduce vehicle weight, and improve EV affordability. Early-stage demonstrations in Sweden (eRoadArlanda), Germany (eCharge), and South Korea (OLEV) are testing the technical viability, power transfer efficiency, and user acceptance of such systems in diverse traffic and environmental conditions. However, WPT systems face non-trivial engineering and regulatory challenges: standardising coil design and frequency ranges, mitigating electromagnetic interference (EMI), ensuring road surface durability, and achieving economic scalability. Addressing these issues will be crucial before large-scale deployment becomes viable, particularly for long-haul logistics and public transportation networks [114]. The eRoadArlanda pilot in Sweden demonstrated dynamic inductive charging for heavy trucks over a 2 km electrified lane, achieving an average transfer efficiency of 85–90% at 60 km/h. In South Korea's OLEV bus trials, continuous in-motion charging extended battery lifespan by reducing deep discharge cycles, while enabling a 40% reduction in on-board battery capacity. These studies highlight WPT's potential to reshape EV design and charging infrastructure, although they also underscore the importance of standardisation and electromagnetic compatibility assessments before scaling to urban road networks.

Feasibility Assessment: While pilot projects such as *eRoadArlanda* in Sweden and *OLEV* in South Korea have demonstrated the technical viability of dynamic inductive charging, large-scale deployment faces substantial challenges. Capital costs for road electrification are orders of magnitude higher than conventional charging infrastructure, with unclear return on investment in most urban contexts. Additionally, regulatory hurdles—including electromagnetic field exposure limits, safety certification for in-motion charging, and long-term maintenance obligations—remain unresolved in many jurisdictions. As such, dynamic inductive roads are more likely to see near-term adoption in niche freight corridors or specialised shuttle routes rather than as a universal charging solution in the next decade.

6.5. Quantum-Safe Security Approaches

With the emergence of quantum computing, current cryptographic algorithms, particularly RSA and elliptic curve cryptography (ECC), face obsolescence, as quantum algorithms such as Shor's algorithm can theoretically break these protocols in polynomial time. This poses a serious risk to secure EV–SG communications, including billing, authentication, firmware updates, and control signals. To address this, the field of post-quantum cryptography (PQC) is advancing rapidly [88]. NIST's ongoing standardisation process is evaluating a suite of quantum-resistant algorithms based on lattice problems, hash functions, error-correcting codes, and multivariate polynomial systems. These approaches are being trialled in experimental EV communication stacks to future-proof charge authorisation and over-the-air update mechanisms [115].

Simultaneously, quantum key distribution (QKD) offers a fundamentally new approach to securing data, utilising quantum entanglement and Heisenberg's uncertainty principle to detect eavesdropping attempts in real-time. Though currently constrained by range and hardware cost, QKD has already been deployed in high-security utility networks and could eventually secure control links between grid substations and EV aggregation platforms. Embedding quantum-safe protocols within EV infrastructure is essential for long-term data security and system trustworthiness. This is particularly important as the number of decentralised actors increases and as EVs assume more autonomous roles in grid regulation and energy exchange [95].

Feasibility Assessment: Despite strong theoretical benefits, blockchain applications in high-frequency V2G transaction environments face performance bottlenecks. Consensus mechanisms, even in energy-efficient configurations, can introduce latencies exceeding sub-second requirements for real-time grid balancing. Scalability concerns also persist, particularly in public or consortium chains, where throughput limitations could restrict market participation. Moreover, ensuring regulatory compliance for blockchain-mediated settlements, especially in cross-border energy markets, remains complex. Consequently, blockchain is best positioned as a medium-term solution for settlement and audit functions, rather than for continuous operational control in the near term.

7. Challenges and Research Gaps

While the integration of EVs into SGs presents transformative opportunities, several persistent barriers must be addressed to achieve reliable, scalable, and sustainable deployment. These challenges span technical, economic, regulatory, and standardisation domains, and they also highlight rich opportunities for future research.

7.1. Technical Barriers

One of the primary technical challenges is achieving system-wide interoperability among heterogeneous devices, platforms, and communication protocols. EVs from different manufacturers, smart chargers, EMSs, and grid controllers are often proprietary technologies that limit seamless integration. The absence of unified data formats and interface standards hinders real-time coordination, especially in multi-vendor deployments [116].

Latency also presents a significant bottleneck, particularly for real-time services such as V2G balancing, DR, and DER aggregation. Current cloud-based architectures often struggle to deliver sub-second response times, making them unsuitable for time-sensitive grid services. This necessitates the adoption of edge computing and time-synchronised communication protocols along with IEEE 1588 [117,118].

System resilience is another pressing concern. EV-grid networks must be capable of withstanding failures, cyberattacks, or unexpected fluctuations in load and generation. However, existing grid infrastructure was not designed for bi-directional energy flows or rapid reconfiguration, making it vulnerable to instabilities when EV penetration increases. Predictive fault detection and decentralised control systems remain underdeveloped in real-world applications [119].

Additionally, high penetration of EVs and bidirectional power converters can introduce voltage instability within distribution networks, particularly during rapid transitions between charging and discharging modes. The switching characteristics of bidirectional chargers may also generate harmonic distortion, affect power quality and potentially lead to overheating or malfunction in sensitive equipment. In decentralised control schemes, response latency, arising from communication delays between distributed nodes, can degrade system performance in fast-changing operational conditions, limiting the effectiveness of real-time V2G balancing. Furthermore, the large-scale replacement of synchronous generators with inverter-based resources, including EV chargers, contributes to a reduction in system inertia, which can impair frequency stability during disturbances. Addressing these challenges will require coordinated hardware–software solutions, such as advanced converter designs with active filtering, low-latency communication protocols, and synthetic inertia provision through inverter controls [119].

7.2. Economic and Regulatory Hurdles

From an economic perspective, the high capital cost of bidirectional chargers, concerns about battery degradation, and uncertain returns from V2G participation have limited their commercial deployment. Many EV owners remain reluctant to allow grid access to their vehicles due to a lack of clear incentives or guarantees regarding battery health and compensation structures [120].

On the regulatory side, policies often lag behind technological innovation. For example, in many jurisdictions, EVs are not yet formally recognised as dispatchable grid assets, complicating their participation in ancillary service markets. Moreover, regulatory frameworks governing data sharing, cybersecurity, and grid access rights are fragmented, creating uncertainty for investors and service providers [30]. A further barrier is the complexity of tariff design. ToU and dynamic pricing schemes, while conceptually ideal for demand-side management, often lack transparency or consistency across regions, discouraging user engagement and automated response [117].

7.3. Lack of Standardisation

The lack of global technical standards for communication protocols, cybersecurity frameworks, and charging interface specifications remains a significant bottleneck. Competing standards, such as ISO 15118, CHAdeMO, and GB/T, limit cross-border interoperability, making it challenging to scale hardware production and software integration across markets [121]. Standardisation is particularly critical in areas such as secure firmware updates, identity management, and blockchain interoperability, where inconsistency can lead to security loopholes or vendor lock-in. The absence of harmonised data governance policies also hinders collaborative innovation, especially in shared mobility and public charging networks [122]. Furthermore, while some organisations (e.g., IEC, NIST, and SAE) are developing guidelines, adoption is inconsistent, and coordination between regions remains limited. A centralised global effort is required to accelerate standards that facilitate secure, interoperable, and scalable EV–SG systems [123].

A comparative assessment of these major protocols reveals significant differences in their security provisions and interoperability scope. *IEC 63110* focuses on EV charging station management and supports secure authentication mechanisms, but its cybersecurity guidance remains high-level and relies on complementary standards for cryptographic detail [124]. ISO/SAE 15118 offers more prescriptive measures, including TLS-based encrypted communication, mutual authentication, and plug-and-charge functionality, enhancing user convenience while reducing exposure to credential theft [125]. However, its adoption is still limited outside Europe, and interoperability with legacy charging equipment is partial at best. IEEE 2030.5, designed primarily for distributed energy resource integration, includes message-level security through XML signatures and TLS, but its broader scope can lead to implementation variability across vendors, affecting interoperability with EV-specific services [126].

Common to all these standards is the use of public key infrastructure (PKI) for authentication, yet differences in certificate formats, revocation processes, and key management policies complicate cross-standard harmonisation. Interoperability challenges are also exacerbated by variations in supported communication stacks, message encoding formats, and optional feature sets.

Recommendations for advancement include:

1. Establishing a global interoperability framework that defines mandatory core features across IEC, ISO/SAE, and IEEE protocols, with certification processes for multi-standard compliance.
2. Developing a harmonised security profile that unifies PKI structures, certificate handling, and encryption protocols, reducing integration complexity for manufacturers and operators.
3. Encouraging testbeds and pilot projects involving cross-vendor, cross-protocol deployments to validate interoperability under real-world operational conditions.

4. Introducing version alignment roadmaps to synchronise updates across standards bodies, avoiding fragmentation caused by asynchronous revisions.

By explicitly comparing security and interoperability features and aligning standard evolution with operational needs, stakeholders can accelerate the deployment of truly global, secure, and interoperable EV–SG ecosystems.

7.4. Research Opportunities

Despite these challenges, the field presents a wide array of open research problems that are both urgent and multidisciplinary:

- Scalable optimisation algorithms for real-time EV fleet coordination across diverse grid topologies and market structures.
- Battery-aware V2G scheduling models that incorporate ageing characteristics, state-of-health estimation, and economic cost functions.
- Federated learning approaches for training AI models across distributed charging infrastructure without compromising data privacy.
- Resilient communication architectures that maintain QoS (Quality of Service) under adversarial conditions, power outages, or DoS attacks.
- Techno-economic analysis frameworks to evaluate the lifecycle value of EV-grid services, accounting for grid deferral, carbon savings, and ancillary services.
- Cross-domain DT models that fuse cyber-physical simulations with real-world telemetry for control system verification and anomaly detection.

In addition, social and behavioural studies are needed to understand user acceptance of V2G better, privacy-preserving data sharing, and the influence of incentive schemes on charging behaviours.

7.5. Human Factors and Socio-Economic Dimensions

While the technical integration of EVs into SGs is essential, human factors and socio-economic considerations are equally critical to ensuring large-scale, equitable adoption of V2G technologies. User behaviour models indicate that participation in V2G programmes is influenced not only by financial incentives but also by perceptions of convenience, trust in the technology, and concerns over battery degradation. Studies show that even attractive tariffs may fail to generate engagement if users anticipate excessive battery wear or perceive scheduling complexity as burdensome. Incorporating behavioural modelling into V2G optimisation, factoring in time-of-use flexibility, interface simplicity, and feedback mechanisms, can improve user enrolment and retention.

From an economic perspective, the commercial viability of V2G depends on the balance between capital expenditure, operational costs, and revenue streams from flexibility markets. Comparative evaluations of business models—such as aggregator-led participation, peer-to-peer energy trading, and utility-managed programmes—highlight trade-offs in revenue predictability, scalability, and risk allocation between users, grid operators, and third-party aggregators [127]. Business model design must account for stakeholder-specific priorities: for example, aggregators may favour low transaction costs and high participation rates, while users value transparent degradation compensation and predictable bill savings.

Social equity considerations further complicate deployment. Populations without access to private charging, such as residents of multi-unit dwellings or low-income households, risk exclusion from the benefits of V2G and smart charging. This "digital divide in energy" can reinforce existing inequalities, as early adopters capture disproportionate economic and environmental benefits [128]. Equitable deployment strategies could include community charging hubs, targeted subsidies, and regulatory mandates for inclusive infrastructure planning. Addressing these human and socio-economic factors is essential for bridging the gap between technical feasibility and sustainable, socially responsible adoption of EV–SG ecosystems.

8. Conclusions and Future Directions

The integration of EVs into SGs represents a pivotal step in the transition towards low-carbon, intelligent energy systems. This review has examined the critical dimensions of EV–SG interaction, spanning energy storage technologies, optimisation strategies, cybersecurity challenges, and emerging innovations. The evidence shows that EVs, when supported by intelligent control architectures and bidirectional communication infrastructure, can evolve from mere transport assets into decentralised energy resources, enhancing grid stability, enabling greater renewable penetration, and improving system flexibility. However, persistent barriers remain. Technical limitations, including interoperability gaps, communication latency, and insufficient system resilience, restrict real-time coordination between EVs and grid infrastructure. Economic and regulatory hurdles, such as unclear market participation frameworks and opaque tariff structures, impede large-scale adoption. Fragmented standardisation and uneven cybersecurity readiness further expose the ecosystem to operational disruptions and privacy risks.

Looking forward, a strategic and prioritised roadmap should focus on:

1. Interoperability and Standards—Establishing harmonised communication protocols and unified data models to integrate multi-vendor EVs, chargers, and grid systems seamlessly.
2. Edge-Enabled AI Optimisation—Deploying low-latency, AI-driven control systems at the network edge to enable localised optimisation of energy flows and demand-side response.
3. Secure, Scalable Energy Trading—Advancing blockchain- and post-quantum cryptography-based platforms that enable secure, real-time, and regulation-compliant energy transactions.
4. Second-Life Battery Deployment—Investing in infrastructure to repurpose decommissioned EV batteries for cost-effective grid-scale storage and ancillary services.

The success of this transition will depend on interdisciplinary collaboration across electrical engineering, computer science, economics, and public policy. Coordinated cross-sector action is essential to develop robust regulations, harmonise international standards, and design user-centric systems that foster participation without compromising privacy, security, or reliability.

In conclusion, while the pathway to fully integrated EV–SG systems is complex, it offers substantial innovation potential. By systematically addressing the identified challenges and advancing the key priorities outlined above, the EV–SG ecosystem can become a cornerstone of the smart, sustainable cities of the future.

Author Contributions: Conceptualisation, M.C., and H.A.; methodology, M.C., and H.A.; validation, M.C.; formal analysis, M.C.; investigation, M.C., and H.A.; resources, M.C.; writing—original draft preparation, M.C., and H.A.; writing—review and editing, M.C., H.A., M.B. and D.D.; visualisation, M.C. All authors have read and agreed to the published version of the manuscript.

Funding: This research received no external funding.

Conflicts of Interest: The authors declare no conflicts of interest.

Abbreviations

The following abbreviations are used in this manuscript:

EV	Electric vehicle
SG	Smart grid
G2V	Grid-to-vehicle
V2G	Vehicle-to-grid
AI	Artificial intelligence
ESS	Energy storage system
ML	Machine learning
RES	Renewable energy source
DT	Digital twin
WCS	Wireless charging system
EMS	Energy management system
BEV	Battery electric vehicle
PHEV	Plug-hybrid electric vehicle
HEV	Hybrid electric vehicle
EV	Multidisciplinary Digital Publishing Institute
FCEV	Fuel cell electric vehicle
Li-ion	Lithium-ion
BMS	Battery management system
HESS	Hybrid energy storage system
DL	Deep learning
IoT	Internet-of-thing
LSTM	Long short-term memory
RNN	Recurrent neural network
RL	Reinforcement learning
GA	Genetic algorithm
GT	Game theory
MPC	Model predictive control
MAS	Multi-agent system
FL	Federated learning
ToU	Time-of-Use
CNN	Convolutional neural network
DR	Demand response
DSM	Demand-side management
OPF	Optimal power flow
PSO	Particle swarm optimisation
NSGA II	Non-dominated Sorting Genetic Algorithm II
DER	Distributed energy resource

References

1. Amini, M.H.; Mohammadi, J.; Kar, S. Distributed Holistic Framework for Smart City Infrastructures: Tale of Interdependent Electrified Transportation Network and Power Grid. *IEEE Access* **2019**, *7*, 157535–157554. [CrossRef]
2. Kobashi, T.; Yamagata, Y.; Yoshida, T.; Chang, S.; Mochizuki, Y.; Ahl, A.; Aleksejeva, J. Smart City and ICT Infrastructure with Vehicle to X Applications toward Urban Decarbonization. In *Urban Systems Design*; Elsevier: Amsterdam, The Netherlands, 2020; pp. 289–333.
3. Li, T.T.; Zhao, A.P.; Wang, Y.; Li, S.; Fei, J.; Wang, Z.; Xiang, Y. Integrating Solar-Powered Electric Vehicles into Sustainable Energy Systems. *Nat. Rev. Electr. Eng.* **2025**, *2*, 467–479. [CrossRef]
4. Dahmani, S. Energy Optimization and Smart Grids. In *Harnessing High-Performance Computing and AI for Environmental Sustainability*; IGI Global: Hershey, PA, USA, 2024; pp. 278–304.
5. Arfeen, Z.A.; Khairuddin, A.B.; Munir, A.; Azam, M.K.; Faisal, M.; Arif, M.S. Bin En Route of Electric Vehicles with the Vehicle to Grid Technique in Distribution Networks: Status and Technological Review. *Energy Storage* **2020**, *2*, e115. [CrossRef]

6. Ullah, Z.; Hussain, I.; Mahrouch, A.; Ullah, K.; Asghar, R.; Ejaz, M.T.; Aziz, M.M.; Naqvi, S.F.M. A Survey on Enhancing Grid Flexibility through Bidirectional Interactive Electric Vehicle Operations. *Energy Rep.* **2024**, *11*, 5149–5162. [CrossRef]
7. Putera, D.A.; Fajri, N.; Alda, T. Advancing Electric Vehicle Safety and Adoption in Indonesia: Insights from Global and Local Perspectives. *Eng. Proc.* **2025**, *84*, 52.
8. Esfandi, S.; Tayebi, S.; Byrne, J.; Taminiau, J.; Giyahchi, G.; Alavi, S.A. Smart Cities and Urban Energy Planning: An Advanced Review of Promises and Challenges. *Smart Cities* **2024**, *7*, 414–444. [CrossRef]
9. Pal, K. Cybersecurity of Electric Vehicle Smart Charging Management Systems. In *Solving Fundamental Challenges of Electric Vehicles*; IGI Global: Hershey, PA, USA, 2024; pp. 84–98.
10. Alqahtani, H.; Kumar, G. Cybersecurity in Electric and Flying Vehicles: Threats, Challenges, AI Solutions & Future Directions. *ACM Comput. Surv.* **2025**, *57*, 1–34. [CrossRef]
11. Cavus, M. Advancing Power Systems with Renewable Energy and Intelligent Technologies: A Comprehensive Review on Grid Transformation and Integration. *Electronics* **2025**, *14*, 1159. [CrossRef]
12. Cavus, M.; Dissanayake, D.; Bell, M. Next Generation of Electric Vehicles: AI-Driven Approaches for Predictive Maintenance and Battery Management. *Energies* **2025**, *18*, 1041. [CrossRef]
13. Divan, D.; Sharma, S. Aligning Innovation, Economics, and Decarbonization. In *ENERGY 2040*; Springer Nature: Cham, Switzerland, 2024; pp. 193–225.
14. Wong, E.Y.C.; Ho, D.C.K.; So, S.; Tsang, C.-W.; Chan, E.M.H. Life Cycle Assessment of Electric Vehicles and Hydrogen Fuel Cell Vehicles Using the GREET Model—A Comparative Study. *Sustainability* **2021**, *13*, 4872. [CrossRef]
15. Kannangara, M.; Bensebaa, F.; Vasudev, M. An Adaptable Life Cycle Greenhouse Gas Emissions Assessment Framework for Electric, Hybrid, Fuel Cell and Conventional Vehicles: Effect of Electricity Mix, Mileage, Battery Capacity and Battery Chemistry in the Context of Canada. *J. Clean. Prod.* **2021**, *317*, 128394. [CrossRef]
16. Teixeira, A.C.R.; da Silva, D.L.; Neto, L.d.V.B.M.; Diniz, A.S.A.C.; Sodré, J.R. A Review on Electric Vehicles and Their Interaction with Smart Grids: The Case of Brazil. *Clean Technol. Environ. Policy* **2015**, *17*, 841–857. [CrossRef]
17. Parikh, A.; Shah, M.; Prajapati, M. Fuelling the Sustainable Future: A Comparative Analysis between Battery Electrical Vehicles (BEV) and Fuel Cell Electrical Vehicles (FCEV). *Environ. Sci. Pollut. Res.* **2023**, *30*, 57236–57252. [CrossRef]
18. Jia, C.; Liu, W.; He, H.; Chau, K.T. Superior Energy Management for Fuel Cell Vehicles Guided by Improved DDPG Algorithm: Integrating Driving Intention Speed Prediction and Health-Aware Control. *Appl. Energy* **2025**, *394*, 126195. [CrossRef]
19. Jia, C.; Liu, W.; He, H.; Chau, K.T. Deep Reinforcement Learning-Based Energy Management Strategy for Fuel Cell Buses Integrating Future Road Information and Cabin Comfort Control. *Energy Convers. Manag.* **2024**, *321*, 119032. [CrossRef]
20. Mwasilu, F.; Justo, J.J.; Kim, E.K.; Do, T.D.; Jung, J.W. Electric Vehicles and Smart Grid Interaction: A Review on Vehicle to Grid and Renewable Energy Sources Integration. *Renew. Sustain. Energy Rev.* **2014**, *34*, 501–516. [CrossRef]
21. Al-Shetwi, A.Q.; Hannan, M.A.; Al-Masri, H.M.K.; Sujod, M.Z. Latest Advancements in Smart Grid Technologies and Their Transformative Role in Shaping the Power Systems of Tomorrow: An Overview. *Prog. Energy* **2025**, *7*, 012004. [CrossRef]
22. Wang, Q. *Renewable Energy and the Smart Grid: Architecture Modelling, Communication Technologies and Electric Vehicles Integration*; University of Trento: Trento, Italy, 2015.
23. Ntombela, M.; Musasa, K.; Moloi, K. A Comprehensive Review of the Incorporation of Electric Vehicles and Renewable Energy Distributed Generation Regarding Smart Grids. *World Electr. Veh. J.* **2023**, *14*, 176. [CrossRef]
24. Mohanty, S.; Panda, S.; Parida, S.M.; Rout, P.K.; Sahu, B.K.; Bajaj, M.; Zawbaa, H.M.; Kumar, N.M.; Kamel, S. Demand Side Management of Electric Vehicles in Smart Grids: A Survey on Strategies, Challenges, Modeling, and Optimization. *Energy Rep.* **2022**, *8*, 12466–12490. [CrossRef]
25. Abdullah, A.A.; Hassan, T.M. Smart Grid (SG) Properties and Challenges: An Overview. *Discov. Energy* **2022**, *2*, 8. [CrossRef]
26. Panda, S.; Mohanty, S.; Rout, P.K.; Sahu, B.K.; Parida, S.M.; Kotb, H.; Flah, A.; Tostado-Véliz, M.; Abdul Samad, B.; Shouran, M. An Insight into the Integration of Distributed Energy Resources and Energy Storage Systems with Smart Distribution Networks Using Demand-Side Management. *Appl. Sci.* **2022**, *12*, 8914. [CrossRef]
27. Brenna, M.; Foiadelli, F.; Zaninelli, D.; Graditi, G.; Di Somma, M. The Integration of Electric Vehicles in Smart Distribution Grids with Other Distributed Resources. In *Distributed Energy Resources in Local Integrated Energy Systems*; Elsevier: Amsterdam, The Netherlands, 2021; pp. 315–345.
28. Rajapandiyan, A.; Thangavel, S. Energy Management in Distribution System Due to the Integration of Renewable Energy Powered EV Charging Infrastructures. In *Power Systems Operation with 100% Renewable Energy Sources*; Elsevier: Amsterdam, The Netherlands, 2024; pp. 281–297.
29. Rathor, S.K.; Saxena, D. Energy Management System for Smart Grid: An Overview and Key Issues. *Int. J. Energy Res.* **2020**, *44*, 4067–4109. [CrossRef]
30. Tirunagari, S.; Gu, M.; Meegahapola, L. Reaping the Benefits of Smart Electric Vehicle Charging and Vehicle-to-Grid Technologies: Regulatory, Policy and Technical Aspects. *IEEE Access* **2022**, *10*, 114657–114672. [CrossRef]

31. Eltamaly, A.M. Smart Decentralized Electric Vehicle Aggregators for Optimal Dispatch Technologies. *Energies* **2023**, *16*, 8112. [CrossRef]
32. Zafar, U.; Bayhan, S.; Sanfilippo, A. Home Energy Management System Concepts, Configurations, and Technologies for the Smart Grid. *IEEE Access* **2020**, *8*, 119271–119286. [CrossRef]
33. Adegbohun, F.; von Jouanne, A.; Agamloh, E.; Yokochi, A. A Review of Bidirectional Charging Grid Support Applications and Battery Degradation Considerations. *Energies* **2024**, *17*, 1320. [CrossRef]
34. Khan, M. Innovations in Battery Technology: Enabling the Revolution in Electric Vehicles and Energy Storage. *Br. J. Multidiscip. Adv. Stud.* **2024**, *5*, 23–41. [CrossRef]
35. Ali, E.; Rai, R.; Aleem, M.; Asghar, R.; Maity, R. Energy Storage Innovations in the Context of Electric Vehicles and Smart Grid Integration: A Review. *Int. J. Econ. Environ. Geol.* **2024**, *15*, 32–40. [CrossRef]
36. Ratshitanga, M.; Ayeleso, A.; Krishnamurthy, S.; Rose, G.; Aminou Moussavou, A.A.; Adonis, M. Battery Storage Use in the Value Chain of Power Systems. *Energies* **2024**, *17*, 921. [CrossRef]
37. Eagon, M.J.; Kindem, D.K.; Panneer Selvam, H.; Northrop, W.F. Neural Network-Based Electric Vehicle Range Prediction for Smart Charging Optimization. *J. Dyn. Syst. Meas. Control* **2022**, *144*, 011110. [CrossRef]
38. Adebayo, D.H.; Ajiboye, J.A.; Okwor, U.D.; Muhammad, A.L.; Ugwuijem, C.D.; Agbo, E.K.; Stephen, V.I. Optimizing Energy Storage for Electric Grids: Advances in Hybrid Technologies. *World J. Adv. Eng. Technol. Sci.* **2025**, *14*, 138–172. [CrossRef]
39. Bai, H.; Song, Z. Lithium-Ion Battery, Sodium-Ion Battery, or Redox-Flow Battery: A Comprehensive Comparison in Renewable Energy Systems. *J. Power Sources* **2023**, *580*, 233426. [CrossRef]
40. Thakur, J.; Martins Leite de Almeida, C.; Baskar, A.G. Electric Vehicle Batteries for a Circular Economy: Second Life Batteries as Residential Stationary Storage. *J. Clean. Prod.* **2022**, *375*, 134066. [CrossRef]
41. Colarullo, L.; Thakur, J. Second-Life EV Batteries for Stationary Storage Applications in Local Energy Communities. *Renew. Sustain. Energy Rev.* **2022**, *169*, 112913. [CrossRef]
42. Sarker, M.T.; Haram, M.H.S.M.; Shern, S.J.; Ramasamy, G.; Al Farid, F. Second-Life Electric Vehicle Batteries for Home Photovoltaic Systems: Transforming Energy Storage and Sustainability. *Energies* **2024**, *17*, 2345. [CrossRef]
43. Faessler, B. Stationary, Second Use Battery Energy Storage Systems and Their Applications: A Research Review. *Energies* **2021**, *14*, 2335. [CrossRef]
44. Akram, M.N.; Abdul-Kader, W. Repurposing Second-Life EV Batteries to Advance Sustainable Development: A Comprehensive Review. *Batteries* **2024**, *10*, 452. [CrossRef]
45. Ferrara, C.; Ruffo, R.; Quartarone, E.; Mustarelli, P. Circular Economy and the Fate of Lithium Batteries: Second Life and Recycling. *Adv. Energy Sustain. Res.* **2021**, *2*, 2100047. [CrossRef]
46. John, J.; Kudva, G.; Jayalakshmi, N.S. Secondary Life of Electric Vehicle Batteries: Degradation, State of Health Estimation Using Incremental Capacity Analysis, Applications and Challenges. *IEEE Access* **2024**, *12*, 63735–63753. [CrossRef]
47. Bhatt, A.; Ongsakul, W.; Madhu, M.N. Optimal Techno-Economic Feasibility Study of Net-Zero Carbon Emission Microgrid Integrating Second-Life Battery Energy Storage System. *Energy Convers. Manag.* **2022**, *266*, 115825. [CrossRef]
48. Farrag, M. *Development of Battery-Supercapacitor Management Systems for Electric Vehicles*; Brunel University: London, UK, 2025.
49. Rajnarayanan, P.N.; Kathiravan, K. Supercapacitors as Distributed Energy Storage Systems for EV Charging Infrastructure. In *Distributed Energy Storage Systems for Digital Power Systems*; Elsevier: Amsterdam, The Netherlands, 2025; pp. 149–166.
50. Ali, E.S.; Elkholy, M.H.; Senjyu, T.; Elazim, S.M.A.; Hassan, E.S.; Alaas, Z.; Lotfy, M.E. A Flexible Multi-Agent System for Managing Demand and Variability in Hybrid Energy Systems for Rural Communities. *Sci. Rep.* **2025**, *15*, 16255. [CrossRef]
51. Mohseni, S.; Khalid, R.; Brent, A.C. Stochastic, Resilience-Oriented Optimal Sizing of off-Grid Microgrids Considering EV-Charging Demand Response: An Efficiency Comparison of State-of-the-Art Metaheuristics. *Appl. Energy* **2023**, *341*, 121007. [CrossRef]
52. Zhou, Y. Renewable-Storage Sizing Approaches for Centralized and Distributed Renewable Energy—A State-of-the-Art Review. *J. Energy Storage* **2024**, *100*, 113688. [CrossRef]
53. Zabihi, A.; Parhamfar, M. Decentralized Energy Solutions: The Impact of Smart Grid-Enabled EV Charging Stations. *Heliyon* **2025**, e41815. [CrossRef]
54. Ouramdane, O.; Elbouchikhi, E.; Amirat, Y.; Gooya, E.S. Optimal Sizing and Energy Management of Microgrids with Vehicle-to-Grid Technology: A Critical Review and Future Trends. *Energies* **2021**, *14*, 4166. [CrossRef]
55. Dai, X.; Batool, K. Optimizing Multi-Objective Design, Planning, and Operation for Sustainable Energy Sharing Districts Considering Electrochemical Battery Longevity. *Renew. Energy* **2024**, *229*, 120705. [CrossRef]
56. Li, H.; Bin Kaleem, M.; Liu, Z.; Wu, Y.; Liu, W.; Huang, Z. IoB: Internet-of-Batteries for Electric Vehicles–Architectures, Opportunities, and Challenges. *Green Energy Intell. Transp.* **2023**, *2*, 100128. [CrossRef]
57. Pandiyan, P.; Saravanan, S.; Kannadasan, R.; Krishnaveni, S.; Alsharif, M.H.; Kim, M.-K. A Comprehensive Review of Advancements in Green IoT for Smart Grids: Paving the Path to Sustainability. *Energy Rep.* **2024**, *11*, 5504–5531. [CrossRef]

58. Tajjour, S.; Singh Chandel, S. A Comprehensive Review on Sustainable Energy Management Systems for Optimal Operation of Future-Generation of Solar Microgrids. *Sustain. Energy Technol. Assess.* **2023**, *58*, 103377. [CrossRef]
59. Arévalo, P.; Ochoa-Correa, D.; Villa-Ávila, E. A Systematic Review on the Integration of Artificial Intelligence into Energy Management Systems for Electric Vehicles: Recent Advances and Future Perspectives. *World Electr. Veh. J.* **2024**, *15*, 364. [CrossRef]
60. Barja Martínez, S. *Energy Management Systems for Smart Homes and Local Energy Communities Based on Optimization and Artificial Intelligence Techniques*; Universitat Politècnica de Catalunya: Barcelona, Spain, 2023.
61. Smith, E.J.; Robinson, D.A.; Elphick, S. DER Control and Management Strategies for Distribution Networks: A Review of Current Practices and Future Directions. *Energies* **2024**, *17*, 2636. [CrossRef]
62. Trivedi, R.; Khadem, S. Implementation of Artificial Intelligence Techniques in Microgrid Control Environment: Current Progress and Future Scopes. *Energy AI* **2022**, *8*, 100147. [CrossRef]
63. Yerasimou, Y.; Kynigos, M.; Efthymiou, V.; Georghiou, G.E. Design of a Smart Nanogrid for Increasing Energy Efficiency of Buildings. *Energies* **2021**, *14*, 3683. [CrossRef]
64. Mansour, S.; Harrabi, I.; Maier, M.; Joós, G. Co-Simulation Study of Performance Trade-Offs between Centralised, Distributed, and Hybrid Adaptive PEV Charging Algorithms. *Comput. Netw.* **2015**, *93*, 153–165. [CrossRef]
65. Cavus, M.; Allahham, A. Enhanced Microgrid Control through Genetic Predictive Control: Integrating Genetic Algorithms with Model Predictive Control for Improved Non-Linearity and Non-Convexity Handling. *Energies* **2024**, *17*, 4458. [CrossRef]
66. Cavus, M.; Dissanayake, D.; Bell, M. Deep-Fuzzy Logic Control for Optimal Energy Management: A Predictive and Adaptive Framework for Grid-Connected Microgrids. *Energies* **2025**, *18*, 995. [CrossRef]
67. Zhang, J. *Middleware-Based Energy Resource Management in Smart Grid*; Auckland University of Technology: Auckland, New Zealand, 2017.
68. Aljehane, N.O.; Mansour, R.F. Optimal Allocation of Renewable Energy Source and Charging Station for PHEVs. *Sustain. Energy Technol. Assess.* **2022**, *49*, 101669. [CrossRef]
69. Cavus, M.; Allahham, A.; Adhikari, K.; Zangiabadia, M.; Giaouris, D. Control of Microgrids Using an Enhanced Model Predictive Controller. In Proceedings of the 11th International Conference on Power Electronics, Machines and Drives (PEMD 2022), Newcastle, UK, 21–23 June 2022; Institution of Engineering and Technology: Stevenage, UK, 2022; pp. 660–665.
70. Chen, Q.; Folly, K.A. Application of Artificial Intelligence for EV Charging and Discharging Scheduling and Dynamic Pricing: A Review. *Energies* **2022**, *16*, 146. [CrossRef]
71. Cui, L.; Wang, Q.; Qu, H.; Wang, M.; Wu, Y.; Ge, L. Dynamic Pricing for Fast Charging Stations with Deep Reinforcement Learning. *Appl. Energy* **2023**, *346*, 121334. [CrossRef]
72. Zhao, Z.; Lee, C.K.M. Dynamic Pricing for EV Charging Stations: A Deep Reinforcement Learning Approach. *IEEE Trans. Transp. Electrif.* **2022**, *8*, 2456–2468. [CrossRef]
73. Narayan, A.; Krishna, A.; Misra, P.; Vasan, A.; Sarangan, V. A Dynamic Pricing System for Electric Vehicle Charging Management Using Reinforcement Learning. *IEEE Intell. Transp. Syst. Mag.* **2022**, *14*, 122–134. [CrossRef]
74. Moghaddam, V.; Yazdani, A.; Wang, H.; Parlevliet, D.; Shahnia, F. An Online Reinforcement Learning Approach for Dynamic Pricing of Electric Vehicle Charging Stations. *IEEE Access* **2020**, *8*, 130305–130313. [CrossRef]
75. Palaniyappan, B.; Kumar R, S.; T, V. Dynamic Pricing for Load Shifting: Reducing Electric Vehicle Charging Impacts on the Grid through Machine Learning-Based Demand Response. *Sustain. Cities Soc.* **2024**, *103*, 105256. [CrossRef]
76. Lee, J.; Lee, E.; Kim, J. Electric Vehicle Charging and Discharging Algorithm Based on Reinforcement Learning with Data-Driven Approach in Dynamic Pricing Scheme. *Energies* **2020**, *13*, 1950. [CrossRef]
77. Liceaga-Ortiz-De-La-Peña, J.M.; Ruiz-Vanoye, J.A.; Xicoténcatl-Pérez, J.M.; Díaz-Parra, O.; Fuentes-Penna, A.; Barrera-Cámara, R.A.; Robles-Camarillo, D.; Márquez-Vera, M.A.; Trejo-Macotela, F.R.; Ortiz-Suárez, L.A. Advancing Smart Energy: A Review for Algorithms Enhancing Power Grid Reliability and Efficiency Through Advanced Quality of Energy Services. *Energies* **2025**, *18*, 3094. [CrossRef]
78. Khan, A.; Bressel, M.; Davigny, A.; Abbes, D.; Ould Bouamama, B. Comprehensive Review of Hybrid Energy Systems: Challenges, Applications, and Optimization Strategies. *Energies* **2025**, *18*, 2612. [CrossRef]
79. Behera, S.; Dev Choudhury, N.B. A Systematic Review of Energy Management System Based on Various Adaptive Controllers with Optimization Algorithm on a Smart Microgrid. *Int. Trans. Electr. Energy Syst.* **2021**, *31*, e13132. [CrossRef]
80. Teo, T.T.; Logenthiran, T.; Woo, W.L.; Abidi, K.; John, T.; Wade, N.S.; Greenwood, D.M.; Patsios, C.; Taylor, P.C. Optimization of Fuzzy Energy-Management System for Grid-Connected Microgrid Using NSGA-II. *IEEE Trans. Cybern.* **2021**, *51*, 5375–5386. [CrossRef]
81. Alzahrani, A.; Rahman, M.U.; Hafeez, G.; Rukh, G.; Ali, S.; Murawwat, S.; Iftikhar, F.; Haider, S.I.; Khan, M.I.; Abed, A.M. A Strategy for Multi-Objective Energy Optimization in Smart Grid Considering Renewable Energy and Batteries Energy Storage System. *IEEE Access* **2023**, *11*, 33872–33886. [CrossRef]

82. Cavus, M.; Ayan, H.; Sari, M.; Akbulut, O.; Dissanayake, D.; Bell, M. Enhancing Smart Grid Reliability Through Data-Driven Optimisation and Cyber-Resilient EV Integration. *Energies* **2025**, *18*, 4510. [CrossRef]
83. Khalaf, M.; Ayad, A.; Tushar, M.H.K.; Kassouf, M.; Kundur, D. A Survey on Cyber-Physical Security of Active Distribution Networks in Smart Grids. *IEEE Access* **2024**, *12*, 29414–29444. [CrossRef]
84. Manias, D.M.; Saber, A.M.; Radaideh, M.I.; Gaber, A.T.; Maniatakos, M.; Zeineldin, H.; Svetinovic, D.; El-Saadany, E.F. Trends in Smart Grid Cyber-Physical Security: Components, Threats, and Solutions. *IEEE Access* **2024**, *12*, 161329–161356. [CrossRef]
85. Kaur, K.; Kaddoum, G.; Zeadally, S. Blockchain-Based Cyber-Physical Security for Electrical Vehicle Aided Smart Grid Ecosystem. *IEEE Trans. Intell. Transp. Syst.* **2021**, *22*, 5178–5189. [CrossRef]
86. Ye, J.; Guo, L.; Yang, B.; Li, F.; Du, L.; Guan, L.; Song, W. Cyber–Physical Security of Powertrain Systems in Modern Electric Vehicles: Vulnerabilities, Challenges, and Future Visions. *IEEE J. Emerg. Sel. Top. Power Electron.* **2021**, *9*, 4639–4657. [CrossRef]
87. Cavus, M.; Ayan, H.; Bell, M.; Oyebamiji, O.K.; Dissanayake, D. Deep Charge-Fusion Model: Advanced Hybrid Modelling for Predicting Electric Vehicle Charging Patterns with Socio-Demographic Considerations. *Int. J. Transp. Sci. Technol.* **2025**. [CrossRef]
88. Shekhawat, H.; Gupta, D.S. A Survey on Lattice-based Security and Authentication Schemes for Smart-grid Networks in the Post-quantum Era. *Concurr. Comput.* **2024**, *36*, e8080. [CrossRef]
89. Aljohani, T.; Almutairi, A. A Comprehensive Survey of Cyberattacks on EVs: Research Domains, Attacks, Defensive Mechanisms, and Verification Methods. *Def. Technol.* **2024**, *42*, 31–58. [CrossRef]
90. Nguyen, L.-H.; Nguyen, V.-L.; Hwang, R.-H.; Kuo, J.-J.; Chen, Y.-W.; Huang, C.-C.; Pan, P.-I. Towards Secured Smart Grid 2.0: Exploring Security Threats, Protection Models, and Challenges. *IEEE Commun. Surv. Tutor.* **2024**, *27*, 2581–2620. [CrossRef]
91. Qureshi, M.A.; Zafar, L.; Shaheen, R. Cybersecurity and Privacy Perspectives on the Green Internet of Vehicles. In *Driving Innovation at the Intersection of Renewable Energy and the Internet of Vehicles*; IGI Global: Hershey, PA, USA, 2025; pp. 383–416.
92. Cao, Y.; Li, S.; Lv, C.; Wang, D.; Sun, H.; Jiang, J.; Meng, F.; Xu, L.; Cheng, X. Towards Cyber Security for Low-Carbon Transportation: Overview, Challenges and Future Directions. *Renew. Sustain. Energy Rev.* **2023**, *183*, 113401. [CrossRef]
93. Ataullah, M.; Chauhan, N. Exploring Security and Privacy Enhancement Technologies in the Internet of Things: A Comprehensive Review. *Secur. Priv.* **2024**, *7*, e448. [CrossRef]
94. Acharya, S.; Dvorkin, Y.; Pandzic, H.; Karri, R. Cybersecurity of Smart Electric Vehicle Charging: A Power Grid Perspective. *IEEE Access* **2020**, *8*, 214434–214453. [CrossRef]
95. Denis, A.; Thomas, A.; Robert, W.; Samuel, A.; Kabiito, S.P.; Morish, Z.; Sallam, M.; Ali, G.; Mijwil, M.M. A Survey on Artificial Intelligence and Blockchain Applications in Cybersecurity for Smart Cities. *SHIFRA* **2025**, *2025*, 1–45. [CrossRef]
96. Cavus, M.; Allahham, A.; Adhikari, K.; Giaouris, D. A Hybrid Method Based on Logic Predictive Controller for Flexible Hybrid Microgrid with Plug-and-Play Capabilities. *Appl. Energy* **2024**, *359*, 122752. [CrossRef]
97. Zhang, C.; Wu, J.; Zhou, Y.; Cheng, M.; Long, C. Peer-to-Peer Energy Trading in a Microgrid. *Appl. Energy* **2018**, *220*, 1–12. [CrossRef]
98. Dutta, A.; Ganguly, S.; Kumar, C. Coordinated Control Scheme for EV Charging and Volt/Var Devices Scheduling to Regulate Voltages of Active Distribution Networks. *Sustain. Energy Grids Netw.* **2022**, *31*, 100761. [CrossRef]
99. Rasheed, A.; San, O.; Kvamsdal, T. Digital Twin: Values, Challenges and Enablers From a Modeling Perspective. *IEEE Access* **2020**, *8*, 21980–22012. [CrossRef]
100. Feliziani, M.; Campi, T.; Cruciani, S.; Maradei, F. Introduction to Wireless Power Transfer for E-Mobility. In *Wireless Power Transfer for E-Mobility*; Elsevier: Amsterdam, The Netherlands, 2024; pp. 1–24.
101. Lukic, S.; Pantic, Z. Cutting the Cord: Static and Dynamic Inductive Wireless Charging of Electric Vehicles. *IEEE Electrif. Mag.* **2013**, *1*, 57–64. [CrossRef]
102. Chen, L.; Jordan, S.; Liu, Y.-K.; Moody, D.; Peralta, R.; Perlner, R.; Smith-Tone, D. *Report on Post-Quantum Cryptography*; NIST: Gaithersburg, MD, USA, 2016.
103. Kang, J.; Yu, R.; Huang, X.; Maharjan, S.; Zhang, Y.; Hossain, E. Enabling Localized Peer-to-Peer Electricity Trading Among Plug-in Hybrid Electric Vehicles Using Consortium Blockchains. *IEEE Trans. Ind. Inf.* **2017**, *13*, 3154–3164. [CrossRef]
104. Shi, W.; Cao, J.; Zhang, Q.; Li, Y.; Xu, L. Edge Computing: Vision and Challenges. *IEEE Internet Things J.* **2016**, *3*, 637–646. [CrossRef]
105. Singh, P.; Singh, M.K.; Singh, R.; Singh, N. Federated Learning: Challenges, Methods, and Future Directions. In *Federated Learning for IoT Applications*; Springer: Cham, Switzerland, 2022; pp. 199–214.
106. Dorokhova, M.; Martinson, Y.; Ballif, C.; Wyrsch, N. Deep Reinforcement Learning Control of Electric Vehicle Charging in the Presence of Photovoltaic Generation. *Appl. Energy* **2021**, *301*, 117504. [CrossRef]
107. Jauhar, S.K.; Sethi, S.; Kamble, S.S.; Mathew, S.; Belhadi, A. Artificial Intelligence and Machine Learning-Based Decision Support System for Forecasting Electric Vehicles' Power Requirement. *Technol. Forecast. Soc. Change* **2024**, *204*, 123396. [CrossRef]
108. Kong, W.; Dong, Z.Y.; Jia, Y.; Hill, D.J.; Xu, Y.; Zhang, Y. Short-Term Residential Load Forecasting Based on LSTM Recurrent Neural Network. *IEEE Trans. Smart Grid* **2019**, *10*, 841–851. [CrossRef]

109. Lopes, J.A.P.; Soares, F.J.; Almeida, P.M.R. Integration of Electric Vehicles in the Electric Power System. *Proc. IEEE* **2011**, *99*, 168–183. [CrossRef]
110. Yan, J.; Menghwar, M.; Asghar, E.; Kumar Panjwani, M.; Liu, Y. Real-Time Energy Management for a Smart-Community Microgrid with Battery Swapping and Renewables. *Appl. Energy* **2019**, *238*, 180–194. [CrossRef]
111. CH, H.B.; Reddy K, R.; C, D.; Kamwa, I.; Muyeen, S.M. A Novel on Intelligent Energy Control Strategy for Micro Grids with Renewables and EVs. *Energy Strategy Rev.* **2024**, *52*, 101306. [CrossRef]
112. Nasrinasrabadi, M.; Hejazi, M.A.; Chaharmahali, E.; Hussein, M. A Comprehensive Review of Blockchain Integration in Smart Grid with a Special Focus on Internet of Things. *Energy Convers. Manag. X* **2025**, *27*, 101196. [CrossRef]
113. Ukkusuri, S.; Hamim, O.F.; Lei, Z.; Ka, E.; Salek, M.S.; Chowdhury, M.; Amini, M.H.; Cardenas, A.; Thuraisingham, B. Cybersecurity for Next-Generation Road Transportation: A Review. *ACM J. Auton. Transp. Syst.* **2025**. [CrossRef]
114. Bhat, J.R.; Alqahtani, S.A. 6G Ecosystem: Current Status and Future Perspective. *IEEE Access* **2021**, *9*, 43134–43167. [CrossRef]
115. Fu, R.; Lichtenwalner, M.E.; Johnson, T.J. A Review of Cybersecurity in Grid-Connected Power Electronics Converters: Vulnerabilities, Countermeasures, and Testbeds. *IEEE Access* **2023**, *11*, 113543–113559. [CrossRef]
116. Gupta, S.; Varshney, T.; Saini, P.; Hussain, M.M. Standards and Protocols for Interoperability in Smart EV Systems. In *Modern Computing Technologies for EV Efficiency and Sustainable Energy Integration*; IGI Global: Hershey, PA, USA, 2025; pp. 115–150.
117. Štogl, O.; Miltner, M.; Zanocco, C.; Traverso, M.; Starý, O. Electric Vehicles as Facilitators of Grid Stability and Flexibility: A Multidisciplinary Overview. *WIREs Energy Environ.* **2024**, *13*, e536. [CrossRef]
118. *IEEE 1588-2008*; IEEE Standard for a Precision Clock Synchronization Protocol for Networked Measurement and Control Systems. IEEE Standards Association: New York, NY, USA, 2008.
119. Harris, R. *Interoperability of Distributed Energy Resources: Benefits, Challenges, and Solutions*; Carbon Trust: London, UK, 2023.
120. Van Eijk, M.W.; Annema, J.A.; Van der Koogh, M.; Lukszo, Z. Institutional Barriers to Vehicle-to-Grid Implementation in Europe. *Renew. Sustain. Energy Rev.* **2025**, *217*, 115653. [CrossRef]
121. Kumar, G.; Mikkili, S. Advancements in EV International Standards: Charging, Safety and Grid Integration with Challenges and Impacts. *Int. J. Green Energy* **2024**, *21*, 2672–2698. [CrossRef]
122. Das, H.S.; Rahman, M.M.; Li, S.; Tan, C.W. Electric Vehicles Standards, Charging Infrastructure, and Impact on Grid Integration: A Technological Review. *Renew. Sustain. Energy Rev.* **2020**, *120*, 109618. [CrossRef]
123. Micari, S.; Napoli, G. Electric Vehicles for a Flexible Energy System: Challenges and Opportunities. *Energies* **2024**, *17*, 5614. [CrossRef]
124. *IEC 63110-1:2022*; Protocol for Management of Electric Vehicles Charging and Discharging Infrastructure—Part 1: Basic Definitions, Use Cases and Architectures. International Electrotechnical Commission (IEC): Geneva, Switzerland, 2022.
125. *ISO/SAE 15118-1:2019*; Road Vehicles—Vehicle to Grid Communication Interface—Part 1: General Information and Use-Case Definition. International Organization for Standardization (ISO): Geneva, Switzerland; Society of Automotive Engineers (SAE): Warrendale, PA, USA, 2019.
126. *IEEE Std 2030.5-2018*; IEEE Standard for Smart Energy Profile Application Protocol. IEEE Standards Association: New York, NY, USA, 2018.
127. Openshaw, S.; Etta, D.; Maji, S.; Ruan, T.; Afridi, K.K. Investigation of Commercial Viability and Public Perception of Electrified Roadways with Dynamic Wireless Charging. In Proceedings of the 2023 IEEE Wireless Power Technology Conference and Expo (WPTCE), San Diego, CA, USA, 4–8 June 2023; IEEE: New York, NY, USA, 2023; pp. 1–6.
128. Ruan, T.; Lv, Q. Exploring Equity Perception of Electric Vehicles from a Social Media Perspective. *Transp. Res. Interdiscip. Perspect.* **2024**, *25*, 101103. [CrossRef]

energies

MDPI

Article

Sensitivity Analysis of Distribution Network Reconfiguration Optimization for Electric Vehicle and Renewable Distributed Generator Integration

Mahmoud Ghofrani

Division of Engineering and Mathematics, School of STEM, University of Washington, Bothell, WA 98011, USA; mrani@uw.edu

Abstract: Distribution networks have faced significant efficiency and reliability challenges, balancing the recent integration of electric vehicles (EVs) and renewable distributed generators (DGs). This study proposes a reconfiguration optimization of the distribution system by adjusting the status of switches within the network. This approach aims to minimize power losses and enhance overall operational efficiency. To model the variability of wind and solar DGs, probability distribution functions (PDFs) are employed, which allow for a more accurate representation of their performance. Additionally, stochastic models and Monte Carlo Simulation (MCS) are utilized to generate various scenarios that reflect real-world conditions, including the charging and discharging behaviors of EVs. A sensitivity analysis is conducted to evaluate the effectiveness of our proposed reconfiguration strategy across different levels of EV and DG penetration.

Keywords: distribution system reconfiguration; electric vehicle; optimization; renewable distributed generation; probability distribution function; sensitivity analysis

1. Introduction

The increasing demand for electric vehicles (EVs) and renewable distributed generation (DG) sources is reshaping the landscape of modern power distribution networks. This transformation is primarily driven by a global push for sustainability, decarbonization, and energy efficiency. However, while these technologies bring significant benefits in terms of reducing carbon emissions and reliance on fossil fuels, their integration into existing power grids presents a series of new challenges. The key issues are largely due to the intermittent nature of renewable energy sources like wind and solar and the variable charging behaviors of electric vehicles, which can significantly affect the overall reliability and operational efficiency of the grid.

If the grid is not able to effectively manage the additional load from both EVs and DGs, certain sections of the network may experience overloading. This can lead to excessive wear on equipment, reduce the lifespan of infrastructure, and increase maintenance costs.

Overloading or incorrect power flow due to inconsistent generation or unpredictable demand can cause voltage deviations. These fluctuations can lead to poor power quality, affecting sensitive equipment and potentially damaging grid components.

Extended periods of overloading or poor voltage regulation can eventually result in equipment failures, requiring costly repairs or replacements and potentially leading to widespread outages.

Inefficiencies in power flow, resulting from the misalignment of supply and demand or the additional power required for EV charging and discharging, can lead to increased energy losses, thereby reducing the overall efficiency of the system.

To effectively address these challenges, the transition to an eco-friendly and sustainable energy system must integrate both transportation (EVs) and generation systems (renewable DGs) in a way that mitigates the adverse impacts on the distribution network. An innovative solution is required that can manage the uncertainties introduced by these technologies while ensuring the reliability, efficiency, and stability of the grid.

One promising approach to tackle these challenges is distribution network reconfiguration (DNR). DNR refers to the process of optimizing the topology of the distribution network by adjusting the status of switches within the grid. By strategically opening or closing certain switches, the flow of electricity can be altered to minimize losses, reduce congestion, and optimize voltage levels across the network. This approach can help mitigate the impacts of EV charging and renewable energy fluctuations by dynamically adapting the grid to changing conditions.

DNR in the presence of DGs has been investigated in several studies [1–4]. In these studies, the primary focus was on optimizing the topology of the distribution network by adjusting switch positions to enhance the network's performance, particularly in the context of the integration of DG sources. The addition of DGs to the system typically requires changes in the topology of the distribution network to ensure that power from the DGs is optimally delivered to the loads and that the network's performance is not negatively affected by fluctuations in power output from renewable sources.

The authors of [1] proposed an approach to optimize DNR in the presence of DG sources using a fast heuristic-based method. The practical effectiveness of the approach in achieving fast and reliable configuration was validated through case studies and simulations. However, the study did not address how the reconfiguration method adapts to dynamic changes such as varying loads, changing generation from renewable sources, and fluctuations in demand. Additionally, the impacts of DG on network losses, voltage profiles, and power quality were not explored.

The authors of [2] proposed a hybrid optimization model, termed the "hybrid Ttloa model", for simultaneously performing DNR and optimal placement of DGs in order to reduce power losses. The key objective was to minimize active power losses while maintaining the operational reliability of the network, especially under contingencies. The proposed model was validated using multiple test systems. However, the study does not adequately discuss the computational complexity of the proposed method. Furthermore, the study did not consider uncertainties in the input parameters, such as variations in renewable generation, load fluctuations, or potential faults in the network.

The authors of [3] developed an innovative approach for optimizing the reconfiguration of distribution networks by incorporating DG systems using a parallel slime mold algorithm (PSMA). The results indicated that using PSMA significantly reduces power losses, improves the voltage profile, and enhances overall network efficiency when compared to other optimization techniques, such as genetic algorithms and particle swarm optimization.

The authors of [4] employed the Whale Optimization Algorithm (WOA) to optimize the reconfiguration of distribution networks while simultaneously determining the optimal placement of DG units. The study formulated the optimization problem as a multi-objective function that includes power loss reduction, voltage profile improvement, and cost-effectiveness in terms of both reconfiguration and DG installation. The method was tested on standard distribution network systems, including the IEEE 33-bus and IEEE 69-bus systems. The results showed that using WOA for simultaneous reconfiguration and

DG allocation achieves significant reductions in power losses and improves the system's voltage stability. The proposed method also demonstrated better convergence speed and solution quality compared to other traditional optimization algorithms.

Several studies tackled the DNR problem to support the integration of EVs [5–12]. DNR optimization studies for EVs typically aim to manage the impact of the EV charging demand on the grid by altering network configurations (e.g., switching operations) to prevent overloads, minimize losses, and maintain voltage stability.

The authors of [5] developed an optimization model that simultaneously considers the network reconfiguration and the integration of EVs. The model takes into account the charging stations' locations, the charging and discharging patterns of EVs, and the overall energy demand. A genetic algorithm (GA) was used to explore different network configurations and determine the optimal placement of charging stations while considering the variable load introduced by EVs. The proposed strategy was tested on a test distribution system (such as the IEEE 33-bus system). By optimally placing charging stations and adjusting the network topology, the strategy effectively reduces power losses, improves voltage profiles, and enhances the system's ability to handle EV charging demand. The results demonstrated that the integration of EVs can lead to significant improvements in the system's performance when the network is properly reconfigured.

The authors of [6] explored an optimization approach for active distribution system planning that integrates non-utility-owned EV charging stations (i.e., privately owned or third-party charging stations) and network reconfiguration. The GA was used to determine the best locations for EV charging stations and the optimal network configuration that minimizes system losses, ensures voltage stability, and accommodates EV charging demands. The optimization approach was tested on a real distribution network. The results demonstrated that the proposed approach significantly reduces power losses and improves voltage stability compared to traditional distribution planning approaches.

The authors of [7] focused on reconfiguring the distribution network to maximize the efficiency and reliability of EV integration. The study utilized Binary Particle Swarm Optimization (BPSO) to minimize several key factors, including the cost of power exchange, active power loss, and operational costs associated with EVs. The simulation results highlighted the effectiveness of BPSO in determining the optimal status of switches within the distribution network, as well as in creating an optimal schedule for EVs. These findings underscore the potential of BPSO as a valuable tool for enhancing the integration of EVs into our distribution networks.

The authors of [8] presented an optimization framework aimed at determining the optimal locations for EV charging stations within a radial distribution network. Network reconfiguration was considered alongside the placement of EV charging stations to minimize power losses and improve voltage profiles under the additional load brought on by EV charging. The optimization method was applied to a test case of a typical radial distribution system. The results showed that the optimal placement of EV charging stations and the reconfiguration of the distribution network led to a significant reduction in power losses and an improvement in voltage profiles.

The authors of [9] focused on optimizing the reconfiguration of distribution networks that integrate EVs while also considering the environmental benefit of carbon reduction. The optimization model developed in the paper considers multiple objectives, including minimizing power losses, improving voltage stability, and maximizing carbon reduction. The optimization framework was applied to a case study of a distribution system with integrated EVs. The results demonstrated that the integration of EVs with proper planning and reconfiguration can contribute to both improved network performance and environmental sustainability.

The authors of [10] proposed a distribution system reconfiguration method that coordinates with maintenance scheduling while accounting for the uncertainty of EV charging loads. In this study, the EV charging loads were modeled stochastically using various scenarios. The distribution network configuration was then optimized to minimize both power loss and maintenance costs based on these EV charging load models. The application of this proposed method on the IEEE 33-bus network demonstrated its feasibility in optimizing network topology both before and after maintenance overhauls, adapting effectively to different EV charging scenarios.

The authors of [11] proposed an optimization-based approach to address the network reconfiguration problem, particularly in managing the unpredictability associated with EV charging patterns. A case study was presented to demonstrate the effectiveness of the proposed reconfiguration strategies conducted on the IEEE 33-bus distribution network with varying levels of EV penetration. The results showed that optimizing the network's configuration can effectively reduce power losses, maintain voltage stability, and prevent overloading.

The authors of [12] developed a fuzzy Pareto heuristic network reconfiguration to balance multiple objectives, including performance improvement and grid-side emission reduction, while optimizing the placement and operation of EV charging stations within the distribution network. However, the study did not perform a sensitivity analysis to assess how changes in key input parameters such as EV penetration rates, grid-side emission factors, or load forecasting errors affect the optimal solutions. Additionally, the study assumed a generalized EV charging pattern, which may not reflect the diverse charging behaviors of different EV owners or geographical variations.

The integration of both renewable DGs and EVs creates new opportunities but also introduces additional complexities. Renewable DGs, with their intermittent and variable generation, combined with the dynamic and often unpredictable behavior of EVs, can result in power imbalances, voltage fluctuations, and inefficient use of available generation resources. To address this, many studies have explored how DNR strategies can be adjusted to support both renewable DGs and EV integration simultaneously [13–18]. These strategies involve a coordinated approach where the network's topology is reconfigured to optimize the flow of energy from both renewable sources and EVs, ensuring minimal losses, stable voltage profiles, and reliable service for consumers. Moreover, advanced optimization techniques like multi-objective optimization, genetic algorithms, particle swarm optimization, and other heuristic methods are often employed to address the multiple conflicting objectives of reducing losses, improving voltage quality, and maximizing the use of renewable energy while managing the varying load from EVs. Such integrated DNR strategies often need to consider the time-varying nature of EV charging demand and the stochastic behavior of renewable generation to achieve optimal solutions.

The authors of [13] proposed a multi-objective optimization method for reconfiguring distribution networks in the presence of DG sources and EVs. The study developed a dynamic reconfiguration model aimed at minimizing active power loss, improving the load balancing index, and reducing voltage deviations within the network. To achieve this, the researchers employed the Lévy flight and chaos distributed beetle antennae search (LDBAS) algorithm. This innovative approach not only enhanced the computing speed but also improved the convergence rate of the optimization process. The proposed method was evaluated through simulations of the IEEE 33-bus and 118-bus systems. The results demonstrated the computational efficiency of the LDBAS algorithm and its effectiveness in finding global optimum solutions. Furthermore, the findings indicated that the reconfiguration method significantly enhances both the economic performance and power quality of distribution networks, particularly those with high penetrations of EVs and DGs.

The authors of [14] developed a novel hybrid optimization method that combines GA and the Shuffled Frog Leaping Algorithm (SFLA) to optimize network reconfiguration, considering the placement and displacement of DGs and EVs. The hybrid GA–SFLA algorithm demonstrated significant improvements in system performance, including reduced power losses and improved voltage profiles, when compared to conventional methods. The study also highlighted that the optimal placement and displacement of DGs and EVs effectively balance the load in the distribution system, reduce losses, and enhance voltage stability. Additionally, the hybrid GA–SFLA algorithm proved to be highly effective in solving the complex multi-objective optimization problem, yielding better results than using GA or SFLA alone.

The authors of [15] developed an optimal planning framework that simultaneously considers the placement and sizing of EV charging stations and DGs within a distribution network while also accounting for network reconfiguration and the uncertainties in renewable energy generation and EV charging behavior. The goal was to minimize the total cost of energy losses, improve the network's efficiency, and enhance its reliability. The proposed model was applied to a case study of a smart distribution network with high levels of renewable energy integration and EV penetration. The results showed that optimal placement and sizing of EV charging stations and DGs can significantly reduce energy losses and operational costs. Additionally, network reconfiguration improved voltage stability and reduced the impact of uncertainties on the network.

The authors of [16] addressed the optimization problem of allocating controllable power factor DG in distribution networks, with a focus on networks impacted by EV charging stations. The study introduced a methodology that integrates DG placement with power factor control while considering network reconfiguration to improve power quality, reduce losses, and increase the system's overall efficiency. However, the proposed method does not account for uncertainties in the system, particularly the variability in EV charging demand and renewable DG output.

The authors of [17] proposed a novel multi-objective optimization of DNR in the presence of DGs, EV charging stations, and demand response programs. The method was validated through case studies and simulation results. However, the study did not explicitly address the uncertainties in key parameters, such as renewable generation (e.g., solar or wind variability), EV charging demand fluctuations, or load forecasting errors. Additionally, the proposed approach assumes that EV charging is flexible and can be easily adjusted based on the demand response signal. However, real-world charging behavior may not always be as flexible or controllable due to user preferences, battery capacity constraints, and other factors. Furthermore, the study lacks an economic analysis to assess the feasibility of the method.

The authors of [18] developed an optimization method combining DNR with renewable energy integration to improve the system's resilience under various contingencies. While the study highlighted the importance of robustness in DNR, particularly in terms of handling uncertainties (e.g., fluctuating renewable generation, load variations, and extreme weather events), the proposed approach did not clearly explain how these uncertainties are incorporated, which may limit the practical applicability of the proposed methodology.

Despite a large body of research on distribution network reconfiguration (DNR) and the integration of distributed generation (DG) sources, including renewable energy and electric vehicles (EVs), a critical gap persists in understanding the sensitivity of network performance to varying penetration levels of EVs and renewable DGs. As the penetration of EVs and renewable DGs increases, the impacts on network performance, such as system losses, voltage stability, and the risk of system overload, can vary significantly. These factors are highly dependent on the level of integration of these resources into the grid, as well as their intermittency and variability. However, most existing optimization studies fail to provide comprehensive sensitivity analyses that assess how changes in penetration levels of EVs and renewable DGs affect the optimal configuration of the distribution network.

In contrast to existing studies, which primarily focus on static or deterministic approaches to network reconfiguration, the novelty of our work lies in the sensitivity analysis conducted to evaluate the proposed optimization strategy across varying penetration levels of EVs and renewable DGs. This type of analysis is crucial for understanding the dynamic impacts on network performance, such as losses and overall system efficiency, as the penetration of EVs and DGs fluctuates. By systematically analyzing different penetration scenarios, our method provides a more comprehensive and adaptive framework for optimizing network configurations, something that is lacking in many existing investigations.

Without detailed sensitivity analysis, it is challenging to identify the most optimal configuration for a network under different scenarios of EV and DG integration. Existing methods typically focus on optimization under fixed conditions without accounting for how fluctuating penetration levels affect network performance. A static approach to network reconfiguration may not provide the flexibility needed to adapt to the dynamic nature of EV charging and renewable energy production. As penetration levels evolve, grid operators require clear guidance on how to adjust the network configuration to maintain efficiency, minimize losses, avoid overloads, and preserve system stability. Our approach, through sensitivity analysis, offers clear insights into how the network can adapt to these fluctuations, thus filling a critical gap in existing literature.

Furthermore, distribution networks with high levels of EV and renewable DG penetration are inherently intermittent due to the unpredictability of renewable energy generation (e.g., solar and wind) and the fluctuating charging and driving patterns of EVs. These uncertainties pose significant challenges to network operation and optimization. Despite this, many studies on DNR fail to account for or sufficiently model such uncertainties, which significantly limits their practical applicability in real-world settings. Our method specifically addresses this gap by incorporating techniques to handle these uncertainties, ensuring that the network can adapt in real time to dynamic changes.

This study proposes a DNR approach that specifically addresses the challenges of integrating EVs and renewable DGs into the grid. The main objective of this approach is to optimize the network's configuration while managing the additional complexities introduced by these technologies. The key challenges that the proposed method tackles include the following:

- System uncertainties: These arise from the intermittent nature of renewable energy generation and the unpredictable charging behaviors of EVs. The optimization approach incorporates techniques to deal with these uncertainties, ensuring the network can adapt to real-time changes.
- Overloading: By dynamically adjusting the distribution network configuration, the approach aims to prevent overload situations that may arise due to the increased demand from EVs or fluctuating renewable energy outputs.
- Increased losses: The optimization process helps reduce losses by reconfiguring the network to minimize power flow inefficiencies, which can increase with the integration of variable renewable DGs and the additional load from EV charging.

One of the novel aspects of this paper is the sensitivity analysis conducted to evaluate the proposed optimization strategy across different penetration levels of EVs and renewable DGs. This analysis is crucial for understanding how system performance, such as losses and voltage stability, changes as the penetration of EVs and DGs varies. The sensitivity analysis provides a clearer picture of the network's ability to handle different levels of technological integration, which is key for determining the most efficient and reliable configuration for each scenario. Additionally, by identifying the optimal penetration rates for EVs and DGs, the sensitivity analysis offers valuable recommendations for efficiently reconfiguring the distribution network and strategically locating DGs, potentially deferring the need for infrastructure upgrades or additional support.

By conducting this analysis, this study aims to provide more robust insights into how the network should be reconfigured under varying scenarios, ensuring that the optimization approach is adaptive and scalable. This will help grid operators plan more effectively for future scenarios, allowing for better management of grid performance as EVs and renewable DGs continue to be integrated.

This study is organized as follows:

- Section 2: Details the methodology proposed for distribution network reconfiguration, along with models for renewable DGs and EV charging/discharging.
- Section 3: Presents case studies and the test system.
- Section 4: Provides simulation results for various scenarios.
- Section 5: Performs sensitivity analysis and evaluates the distribution system reconfiguration optimization at various EV and renewable DG penetration levels.
- Section 6: Concludes with the findings of the study.

2. Methods

2.1. Distributed Generation Model

This section focuses on the mathematical modeling of variability and randomness in renewable energy sources, specifically wind and solar generation, using probability distribution functions (PDFs). By employing PDFs, we can effectively predict the likelihood of various energy outputs while accounting for the inherent uncertainties associated with solar irradiance and wind speeds.

2.1.1. Wind DG Model (Weibull PDF)

Curve fitting is conducted on historical hourly wind speed data to evaluate different PDFs and identify the most accurate model for wind speed variations. We found that the Weibull PDF most effectively captures the random behavior of wind. The Weibull distribution is governed by the following equation:

$$f_v(v) = (\frac{k}{\lambda})(\frac{v}{\lambda})^{k-1}e^{-(v/\lambda)^k},\ 0 \le v \le \infty \quad (1)$$

$f_v(v)$ is the Weibull PDF for wind speed. k and λ are the shape and scale parameters for the Weibull PDF. v is the wind speed.

Furthermore, we modeled the power generated by the wind renewable DG using the following equation:

$$P_{W-DG} = \begin{cases} 0, & v \le v_{cut-in},\ v \ge v_{cut-out} \\ \frac{v-v_{cut-in}}{v_r-v_{cut-in}} P_{W-DG_{rated}} & v_{cut-in} \le v \le v_r \\ P_{W-DG_{rated}} & v_r \le v \le v_{cut-out} \end{cases} \quad (2)$$

P_{W-DG} and $P_{W-DG_{rated}}$ are the output power and rated power of the wind DG. v_{cut-in}, $v_{cut-out}$, and v_r are the cut-in, cut-out, and rated speeds of the wind turbine.

2.1.2. Solar DG Model (Beta PDF)

PDFs cannot be directly applied to solar DG due to the certainty of zero solar irradiance during night-time, which is influenced by strong diurnal patterns.

To address this, we propose an alternative approach by modeling the clearness index to characterize solar irradiance indirectly. The hourly clearness index (k_t) is calculated using the following formula:

$$k_t = \frac{I_t}{I_o} \tag{3}$$

k_t is the hourly clearness index. I_t is the irradiance on a horizontal plane, and I_o is the extraterrestrial solar irradiance.

We performed curve fitting on our historical hourly clearness index data to evaluate different PDFs and found that the Beta PDF most accurately models our clearness index. The Beta distribution is defined by the following equation:

$$f_{ci}(k_t) = \frac{\gamma(a+b)}{\gamma(a)\gamma(b)} \cdot k_t^{a-1} \cdot (1-k_t)^{b-1} \tag{4}$$

$f_{ci}(k_t)$ is the Beta PDF for clearness index. γ is the gamma function. a and b are the shape parameters of the Beta distribution.

Upon calculating the hourly clearness index, the irradiance on a surface with inclination β to the horizontal plane is calculated using the following formula:

$$I_\beta = T \cdot k_t - T' \cdot k_t^2 \tag{5}$$

Here, T and T' are parameters influenced by various factors, such as inclination, declination, sunset hour angle, and latitude.

The power generated by the solar renewable DG can be expressed as

$$P_{PV-DG} = A_{PV} \cdot \eta_{PV} \cdot I_\beta \tag{6}$$

P_{PV-DG} is the output power of the PV DG. A_{PV} and η_{PV} are the area and efficiency of the solar PV array.

2.2. *EV Charging/Discharging Model*

This section focuses on modeling of EV driving patterns and charging behaviors. Probabilistic methods based on historical driving data are utilized to derive the charging and discharging power of EVs, adhering to specific inequality constraints:

$$\begin{aligned} |P_C(t_C)| &\leq P_C^{max} \quad \forall c \in L \\ |P_d(t_d)| &\leq P_d^{max} \quad \forall d \in L \end{aligned} \tag{7}$$

P_c and P_d are the charging and discharging powers of the EV battery. P_c^{max} and P_d^{max} are the maximum charging and discharging capacities of the EV battery. L is the number of time steps throughout the simulation period.

To enhance our analysis, we cluster EVs with similar driving patterns into separate fleets. This approach allows us to create an aggregate model for EV charging and discharging behaviors, facilitating more accurate predictions and optimizations.

2.3. *Monte Carlo Simulation*

MCS is used to analyze various samples of wind power, solar power, and EV charging/discharging power. This analysis is based on the stochastic models developed in Sections 2.1 and 2.2 of the paper.

The generated samples serve as inputs into our objective function (f) to calculate the MCS outputs (Y), as represented by the following equation:

$$Y = f(P_{W-DG}, P_{PV-DG}, P_c, P_D) \quad (8)$$

These MCS outputs create scenarios that replicate real-world conditions by simulating all possible states. At the conclusion of each simulation, we calculate the average values of the outputs. We ensure that the simulation is repeated sufficiently to achieve convergence to an optimal solution, at which point the average values stabilize and do not change with the inclusion of additional scenarios.

2.4. Distribution Network Reconfiguration (DNR)

Integration of DGs and EVs in distribution networks poses challenges for efficient and reliable operation of the system. The increased charging demands from EVs can elevate system loading, potentially stressing the network. Additionally, the power flow between DGs and EVs may lead to increased distribution system losses.

To tackle these issues, we propose a system reconfiguration optimization strategy aimed at adjusting the physical layout of the network to enhance power flow efficiency. This involves optimally opening normally closed sectionalizing switches and/or closing typically open tie switches.

An unbalanced three-phase power flow (PF) analysis is employed to calculate power loss for both the original and reconfigured systems. This analysis accounts for the inherent heterogeneity of overhead lines and the unbalanced nature of loads within the distribution network. By comparing power losses between the original and reconfigured networks, we can quantify the reduction in power loss achieved through network reconfiguration.

The objective function for minimizing power loss within the distribution grid is modeled as follows:

$$Obj.Function = \min(P_{Loss}) = \min\left(\sum_{l=1}^{N_{br}+N_{ts}} x_l I_l^2 R_l\right) \quad (9)$$

P_{Loss} is the active power loss within the distribution network. N_{br} and N_{ts} are the number of branches and tie-switches. R_l, I_l, and x_l are the resistance, current, and status of the lth branch/tie switch.

The objective function is subject to specific constraints as follows:

2.4.1. Node Voltage Limit

The operating voltage at each bus (i) within the distribution network must adhere to standard limits, defined as follows:

$$v_i^{min} \le v_i \le v_i^{max} \quad (10)$$

v_i is the operating voltage at each bus, with v_i^{min} set at 0.95 pu and v_i^{max} at 1.05 pu.

2.4.2. Feeder Capacity Limit

To ensure safe and reliable operation, the current through each branch must not exceed its maximum capacity:

$$I_l \le I_l^{max} \qquad l\epsilon\{1, 2, 3, \ldots, N_{br} + N_{ts}\} \quad (11)$$

I_l^{max} is the maximum current threshold of the lth branch.

2.4.3. Bus Isolation Limit

To maintain connectivity, only one switch should remain open in a loop, ensuring that each bus is connected to at least one feeder during system reconfiguration. This prevents open loops and guarantees a constant power supply across the network.

2.4.4. Network Configuration Limit

The number of closed switches must equal the number of buses minus one, characteristic of a radial network. This constraint ensures that there is only one path between any two points in the network and is modeled as follows:

$$\sum_{1}^{(N_{Br}+B_{ts})} x_l = N_b - 1 \tag{12}$$

N_b is the number of buses within the distribution network.

2.4.5. Power Balance Limit

Active and reactive power balance must be maintained in the reconfigured distribution system to ensure that all loads are met while compensating for losses. The balance equations are as follows:

$$\begin{aligned} \sum P_i^{Gen} &= P_{Loss} + \sum P_i^{Load} \\ \sum Q_i^{Gen} &= Q_{Loss} + \sum Q_i^{Load} \end{aligned} \tag{13}$$

P_i^{Gen} and P_i^{Load} are the active power generation and demand at bus i. Q_i^{Gen} and Q_i^{Load} are the reactive power generation and demand at bus i. P_{Loss} and Q_{Loss} are the active and reactive power losses within the distribution network.

GA is used in our analysis to achieve optimal reconfiguration and minimize system losses.

3. Case Study

This section provides an overview of the test system utilized in our study to evaluate the performance of the proposed optimization method. The system comprises a 13.8 kV, 134-bus distribution network, as illustrated in Figure 1, which provides the single-line diagram.

In this setup, we installed a 10 MW renewable DG capacity across six buses within the network. Specifically, buses 67, 84, and 103 are connected to three wind DGs operating at a power factor of 0.85. Additionally, buses 11, 39, and 52 host three solar DGs with a power factor of 0.95.

Furthermore, the network accommodates 484 EVs organized into six fleets. Reference [19] is used to derive the EV driving pattern data and to generate the aggregated charging and discharging power for each fleet. To enhance energy flow between the EVs and the renewable DGs, EV charging stations are strategically located on the same buses as the DGs. Notably, the EVs are equipped with vehicle-to-grid (V2G) functionality, allowing them to discharge power back to the grid when needed.

To simulate the renewable DG generation and the charging/discharging power of the EVs over a 24 h period, we generated 10,000 scenarios using MCS and averaged the results. The Mersenne Twister method is used for generating random numbers in the MCS due to its long period and efficiency. The MCS is run for a total of 10,000 iterations to ensure sufficient statistical sampling and convergence. Convergence is monitored by checking the stability of the mean across successive iterations. Convergence is defined as achieving a relative change of less than 1% in the mean value over the last 1000 iterations. This threshold is chosen to ensure that the simulation results are stable and reliable.

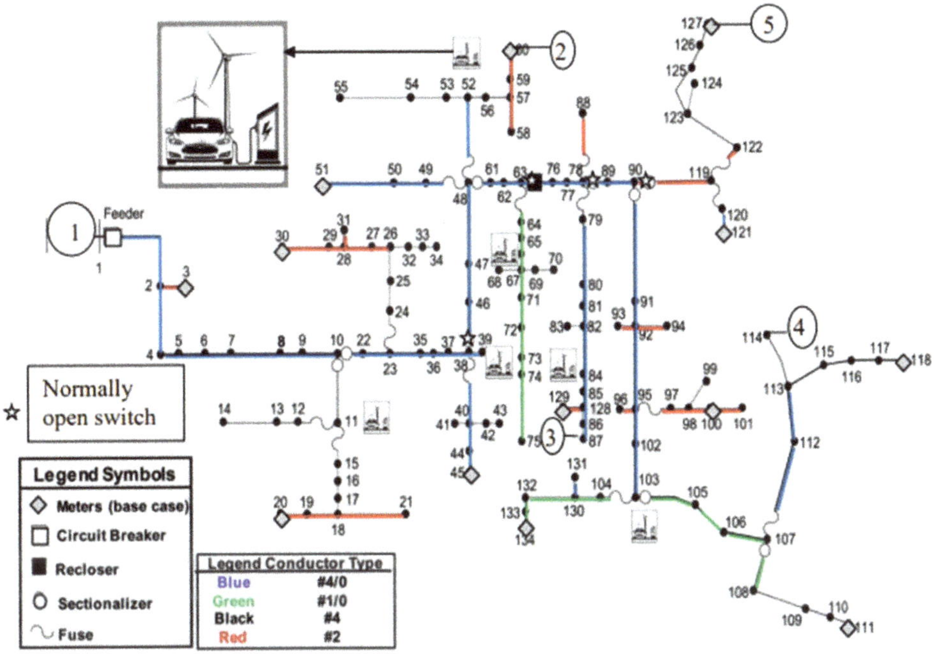

Figure 1. Single-line diagram of the test network.

Table 1 provides the parameters of the GA used for DRN optimization in this study.

Table 1. GA parameters.

Parameter	Population Size	Max Generation	Crossover Rate	Mutation Rate	Elitism Number
Value	200	150	78%	20%	2

4. Simulation Results

The State of Charge (SOC) values for each EV fleet are calculated across 10,000 MCS scenarios. These simulations model a variety of potential real-world conditions, allowing for a comprehensive analysis of the fleet's charging and discharging behaviors. Once the SOC values are calculated for each simulation, they are averaged over time to determine the hourly power consumption or generation profile for each fleet. This averaging helps smooth out fluctuations and provides a reliable representation of the fleet's energy use over time.

To determine the charging and discharging power for each fleet, the SOCs of consecutive hours are subtracted from one another. The difference between these values indicates whether the fleet is charging (positive difference) or discharging (negative difference). This subtraction method enables the tracking of power flows in and out of the EV fleet during the simulation period. Figure 2 visually illustrates the EV charging and discharging power over the 24 h simulation period. In the figure, positive values represent power being supplied from the grid to the EVs (charging), while negative values indicate power being discharged from the EVs back into the grid (discharging).

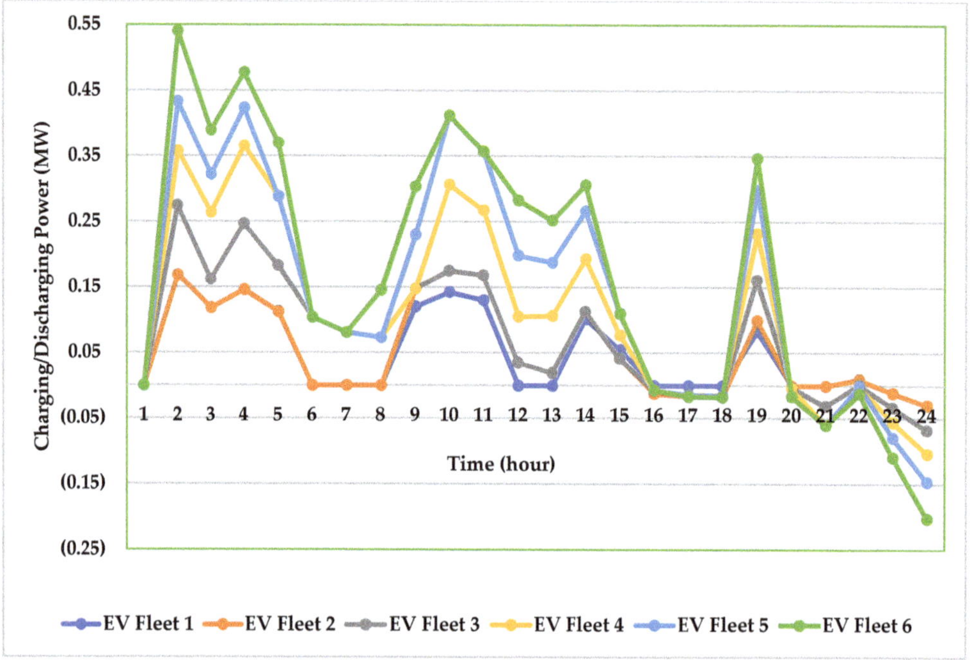

Figure 2. Charging and discharging of the EV fleets for the 24 h simulation period.

In some cases, the values are very close to zero, which corresponds to minimal or negligible power exchange between the grid and the EV fleet. These small fluctuations can be attributed to the relatively steady SOC during certain hours, where the battery power remains constant, and there is little need for charging or discharging.

For the purposes of this study, the EVs are assumed to have a battery power rating of 6 kW, which serves as a baseline for the charging and discharging rates. However, the average power of each fleet varies depending on the number of EVs within the fleet, which is determined through a clustering process. The clustering groups the EVs based on their usage patterns, allowing for the simulation of different fleet sizes and their corresponding power demands or generation capabilities.

The proposed reconfiguration optimization is used in this section to minimize system losses through the optimal operation of switches. This optimization involves strategically closing and opening switches, and the effectiveness of loss reduction is assessed by conducting a three-phase power flow (PF) analysis on the reconfigured system. The power loss is then compared to that of the original network.

The optimization results indicate the following configurations:

- Four tie switches between buses 38–46, 63–76, 78–89, and 90–119 are set to a normally open status.
- Three sectionalizing switches between buses 10–22, 61–48, and 105–106 are maintained in a normally closed status.
- Additionally, new sectionalizing switches have been integrated between buses 62–63, 89–90, 95–102, and 102–103.

Table 2 provides a detailed view of the optimization results, outlining the hourly switching actions over the 24 h simulation period. In this table, green cells indicate switch-closing actions, red cells represent switch-opening operations, and shaded cells denote idle switching operations.

Table 2. DNR optimization results.

Switch Bus#-Bus#/Line# I Hour	1	2	3	4	5	6	7	8	9	10	11	12	13	14	15	16	17	18	19	20	21	22	23	24
Switch10–22/Line 23	o																							
Switch 38–46/Line 47	c																							
Switch 61–48/Line 62		o																						
Switch 62–63/Line 62	o	c																						
Switch 63–76/Line 77	c																							
Switch 78–89/Line 91	c																							
Switch 89–90/Line 92	o	c		o			c									o					c			
Switch 95–102/Line 104	o	c		o			c					c			o		c			o				o
Switch 102–103/Line 105		o		c			o					o			c		o			c				
Switch 105–106/Line 108																								o
Switch 90–119/Line 121	c	o		o			o									c					o			

Figure 3 presents the power flow results, comparing system losses for both the original and reconfigured networks. Figure 4 demonstrates the loss reduction in percentage due to DNR optimization.

The simulation results can be characterized into two main areas: planning optimization and real-time operation optimization.

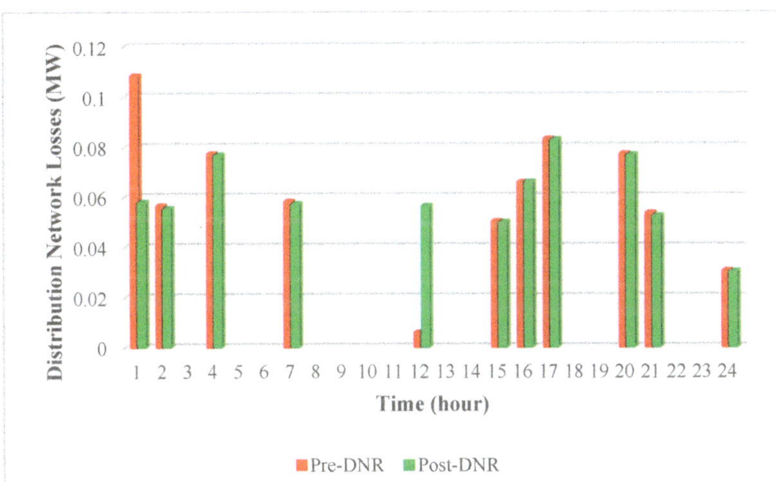

Figure 3. Distribution network losses before and after DNR for the 24 h simulation period.

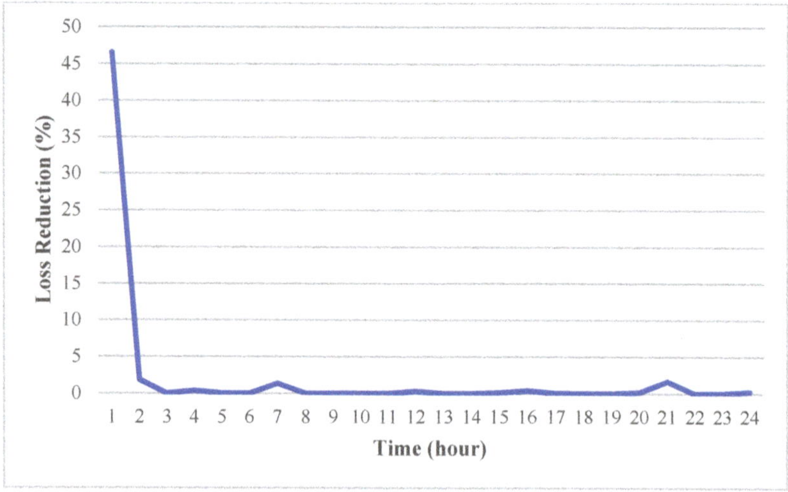

Figure 4. Distribution network loss reduction after DNR optimization for the 24 h simulation period.

4.1. Planning Optimization

The simulation results demonstrate that adjusting the system configuration during hour 1 led to a substantial reduction in power losses, achieving a remarkable decrease of up to 46.52%. This significant improvement was made possible by strategically integrating new sectionalizing switches and executing eight switching actions that involved both the existing and newly added switches. By modifying the system configuration, power flow was more efficiently managed, which drastically minimized power loss. This emphasizes the importance of a well-thought-out planning approach that optimizes the distribution network's structure in advance to reduce operational losses. The drastic reduction in power loss showcases the effectiveness of this planning strategy in enhancing the overall efficiency of the distribution network. These results underscore the effectiveness of the proposed optimization method in a planning context, where advanced system configuration and proactive changes in the distribution network can significantly enhance its efficiency and reduce operational losses. This emphasizes that the primary strength of the proposed method lies in its ability to optimize the network structure well in advance, in line with long-term planning objectives.

4.2. Real-Time Operation Optimization

From hours 2 to 24, the switching actions were largely responsive to dynamic changes in EV charging demand and the output from renewable DGs (such as solar and wind). These switching operations were designed to adapt to fluctuating energy generation from renewable sources and changing demand patterns caused by electric vehicle charging behaviors. Unlike the planning optimization, where substantial changes were made, the power losses during these hours were relatively minimal, resulting in only marginal power reductions. This indicates that the system had already been optimized in the earlier planning stage, and the real-time optimization mainly focused on fine-tuning the network in response to real-world fluctuations. This suggests that the proposed method is not primarily intended for real-time operation but rather as a planning tool designed to set up a robust foundation for the network. While real-time operations are important for adapting to immediate changes, the core advantage of the proposed method lies in its ability to establish an optimized system configuration that minimizes operational losses before these fluctuations even occur.

4.3. Discussion

Given these simulation results, it is clear that the proposed method is most effective as a planning tool, ensuring the distribution network is optimized well in advance to minimize losses during actual operation. While real-time adjustments are important, the proposed method's value is realized primarily through strategic planning, where the significant reduction in power losses from hour 1 illustrates its potential for long-term efficiency improvements.

The real-time system reconfiguration and switching actions pose several challenges, such as communication delays (latency) and limited real-time control and automation [20]. Communication systems can experience delays due to various factors, such as network congestion, limited bandwidth, or the distance between components. These delays can cause a lag between the occurrence of a system event (e.g., a change in energy demand or renewable generation) and the response from the control system. This can result in delayed or suboptimal switching actions, which may affect the system's stability or efficiency, especially when rapid reconfigurations are needed. Additionally, some areas of the grid may still rely on manual intervention or have slower automated response times due to limitations in control algorithms or hardware. This can lead to slower decision-making and may require human operators to intervene in some cases. Real-time switching actions would, therefore, be delayed, reducing the overall effectiveness of the reconfiguration. While the proposed method is designed as a planning tool, it could be adapted for real-time applications, with modifications to account for communication latency and other practical constraints. Future research could focus on integrating faster communication systems, distributed control mechanisms, or more advanced real-time optimization techniques to mitigate these challenges.

5. Sensitivity Analysis

A sensitivity analysis is conducted to evaluate the performance of the distribution system reconfiguration optimization at varying penetration levels of EVs and renewable DGs. The penetration level refers to the proportion of the total system load that is attributed to either EVs or renewable DGs, as outlined in Table 3. For this analysis, the total load of the system is fixed at 35.32 MVA. The penetration levels of both EVs and renewable DGs are expressed as percentages of the total system load and are distributed equally among these two technologies.

Table 3. MVA load for EV fleets and renewable DGs at various penetration levels.

Penetration Level	25%	50%	75%	90%
DG/EV Fleet 1	0.74	1.47	2.21	2.65
DG/EV Fleet 2	0.74	1.47	2.21	2.65
DG/EV Fleet 3	0.74	1.47	2.21	2.65
DG/EV Fleet 4	0.74	1.47	2.21	2.65
DG/EV Fleet 5	0.74	1.47	2.21	2.65
DG/EV Fleet 6	0.74	1.47	2.21	2.65

To effectively conduct the sensitivity analysis, the system is tested under two distinct scenarios to isolate the impacts of each technology individually:

1. For Renewable DGs: The impact of renewable DGs is assessed by setting the penetration level of EVs to 0% (effectively turning off all EVs). In this case, the focus is placed entirely on varying the penetration levels of renewable DGs within the system. This

allows us to understand how the integration of renewable energy sources, such as solar and wind, influences the system's performance, including power loss reduction and efficiency, when there is no additional load from EVs.

2. For EVs: Conversely, to examine the effects of EV penetration, the renewable DGs are set to 0% penetration, effectively turning off the renewable generators. This scenario focuses solely on the effects of EV integration on the distribution system. By doing so, the analysis can identify how varying levels of EV charging demand impact the overall network performance, including any potential overloads or increased losses that may result from the fluctuating charging behaviors of electric vehicles.

As in our previous study, we assume that switches 38–46, 63–76, 78–69, and 90–119 are normally open tie switches connecting five feeders before optimization. These switches serve as critical network components that can be toggled to alter the system configuration and optimize power flow. The status of these switches is important in the optimization process, as their operation directly affects the distribution network's efficiency, especially in terms of power loss reduction and voltage regulation.

For the analysis, hour 1 is selected as the focal point for planning optimization due to the significant power loss reduction observed during this period. Hour 1 is seen as the optimal starting point for reconfiguration because it reflects the most impactful configuration adjustments that yield the highest reduction in power losses. The sensitivity analysis is specifically conducted at different penetration levels of EVs and renewable DGs for this hour. The focus on hour 1 allows for a detailed understanding of how the system responds to changes in penetration levels of EVs and DGs during an optimized planning phase.

However, it is important to note that although the sensitivity analysis is concentrated on hour 1, it can be extended to cover the entire 24 h simulation period.

Tables 4 and 5 present the results of the sensitivity analysis. These tables include the proposed switching actions at various penetration levels of EVs and DGs, as well as the corresponding power loss reductions achieved under these conditions. The results provide a clear picture of how different configurations of the distribution network, influenced by varying EV and DG penetration levels, impact overall network performance. By reviewing these results, we can assess how the proposed optimization strategies perform in different scenarios and determine the most effective configurations for minimizing losses and ensuring the network's efficient operation under various levels of renewable energy and EV integration.

5.1. Key Findings

The analysis revealed several important insights into the relationship between EV penetration, DG penetration, and system losses. These findings provide a deeper understanding of how the optimization process can reduce power losses based on varying levels of these technologies.

5.1.1. Impact of EV Penetration on Loss Reduction

As EV penetration in the system increases, the percentage of loss reduction also rises. This is because an increase in EV penetration corresponds to a higher total system load, which provides more opportunities for the optimization process to minimize losses. When more EVs are integrated, the overall demand on the system increases, allowing the reconfiguration process to adjust the network more efficiently and reduce losses. This phenomenon demonstrates how higher load levels can improve the effectiveness of the optimization in reducing power losses.

Table 4. DNR optimization and loss reduction for variable EV penetration with 0% DG penetration.

Switch Bus#-Bus#/Line# | Action	**25% EV**	**50% EV**	**75% EV**	**90% EV**
Switch 22–23/Line 24	NO ACTION	OPEN	OPEN	OPEN
Switch 23–35/Line 36	OPEN	NO ACTION	NO ACTION	NO ACTION
Switch 89–90/Line 92	OPEN	OPEN	NO ACTION	NO ACTION
Switch 102–103/Line 105	OPEN	OPEN	OPEN	OPEN
Switch 48–61/Line 62	OPEN	OPEN	OPEN	OPEN
Switch 38–46/Line 47	CLOSE	CLOSE	CLOSE	CLOSE
Switch 63–76/Line 77	CLOSE	CLOSE	CLOSE	CLOSE
Switch 78–69/Line 91	CLOSE	CLOSE	NO ACTION	NO ACTION
Switch 90–119/Line 121	CLOSE	CLOSE	CLOSE	CLOSE
System Losses				
Before Optimization (MW)	1.365159	2.041012	2.943356	3.617399
After Optimization (MW)	0.814860	1.038650	1.462397	1.758691
| Difference |	0.550299	1.002362	1.480959	1.858708
Difference (%)	40.310274	49.111023	50.315321	51.382448

Table 5. DNR optimization and loss reduction for variable DG penetration with 0% EV penetration.

Switch Bus#-Bus#/Line# | Action	**25% DG**	**50% DG**	**75% DG**	**90% DG**
Switch 22–23/Line 24	NO ACTION	NO ACTION	OPEN	OPEN
Switch 23–35/Line 36	OPEN	NO ACTION	NO ACTION	NO ACTION
Switch 56–57/Line 58	NO ACTION	OPEN	NO ACTION	NO ACTION
Switch 59–60/Line 61	NO ACTION	NO ACTION	OPEN	NO ACTION
Switch 80–81/Line 83	NO ACTION	NO ACTION	NO ACTION	OPEN
Switch 89–90/Line 92	OPEN	NO ACTION	OPEN	OPEN
Switch 91–92/Line 94	NO ACTION	NO ACTION	NO ACTION	OPEN
Switch 92–95/Line 97	NO ACTION	NO ACTION	OPEN	NO ACTION
Switch 38–46/Line 47	CLOSE	CLOSE	CLOSE	CLOSE
Switch 48–61/Line 62	OPEN	OPEN	NO ACTION	NO ACTION
Switch 63–76/Line 77	CLOSE	CLOSE	CLOSE	CLOSE
Switch 78–69/Line 91	CLOSE	CLOSE	CLOSE	CLOSE
Switch 95–102/Line 104	OPEN	NO ACTION	NO ACTION	NO ACTION
Switch 103–105/Line 107	NO ACTION	OPEN	NO ACTION	NO ACTION
Switch 90–119/Line 121	CLOSE	CLOSE	NO ACTION	CLOSE
Simulation Results				
Before Optimization (MV)	0.094821	0.121754	0.249291	0.490176
After Optimization (MV)	0.068204	0.077211	0.240431	0.365409
| Difference |	0.026617	0.044542	0.008860	0.124767
Difference (%)	28.070539	36.583857	3.554136	25.453480

5.1.2. Non-Linear Correlation Between DG Penetration and Loss Reduction

Unlike EV penetration, the loss reduction percentage is not linearly correlated with the DG penetration level. In scenarios where DG penetration is either significantly lower or higher than the total system load, there tends to be a substantial power flow either from the grid to the distribution system or vice versa. This imbalance in power flow causes additional system losses, but these losses can still be minimized through the optimization process.

On the other hand, when the DG penetration is more closely matched to the system load, the local DG units are able to supply power directly to nearby loads. This results in minimal power flow between the distribution system and the grid, leading to lower system losses. As a result, the optimization process is less impactful in these scenarios because the need for further loss reduction is minimal.

An optimal DG penetration range appears to be one where DG capacity is well-aligned with the distribution network's demand profile, typically in the moderate range of 40–60% of system load, allowing for efficient power distribution and minimal grid reliance.

5.1.3. Effect of DG Proximity to Load Pockets

The effectiveness of loss reduction is also influenced by the proximity of DG units to local load pockets. When large loads are served by nearby DGs, there is a reduced need for power to flow through the main branches of the distribution system. This reduces the amount of power loss in the system, as energy is consumed closer to its point of generation, minimizing transmission and distribution losses. In these situations, the optimization process is less effective because the system is already operating efficiently with minimal loss.

For instance, in the scenario where DG penetration is 75%, the loss reduction is relatively small (only 3%). This indicates that in cases with high DG penetration but limited proximity of DGs to large loads, the power flow remains optimized with little need for further adjustments, making the optimization process less impactful. Therefore, strategically locating DG units closer to major load centers can significantly reduce the need for extensive reconfiguration, and this can be an important consideration in determining the optimal penetration rate.

5.2. Switching Actions

The number of switching actions required in a distribution network can vary significantly depending on the penetration levels of EVs and renewable DG sources. In particular, the number of proposed switching actions for variable DG penetrations at 0% EV penetration is notably higher compared to those needed for varying EV penetrations. This discrepancy is primarily driven by the unique characteristics of renewable DG sources, specifically their intermittent nature and the associated challenges in balancing power generation and load.

As DG penetration increases, the network needs to adjust to the real-time availability of renewable power. The higher the DG penetration, the more frequently the system needs to reconfigure itself to either connect or disconnect specific parts of the network to align power generation with the demand in different areas.

These switching actions are essential to optimize the flow of electricity from renewable sources, ensuring that power generated by wind or solar energy is used efficiently and that the network remains stable. For instance, in times when solar generation is high, certain feeders might be reconfigured to utilize this local generation, while at other times, when solar output drops, the network may need to redirect power from the grid to maintain supply.

On the other hand, when EV penetration increases, the need for frequent switching actions is less pronounced. While EV charging patterns can certainly introduce additional demand and variability in load, they do not have the same level of fluctuation as renewable DG generation. EVs typically charge based on pre-set schedules or demand, which is more predictable than the instantaneous changes in solar or wind power.

Thus, while the increase in EV penetration does require adjustments in load management and potentially some switching, it does not demand as many dynamic and responsive switching actions as high DG penetration. This is because the charging behaviors of EVs are generally more stable and less subject to sudden fluctuations, unlike the intermittent nature of renewable energy generation.

Therefore, while both EV and DG penetration require network adjustments, DG penetration has a higher impact on the frequency of switching actions due to the fluctuating nature of renewable generation. This insight suggests that balancing DG and EV penetration rates is key to minimizing the need for excessive network reconfigurations.

5.3. *Optimal Penetration Rates*

Based on our analysis, this study offers recommendations regarding the most effective penetration rate levels for both technologies, which can significantly enhance network performance.

5.3.1. Optimal Penetration Rates for EVs

As our analysis demonstrates, increased EV penetration in the distribution network leads to a higher total system load, which offers more opportunities for the optimization process to reduce power losses. However, the effectiveness of loss reduction increases with the penetration of EVs up to a certain level, after which the system load may become too high to optimize effectively without additional infrastructure or support.

We suggest that the optimal EV penetration range lies between 20 and 40% of the total system load. This range allows the system to benefit from the additional load provided by EVs while maintaining manageable operational flexibility. Within this range, the optimization process can adjust the network configuration efficiently, leading to noticeable reductions in power losses.

5.3.2. Optimal Penetration Rates for DGs

In the case of DG penetration, the relationship with loss reduction is more complex. At low DG penetration levels, the system still depends heavily on power from the grid, leading to power flow imbalances and increased losses. Conversely, at very high DG penetration, power generation can become highly intermittent, causing potential imbalances in power flow and additional system losses.

Our findings suggest that the most effective loss reduction occurs when DG penetration is between 40 and 60% of the total system load. This range ensures that the local generation from DG units closely matches the system load, thereby reducing the need for power to flow from the grid and minimizing losses. When DG penetration exceeds this range, intermittent generation (e.g., from wind or solar) can cause fluctuations in power generation and demand, which might reduce the effectiveness of the optimization process.

5.4. *Policy Implications*

Based on the findings above, we suggest that grid planners aim for a balanced integration of EVs and DGs within these optimal penetration ranges to achieve the best network performance.

5.4.1. Targeted Incentives for EV Deployment

Policymakers should consider incentivizing EV adoption to reach the 20–40% penetration level within a reasonable timeframe. This will ensure that the additional load from EVs enhances the overall network performance by providing opportunities for optimization while keeping the system load manageable.

5.4.2. Support for DG Integration

Efforts should be made to encourage DG penetration within the 40–60% range. Policies that support the alignment of DG generation with local load demands, such as encouraging the installation of DG systems in areas with high energy consumption, will maximize the benefits of DG integration while minimizing additional losses caused by generation imbalances.

By focusing on these optimal penetration ranges, policymakers can create an efficient and sustainable energy system that balances the benefits of EVs and DGs while minimizing system losses and maximizing the potential for loss reduction.

6. Conclusions

This study presents a multi-objective network reconfiguration strategy aimed at effectively integrating EVs and renewable DGs into distribution grids. To address the variabilities in renewable generation and system uncertainties, we employed probabilistic models for wind and solar energy combined with Monte Carlo simulation techniques. This approach allows us to accurately capture the real-world conditions of such systems. Our application of this method to a 134-bus distribution network demonstrated significant efficacy in reducing power losses.

The optimization results are categorized into two phases: planning optimization and real-time operation optimization. The highest loss reduction was observed at hour 1, driven by integrating new sectionalizing switches and implementing switching actions during the planning phase. The optimization efforts during hours 2 to 24 focused on real-time operational adjustments, where the system continuously adapted to immediate changes in renewable generation and EV charging behaviors. While these adjustments were smaller in scale compared to the planning phase, they played a critical role in maintaining system stability and ensuring that the network could efficiently handle the variable nature of renewable energy and EV charging demands.

The sensitivity analysis conducted as part of this study provides optimal switching actions to minimize power losses while ensuring voltage stability and load balancing as EV and DG penetration levels change. Notably, the results of the sensitivity analysis indicate that more switching actions are required for variable DG penetration compared to variable EV penetration. The increased need for switching actions with renewable DGs reflects the dynamic nature of renewable energy generation. Unlike EVs, which introduce a more predictable load, the intermittent nature of wind and solar power generation necessitates more frequent adjustments to the distribution system to maintain balance and minimize losses. These dynamic adjustments are essential for optimizing system performance, reducing power losses, and ensuring network stability as the penetration of renewable energy sources increases.

However, the increased number of switching actions introduces additional challenges in terms of system complexity and control. Networks with high penetration of renewables require advanced control algorithms and real-time data monitoring to adjust the network configuration efficiently, ensuring that these increased switching actions do not result in undesirable side effects, such as system instability or operational delays.

Additionally, the most effective network performance in terms of loss reduction and switching actions can be expected when EV penetration ranges between 20 and 40% of the total system load and DG penetration is between 40 and 60% of the total system load. This range ensures that the system benefits from the additional load provided by EVs while maintaining a balance between DG generation and local demand, minimizing losses, and optimizing network performance. Policymakers should aim to target these ranges when planning for future grid integration, as they will likely yield the most efficient and sustainable performance while minimizing system losses and the need for frequent switching.

While the proposed strategy shows great promise, real-world implementation will require addressing additional challenges, including policy and regulatory issues, as well as the development of necessary communication and data infrastructure. These considerations are crucial as the adoption of EVs and renewable DGs continues to accelerate.

Funding: This research was funded by the University of Washington Bothell SRCP Seed Grant #06-0398.

Data Availability Statement: The datasets presented in this article are not readily available because the data are part of an ongoing study.

Conflicts of Interest: The authors declare no conflicts of interest.

Nomenclature

$f_v(v)$	Weibull PDF for wind speed.
k	Shape parameter for Weibull PDF.
λ	Scale parameter for Weibull PDF.
v	Wind speed.
P_{W-DG}	Output power of the wind DG.
$P_{W-DG_{rated}}$	Rated power of the wind DG.
v_{cut-in}	Cut-in speed of the wind turbine.
$v_{cut-out}$	Cut-out speed of the wind turbine.
v_r	Rated speed of the wind turbine.
k_t	Hourly clearness index.
I_t	Irradiance on a horizontal plane (kW/m^2).
I_o	Extraterrestrial solar irradiance (kW/m^2).
$f_{ci}(k_t)$	PDF for clearness index.
γ	Gamma function.
a, b	Shape parameters of the Beta distribution.
P_{PV-DG}	Output power of the PV DG.
A_{PV}	Area of the solar PV array.
η_{PV}	Efficiency of the solar PV array.
P_c	Charging power of the EV battery.
P_d	Discharging power of the EV battery.
P_c^{max}	Maximum charging capacity of the EV battery.
P_d^{max}	Maximum discharging capacity of the EV battery.
L	Number of time steps throughout the simulation period.
P_{Loss}	Active power loss within the distribution network.
Q_{Loss}	Reactive power loss within the distribution network.
N_{br}	Number of branches.
N_{ts}	Number of tie switches.
R_l	Resistance of the lth branch.
I_l	Current of the lth branch.

x_l	Status of the lth branch/tie switch (0 for open and 1 for close).
v_i	Operating voltage at bus i.
v_i^{min}	Minimum voltage at bus i.
v_i^{max}	Maximum voltage at bus i.
I_l^{max}	Maximum current threshold of the lth branch.
N_b	Number of buses within the distribution network.
P_i^{Gen}	Active power generation at bus i.
P_i^{Load}	Active power demand at bus i.
Q_i^{Gen}	Reactive power generation at bus i.
Q_i^{Load}	Reactive power demand at bus i.

References

1. Sharma, T.; Chauhan, S. Fast heuristic based distribution network reconfiguration with distributed generations. In Proceedings of the 5th International Conference for Emerging Technology (INCET), Belgaum, India, 24–26 May 2024; pp. 1–6.
2. Chanbasha, S.; Jayakumar, N.; Kumar, N.B. Hybrid Ttloa model for simultaneous distribution network reconfiguration and optimal placement of distributed generators for loss reduction: Validation on multiple test bus systems and outage scenarios. *J. Electr. Syst.* **2024**, *20*, 5146–5168.
3. Wang, H.; Pan, J.-S.; Nguyen, T.-T.; Weng, S. Distribution network reconfiguration with distributed generation based on parallel slime mould algorithm. *Energy* **2022**, *244*, 123011. [CrossRef]
4. Mahdavi, E.; Asadpour, S.; Macedo, L.H.; Romero, R. Reconfiguration of distribution networks with simultaneous allocation of distributed generation using the whale optimization algorithm. *Energies* **2023**, *16*, 4560. [CrossRef]
5. Zhang, J.; Chen, C.; Xu, Z.; Dong, M.; Han, D.; Zhou, X. An optimization strategy for distribution network reconfiguration considering integration of electric vehicles. In Proceedings of the IEEE 3rd China International Youth Conference on Electrical Engineering (CIYCEE), Wuhan, China, 3–5 November 2022; pp. 1–6.
6. Mejia, M.; Macedo, L.H.; Muñoz-Delgado, G.; Contreras, J.; Padilha-Feltrin, A. Active distribution system planning considering non-utility-owned electric vehicle charging stations and network reconfiguration. *Sustain. Energy Grids Netw.* **2023**, *35*, 101101. [CrossRef]
7. Paudel, A.; Supingklad, W.; Ongsakul, W. Distribution network reconfiguration to support electric vehicles integration. In Proceedings of the International Conference on Cogeneration, Small Power Plants and District Energy (ICUE), Bangkok, Thailand, 14–16 September 2016; pp. 1–6.
8. Srinivas, D.; Reddy, M.R. Optimal planning of electric vehicle charging station in radial distribution network along with reconfiguration. In Proceedings of the International Conference on Power Energy, Environment & Intelligent Control (PEEIC), Greater Noida, India, 19–23 December 2023; pp. 1039–1044.
9. Gao, F.; Du, K.; Chen, X.; Xu, Y.; Li, D. Reconfiguration of the electric vehicle integrated distribution system considering carbon-reduction benefit. In Proceedings of the IEEE 5th International Electrical and Energy Conference (CIEEC), Nangjing, China, 27–29 May 2022; pp. 4975–4981.
10. Wang, J.; Di, F.; Meng, H.; Meng, M.; Luo, X.; Wang, Y. Coordinated distribution reconfiguration with maintenance scheduling considering electric vehicles charging uncertainty. In Proceedings of the IEEE 5th International Electrical and Energy Conference (CIEEC), Nangjing, China, 27–29 May 2022; pp. 168–173.
11. Tekeler, E.; Povrazoglu, G. Revisiting network reconfiguration in the era of unpredictable EV charger integration. In Proceedings of the 5th Global Power, Energy and Communication Conference (GPECOM), Nevsehir, Turkiye, 14–16 June 2023; pp. 489–494.
12. Goli, P.; Gampa, S.R.; Kumar, N.M.; Jasthi, K.; Parimi, A.M.; Das, D.; Bansal, R.C.; Siano, P.; Guerrero, J.M. Strategic planning of distribution network integrated with EV charging stations using fuzzy pareto optimality for performance improvement and grid-side emission reduction benefits. *Sustain. Energy Grids Netw.* **2023**, *36*, 101–199. [CrossRef]
13. Wang, J.; Wang, W.; Wang, H.; Zuo, H. Dynamic reconfiguration of multiobjective distribution networks considering DG and EVs based on a novel LDBAS algorithm. *IEEE Access* **2020**, *8*, 216873–216893. [CrossRef]
14. Ghadi, Y.; Kotb, H.; Aboras, K.M.; Alqarni, M.; Yousef, A.; Dashtdar, M.; Alanazi, A. Reconfiguration and displacement of DG and EVs in distribution networks using a hybrid GA-SFLA multi-objective optimization algorithm. *Front. Energy Res.* **2023**, *11*, 1304055. [CrossRef]
15. Pagidipala, S.; Sandeep, V. Optimal planning of electric vehicle charging stations and distributed generators with network reconfiguration in smart distribution networks considering uncertainties. *Meas. Sens.* **2024**, *36*, 101400. [CrossRef]
16. Pratap, A.; Tiwari, P.; Maurya, R. Optimal allocation of controllable power factor distributed generation with network reconfiguration on electric vehicle charging station loaded distribution network. *Int. J. Ambient. Energy* **2024**, *45*, 2326157. [CrossRef]

17. Katanchi, B.; Shojaei, A.A.; Yaghoobi, M. Distribution network reconfiguration in effective presence of DGs, EVCSs and DRP: A novel multi-objective approach. *Electr. Power Compon. Syst.* **2024**, 1–17. [CrossRef]
18. Choobdari, M.; Samiei Moghaddam, M.; Davarzani, R.; Azarfar, A.; Hoseinpour, H. Robust distribution networks reconfiguration considering the improvement of network resilience considering renewable energy sources. *Sci. Rep.* **2024**, *14*, 23041. [CrossRef] [PubMed]
19. Available online: https://github.com/groupoasys/Data_EV_Driving_Patterns?utm_source=chatgpt.com (accessed on 1 March 2025).
20. Lotfi, H.; Hajiabadi, M.E.; Parsadust, H. Power distribution network reconfiguration techniques: A thorough review. *Sustainability* **2024**, *16*, 10307. [CrossRef]

energies

MDPI

Article

Smart Grids and Sustainability: The Impact of Digital Technologies on the Energy Transition

Paola Campana *, Riccardo Censi, Roberto Ruggieri and Carlo Amendola

Department of Management, Sapienza University of Rome, 00185 Rome, Italy; riccardo.censi@uniroma1.it (R.C.); roberto.ruggieri@uniroma1.it (R.R.); carlo.amendola@uniroma1.it (C.A.)

* Correspondence: paola.campana@uniroma1.it; Tel.: +39-335-715-3413

Abstract: Smart Grids (SG) represent a key element in the energy transition, facilitating the integration of renewable and conventional energy sources through the use of advanced digital technologies. This study analyzes the main research trends related to SG, energy efficiency, and the role of Artificial Intelligence (AI) and the Internet of Things (IoT) in smart energy management. Following the PRISMA protocol, 179 relevant academic articles indexed in the Scopus database were selected and analyzed using VOSviewer software, version 1.6.20, to identify the main thematic clusters. The results reveal a converging research focus on energy flow optimization, renewable energy integration, and the adoption of digital technologies—including cybersecurity solutions—to ensure grid efficiency, security, and resilience. The study confirms that digitalization acts as a key enabler for building a more sustainable and reliable energy system, aligned with the objectives of the European Union and the United Nations 2030 Agenda. The contribution of this work lies in its integrated approach to the analysis of digital technologies, linking the themes of efficiency, resilience, and infrastructure security, in order to provide valuable insights for future research and sustainable energy policy development.

Keywords: smart grids; digital energy technologies; artificial intelligence (AI) in energy; energy efficiency; cybersecurity in power systems

1. Introduction

Energy is a fundamental element for economic and social development, influencing both quality of life and the competitiveness of modern economies [1,2]. In recent decades, the increasing energy demand, coupled with the need to reduce greenhouse gas emissions and the growing focus on sustainability, has made the transition toward more efficient and resilient production and consumption models imperative [1,3,4]. This transformation is driven by technological innovation and the implementation of energy policies aimed at decarbonization. The rising energy demand and the urgent need to mitigate greenhouse gas emissions necessitate a radical transformation of energy infrastructures [5]. In this context, SGs represents a pivotal evolution, enabling the optimization of energy production, distribution, and consumption through the deployment of digital technologies and advanced management systems [6,7].

Figure 1 presents the reductions in greenhouse gas emissions compared to 1990 levels within the European Union for the period 2019–2023 [8].

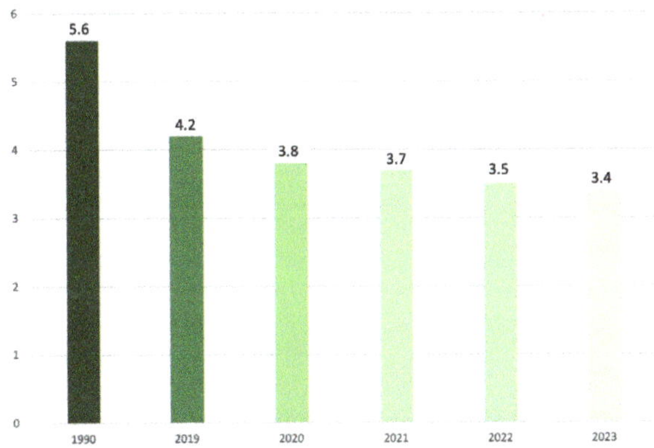

Figure 1. Reduction in greenhouse gas emissions in the European Union (2019–2023) in million tons of CO_2 equivalent. Source: [8].

Furthermore, energy efficiency has become a central pillar in this evolution, not only to minimize consumption but also to ensure the optimization of resources throughout the energy value chain. Technologies such as predictive analytics, real-time data processing, and decentralized control systems contribute to the reduction of transmission losses and enable more sustainable grid operation [6–8]. The integration of renewable sources, such as wind and solar energy, with traditional energy sources presents a complex challenge that requires innovative solutions to ensure the balance between energy supply and demand [9]. In 2022, renewable energy accounted for 23% of the European Union's total energy consumption [10]. By 2024, renewable energy production surpassed fossil fuel-based energy in terms of contribution [11]. Solar energy generated 11% of the EU's electricity, exceeding coal for the first time, which contributed 10% [12]. Table 1 presents the contribution percentages of different energy sources to the EU's electricity consumption in 2024.

Table 1. Contribution of energy sources to electricity consumption in the EU (2024). Source: [8].

Energy Source	Contribution (%)
Nuclear energy	24%
Wind energy	18%
Natural gas	16%
Solar energy	11%
Coal	10%
Hydroelectric	8%
Biomass	7%
Imports and other renewables	6%

A key component of this transformation is the adoption of Artificial Intelligence (AI) and the Internet of Things (IoT), which facilitate real-time monitoring, predictive maintenance, and adaptive demand response strategies [5]. These technologies empower the grid to become more autonomous, intelligent, and capable of self-regulation, aligning supply with fluctuating demand while improving overall system reliability [6].

Cybersecurity also plays a decisive role in this digital transition. As energy infrastructures become increasingly connected and reliant on digital systems, they are exposed to higher risks of cyber threats. Ensuring the effectiveness of cybersecurity frameworks is essential to protect critical assets and guarantee the continuity and safety of energy supply. Measures such as encryption protocols, intrusion detection systems, and AI-based threat analysis are becoming integral components of smart energy systems.

One aspect of SGs is the reduction of energy waste through the use of monitoring tools, predictive analytics, and automation [13]. The application of AI, IoT, and blockchain enhances system efficiency by minimizing energy transmission and distribution losses [13]. In 2023, emissions related to energy generation in the EU decreased by 18%, primarily due to increased renewable energy production and the adoption of advanced energy management technologies [14]. The European Union has established a clear regulatory framework to support the energy transition and emission reduction efforts [15].

The SDGs serve as a strategic guide to addressing these challenges, with a particular focus on SDG 7 (Affordable and Clean Energy), which promotes equitable access to modern, reliable, and sustainable energy services [16], SDG 9 (Industry, Innovation, and Infrastructure), which encourages the development of resilient infrastructure and the adoption of innovative technologies [17], and SDG 13 (Climate Action), which emphasizes the urgency of measures to combat climate change, an objective pursued by the EU through the European Green Deal and Fit for 55, aiming to reduce emissions by 55% by 2030 [18]. The three analyzed SDGs are illustrated in Figure 2.

Figure 2. SDGs 7, 9, and 13: Drivers of Energy Transition and Innovation. Our elaboration.

This manuscript is structured to guide the reader through a clear progression of content. It includes a literature-based Introduction, a detailed Methodology section grounded in PRISMA and bibliometric analysis, followed by the Results and Discussion sections, which present and interpret the findings, and concludes with a summary of key insights and study limitations. This structure is designed to ensure methodological transparency and thematic coherence across all sections of the paper.

This study aims to analyze European distribution networks in relation to the SDGs and EU energy plans, highlighting the role of digital technologies in smart energy management.

2. Methodology

The methodology adopted in this study consists of two main phases: systematic literature selection and bibliometric analysis.

In the first phase, a PRISMA-based approach was employed to identify and filter the most relevant academic literature on SG, the integration of renewable energy sources, and the role of digital technologies in energy management [19]. The selection was conducted using the Scopus database, applying rigorous inclusion and exclusion criteria to ensure the relevance and scientific quality of the analyzed documents [20].

The Scopus database has been used as a source of articles for review [21].

Scopus is among the most comprehensive information resources globally, covering many disciplines and providing scholars with high-quality and reliable academic information, and has gradually become the main source of data for bibliometric analysis and systematic literature review [22].

In the second phase, a bibliometric analysis was performed using VOSviewer software to map the key research trends in the field [23]. The integration of these two phases, along with the analysis of the 20 most cited articles, provided a systematic and updated overview of research dynamics related to SGs and the energy transition.

2.1. Systematic Literature Selection and Inclusion Criteria

To ensure a systematic and reproducible analysis of the scientific literature on SGs and the integration of energy sources, an approach based on the principles of the PRISMA methodology was adopted [19]. The research was conducted using the Scopus database with an advanced search string ("smart grid" OR "intelligent grid" OR "smart energy network") AND ("renewable energy" OR "solar energy" OR "wind energy" OR "hydropower" OR "non-renewable energy") AND ("energy efficiency" OR "energy waste reduction" OR "grid optimization") AND ("digital technology" OR "artificial intelligence" OR "IoT" OR "blockchain" OR "big data").

This string was carefully designed to include the most relevant synonyms and semantically related keywords commonly used in the literature on smart grids and digital energy technologies. The use of Boolean operators (OR, AND) allowed for the inclusion of a wide range of terms while maintaining a coherent and targeted scope. For this reason, a single, comprehensive string was considered sufficient to capture the breadth of existing studies in the field.

The initial query returned a total of 256 documents. To ensure temporal relevance, a publication date filter was applied (2015–2025), reducing the set to 249 documents. Only peer-reviewed scientific articles and conference papers were retained, leading to a refined set of 188 documents [24]. Subsequently, documents were filtered based on disciplinary relevance, focusing on areas such as engineering, computer science, energy, decision sciences, environmental science, and materials science. Studies outside these domains—particularly those focused on purely sociological, economic, or philosophical frameworks—were excluded as they fell outside the technological and infrastructural scope of the present research. The final corpus consisted of 179 documents considered suitable for analysis.

The selection process is illustrated in the PRISMA flow diagram in Figure 3. The diagram follows the standard four-phase structure of PRISMA—Identification, Screening, Eligibility, and Inclusion—providing a transparent overview of how the dataset was progressively refined. In the Identification phase, documents were retrieved using a clearly defined Boolean search string. In the Screening phase, non-peer-reviewed content was excluded. The Eligibility step involved manual screening based on relevance to the research fields. Finally, in the Inclusion phase, only articles meeting all methodological and thematic criteria were selected for analysis [25,26].

This stepwise filtering process ensured methodological rigor and reproducibility, aligning with the best practices of systematic review protocols. It also guaranteed that the final dataset was both current and aligned with the research objective, enabling a robust bibliometric and thematic analysis.

2.2. Bibliometric Analysis

From a bibliometric perspective, VOSviewer is a software widely used in bibliometric analysis for the visualization and analysis of bibliographies and datasets containing bibliographic information, such as title, author, and keywords [27,28].

Thanks to its versatility, it allows for the identification of trends, impacts, and thematic evolutions through the analysis of citation recurrence [29]. In the landscape of scientific research, VOSviewer establishes itself as an essential tool for the representation of bibliometric data, facilitating the identification of research opportunities in specific sectors and the recognition of the most frequently cited sources [30,31].

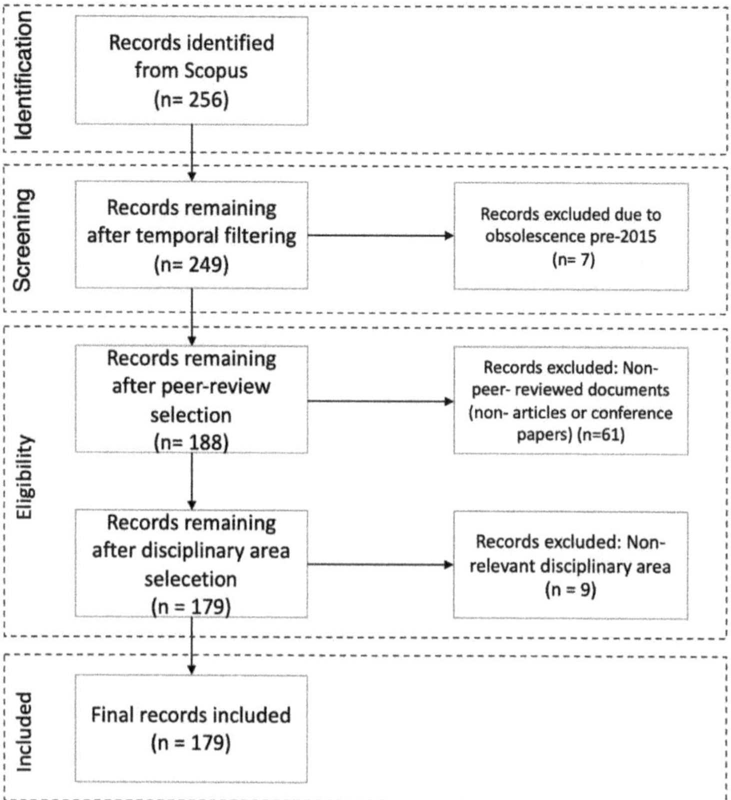

Figure 3. PRISMA flow diagram for document selection. The diagram summarizes the four phases (Identification, Screening, Eligibility, and Inclusion) and shows the number of records excluded at each step. Our elaboration.

The software focuses on analysis at an aggregated level, particularly in the field of cluster analysis [30]. It has been employed to examine and represent connections between keywords, applying the VOS clustering method to the frequency of occurrences and assigning a distinctive color to each group [32].

The interpretation of this methodology assumes that the circle sizes reflect the frequency of keyword usage, while the colors represent the different clusters [33]. It is important to note that the x and y axes do not have a specific meaning; therefore, the generated maps can be freely rotated or flipped without altering their informational content [34].

The analyzed data were extracted from Scopus in CSV format to ensure compatibility with VOSviewer. Subsequently, a map based on bibliographic results was created using both author keywords and indexed keywords. To ensure a meaningful representation, a filter was applied, considering only keywords with a minimum of three occurrences.

This methodology made it possible to highlight the most relevant concepts in the field of study, providing a clear and detailed vision of emerging terms in the scientific literature.

3. Results

The bibliometric analysis conducted using VOSviewer generated the visualization shown in Figure 4.

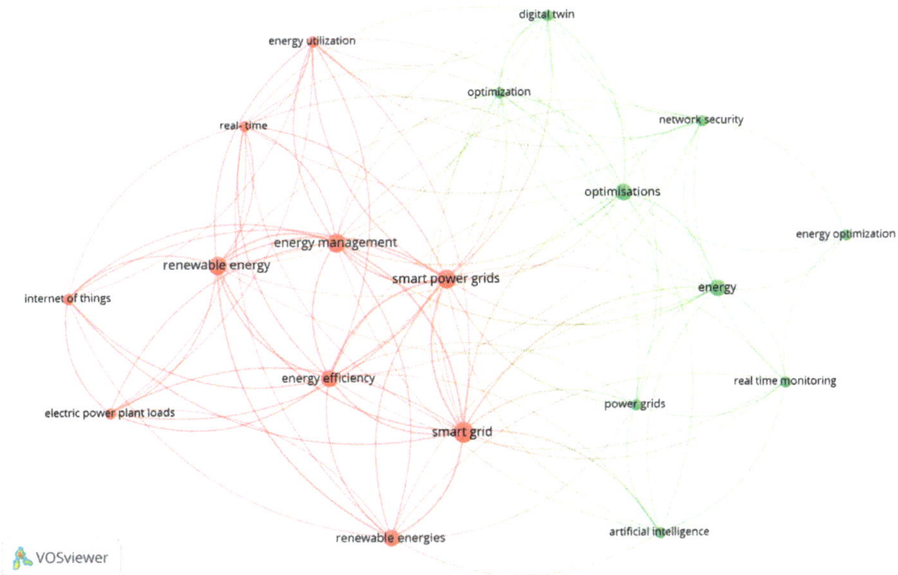

Figure 4. Bibliometric keyword analysis of articles and conference papers. Our elaboration.

In the VOSviewer visualization, concepts and keyword connections are organized into two main clusters, distinguishable by color.

The red cluster, located on the left and central part of the map, is strongly focused on energy management and SGs. The main keywords in this group include smart power grids, energy management, renewable energy, and IoT, indicating a focus on energy transition, renewable energy integration, and the role of digitalization in optimizing energy flows. This cluster also highlights the link between real-time monitoring and energy efficiency, suggesting the increasing use of technologies for dynamic network control.

On the other hand, the green cluster, positioned on the right side of the map, is more focused on optimization, security, and advanced technologies for energy management. Concepts such as optimization, digital twin, network security, and artificial intelligence emerge here, reflecting a growing emphasis on AI-based tools and digital simulation to enhance grid resilience. The importance of cybersecurity is evident, with network security directly linked to energy management, demonstrating the crucial role of protecting digitalized infrastructures.

Visually, the red cluster is denser and more concentrated in the center-left, with many internal connections between renewable energy management concepts and system efficiency. The green cluster, more distributed on the right side, highlights a research trend oriented towards digital simulation, optimization, and the security of modern power grids.

The interconnection between the two clusters suggests that digitalization and smart grid optimization are key elements for enhancing the efficiency and sustainability of renewable energy integration in the global power system.

The bibliographic analysis then examined the 20 most cited articles, shown in Table 2, which revealed the state-of-the-art of the research topic.

Table 2. State-of-the-art analysis. Our elaboration.

Title	Author	Year	Cite	Main Focus	Future Recommendation
Designing microgrid energy markets: A case study: The Brooklyn Microgrid	[35]	2018	1265	This study focuses on the design and evaluation of a blockchain-based microgrid energy market, using the Brooklyn Microgrid as a case study. It introduces a framework of seven key components for efficient market operation, highlighting the economic, environmental, and social benefits of local energy generation and peer-to-peer trading.	Future work should explore more efficient market mechanisms, implement advanced bidding strategies, and address regulatory barriers to commercial peer-to-peer energy markets. Socio-economic research is recommended to better align market design with user preferences and community values.
Artificial intelligence techniques for stability analysis and control in smart grids: Methodologies, applications, challenges and future directions	[36]	2020	227	This paper presents a comprehensive review of artificial intelligence (AI) techniques applied to stability analysis and control in smart grids. It outlines AI methodologies such as machine learning, deep learning, and reinforcement learning, and evaluates their applications in areas including security assessment, fault diagnosis, and various control strategies, highlighting their advantages over traditional methods.	Future research should focus on addressing challenges such as data quality, imbalanced learning, AI interpretability, transfer learning difficulties, and robustness to communication issues or cyber-attacks. Emphasis is also placed on integrating model- and data-driven approaches and developing adaptive, explainable, and scalable AI systems suitable for real-world smart grid operations.
State-of-the-art artificial intelligence techniques for distributed smart grids: A review	[37]	2020	202	This review explores the application of cutting-edge artificial intelligence (AI) techniques in distributed smart grids. It covers AI-enabled solutions for integrating renewable energy sources, managing energy storage systems, enabling demand response, securing grid operations, and supporting home energy management. The paper emphasizes how AI can enhance automation, efficiency, and adaptability in future energy systems.	Future work should focus on developing self-learning, fully automated, and self-healing smart grid systems. There is a need for scalable AI algorithms, advanced cybersecurity protocols, improved user-centric designs, and plug-and-play capabilities. Further, a skilled workforce and better integration of AI with regulatory and market frameworks are critical for real-world deployment.
Blockchain and AI amalgamation for energy cloud management: Challenges, solutions, and future directions	[38]	2020	135	The paper provides a comprehensive survey on the integration of blockchain and artificial intelligence for secure and efficient energy cloud management (ECM). It proposes a decentralized AI-based ECM architecture that enables energy load prediction, secure data transmission, and peer-to-peer energy trading, while addressing security, privacy, and scalability challenges in smart grids.	Future research should address secure data analytics, enhance privacy in public blockchains such as Ethereum, protect against 51% attacks, and carefully develop bug-free smart contracts. There is also a need for standardized frameworks, affordable blockchain data storage solutions, and broader professional expertise to enable scalable and real-world deployment of AI–blockchain ECM systems.
Optimized Day-Ahead Pricing with Renewable Energy Demand-Side Management for Smart Grids	[39]	2017	130	This study proposes a distributed, day-ahead dynamic pricing framework for smart grids that incorporates renewable energy generation and storage at the user level. It models the interaction between electric companies and users as a convex optimization problem, aiming to maximize social welfare. The framework enables privacy-preserving energy trading while balancing grid load and incentivizing carbon emissions reduction through a dynamic buyback pricing strategy.	Future research should explore the integration of additional real-world factors such as uncertainty in renewable generation, user behavior variability, and communication delays. There is also potential in enhancing the scalability of the distributed algorithms and incorporating real-time adaptability into the pricing scheme to better support large-scale deployment in smart grid environments.
Privacy-Preserving Transactive Energy Management for IoT-Aided Smart Homes via Blockchain	[40]	2021	109	This paper proposes a blockchain-based transactive energy management (TEM) system for IoT-enabled smart homes. It enables both vertical (with the grid) and horizontal (peer-to-peer) energy transactions while preserving user privacy. The authors develop a distributed optimization algorithm using smart contracts on a lightweight IoT blockchain, demonstrating improved efficiency, privacy, and feasibility on real IoT devices.	Future work should aim to reduce the computational complexity of decentralized algorithms to support large-scale IoT deployments. The authors also suggest building broader test networks and exploring efficient smart contract execution methods, such as WebAssembly and predefined contracts, to scale and optimize blockchain-based TEM systems.

Table 2. *Cont.*

Title	Author	Year	Cite	Main Focus	Future Recommendation
Developing Novel 5th Generation District Energy Networks	[41]	2020	109	This paper presents a novel concept and design methodology for 5th generation (5G) district energy networks that integrate thermal, power, and mobility sectors. Using a case study from the London Borough of Islington, the authors demonstrate how ultra-low temperature "ambient loop" systems can leverage local renewable and secondary energy sources (e.g., waste heat from data centers and subways), heat pumps, energy storage, and smart control systems to deliver efficient, low-carbon energy to urban communities.	Future research should refine modeling tools for dynamic thermal behavior, optimize ambient loop temperatures, and improve integration of smart control systems. Further exploration is also needed on real-time balancing of heat and cooling demands, stakeholder engagement models, and regulatory adjustments to enable large-scale deployment of 5G smart energy networks.
Environmentally Data-driven Smart Sustainable Cities: Applied Innovative Solutions for Energy Efficiency, Pollution Reduction, and Urban Metabolism	[42]	2020	100	This paper investigates the role of IoT and big data technologies in enhancing environmental sustainability within the framework of smart and sustainable cities. Using case studies of Stockholm and Barcelona, it explores how data-driven solutions—such as smart grids, smart buildings, and urban metabolism monitoring—can reduce pollution, improve energy efficiency, and support real-time urban management.	Future work should aim at advancing integration between data-driven technologies and environmental strategies, improving urban analytics platforms, and addressing city-specific needs through tailored solutions. Research should also focus on scalable infrastructures and interoperable systems to support continuous environmental monitoring and adaptive urban planning.
Green Industrial Internet of Things from a Smart Industry Perspective	[43]	2020	95	This paper explores the evolution of the Green Industrial Internet of Things (GIIoT) as a key enabler of Industry 4.0, Society 5.0, and MIC2025 strategies. It reviews the technological foundations, protocols, and applications of GIIoT across sectors including energy, manufacturing, agriculture, healthcare, and transportation. The study emphasizes how smart technologies—such as IoT, AI, cobots, and digital twins—enhance efficiency, sustainability, and automation in modern industrial ecosystems.	Future work should address challenges in device interoperability, energy efficiency, communication reliability in industrial environments, and cybersecurity. It is recommended to develop robust, scalable, and secure GIIoT architectures tailored to real-time applications. Further integration of AI, machine learning, and decentralized systems (e.g., blockchain) is also crucial for enabling predictive analytics and autonomous industrial processes.
Edge Computing for IoT-Enabled Smart Grid: The Future of Energy	[44]	2022	90	This paper presents a comprehensive survey on integrating edge computing with IoT-enabled smart grids (SGs). It proposes a layered architecture and an all-in-one computing model aimed at reducing latency and improving real-time decision-making in SG systems. Applications span power distribution, micro-grids, and advanced metering, emphasizing benefits such as faster service response, efficient energy management, and scalability across domains such as smart cities and factories.	Future research should address challenges such as scalability, system sustainability, flexible resource allocation, and cybersecurity for edge-enabled SGs. Emphasis is placed on building resilient infrastructures, adopting lightweight security protocols, and integrating advanced technologies such as AI, blockchain, and homomorphic encryption to support real-time, secure, and sustainable smart grid operations.
Intelligent Power Equipment Management Based on Distributed Context-aware Inference in Smart Cities	[45]	2018	83	This paper proposes a context-aware framework for intelligent power equipment management in smart cities. It introduces a distributed inference architecture that utilizes IoT sensor data, context ontologies, and semantic reasoning to monitor and manage power equipment efficiently. The system enhances decision-making by converting low-level data into high-level context, enabling real-time, automated, and intelligent control of smart grid infrastructure.	Future work should explore decentralized decision-making in smart grids, including the handling of fuzzy and uncertain data. Additional research is needed to address privacy and cybersecurity challenges in distributed intelligent control systems, and to improve the scalability and reliability of semantic reasoning frameworks across diverse smart city applications.

Table 2. *Cont.*

Title	Author	Year	Cite	Main Focus	Future Recommendation
Weather Forecasting for Renewable Energy System: A Review	[46]	2022	77	This paper reviews forecasting methods for integrating solar and wind energy into smart grids, emphasizing the need for accurate weather prediction to manage the variability of renewable resources. It covers physical, statistical, artificial intelligence (AI), machine learning, and deep learning models, highlighting hybrid and ensemble methods as effective tools for improving forecasting accuracy.	Future work should focus on enhancing forecast accuracy through hybrid deep learning models (e.g., CNN–LSTM), increasing data availability and resolution, and integrating AI-driven forecasting into real-time smart grid operations. There is also a need to develop lightweight, scalable models that maintain high accuracy while reducing computational complexity and training time.
Investigation and Analysis of Effective Approaches, Opportunities, Bottlenecks and Future Potential Capabilities for Digitalization of Energy Systems and Sustainable Development Goals	[47]	2022	74	This paper analyzes the role of digitalization in transforming energy systems to meet Sustainable Development Goals (SDGs), particularly under the impact of the COVID-19 crisis. It explores how technologies such as blockchain, IoT, smart grids, and Industry 4.0 can improve energy efficiency, resilience, and integration of renewables. The study presents policy strategies and short-, mid-, and long-term planning for achieving net-zero emissions and enhancing environmental sustainability.	Future efforts should focus on overcoming bottlenecks in renewable energy adoption, such as economic, social, and regulatory barriers. Emphasis is placed on enhancing public awareness, developing skilled workforces, and modernizing infrastructures. The integration of digital tools for decentralized energy systems and real-time management is key to accelerating the energy transition and achieving global climate targets.
A Blockchain-enabled Secure Power Trading Mechanism for Smart Grid Employing Wireless Networks	[48]	2020	62	This paper proposes a secure and decentralized power trading mechanism for smart grids that integrates blockchain with wireless networks. It introduces a dual-chain structure—Local Energy Trading Blockchain (LETB) and Renewable Energy Trading Blockchain (RETB)—to facilitate efficient, secure, and trustable energy transactions. Smart contracts manage trading decisions, optimize distribution based on reputation and pricing, and incentivize renewable energy producers through automated feedback mechanisms.	Future work should focus on integrating electric vehicle energy trading into the system and expanding the platform to support a wider range of renewable energy sources. Further research is also recommended on improving blockchain scalability, reducing transaction overhead, and enhancing system adaptability to diverse energy market scenarios.
Internet of Energy (IoE) and High-renewables Electricity System Market Design	[49]	2019	62	This paper explores the integration of the Internet of Energy (IoE) into future high-renewables electricity market design. It analyzes how IoE technologies—including smart grids, smart meters, electric vehicles, and peer-to-peer trading—can support decentralized, flexible, and intelligent energy systems. The study highlights the benefits of IoE in enhancing energy efficiency, real-time coordination, and sustainability through two-way energy and data flows.	Future efforts should focus on fostering social acceptance of IoE technologies, addressing technical challenges such as interoperability and cybersecurity, and incentivizing early adoption through targeted government policies. Additionally, flexible planning frameworks and dynamic market designs are needed to manage demand uncertainty, support prosumers, and scale up smart grid infrastructure in line with decarbonization goals.
Green Energy Aware and Cluster Based Communication for Future Load Prediction in IoT	[50]	2022	49	This paper introduces a green energy-aware clustering and load prediction method (GEQCC-FLP) for IoT networks. It combines a satin bowerbird optimizer (SBO) for energy-efficient cluster head selection with a deep random vector functional link network (DRVFLN) for future load forecasting. The approach aims to optimize energy consumption, communication delay, and network lifetime in energy-constrained IoT environments, proving effective through extensive simulations.	Future work should focus on integrating the proposed GEQCC-FLP technique with real-world IoT infrastructure and exploring further optimization of the model for large-scale, heterogeneous networks. Additionally, addressing security, data privacy, and interoperability issues will be crucial for practical deployment in smart grid and smart city applications.

Table 2. *Cont.*

Title	Author	Year	Cite	Main Focus	Future Recommendation
Design and implementation of an AI-based and IoT-enabled Home Energy Management System: A case study in Benguerir—Morocco	[51]	2021	46	This paper presents the design and implementation of an AI-based and IoT-enabled Home Energy Management System (HEMS), deployed in a smart testbed house in Morocco. The system integrates supply-side and demand-side management using real-time pricing, user preferences, and forecasting data to optimize energy flows. A hybrid control strategy—combining rule-based and PSO-based optimization—enhances energy efficiency, cost savings, and comfort while promoting renewable energy self-consumption.	Future efforts should address the scalability of the HEMS framework, reduce implementation costs, and improve forecasting under uncertainty. Additional research is recommended on enhancing interoperability, user comfort modeling, and adapting pricing models to local contexts. Expanding dynamic pricing and integrating energy storage and electric vehicles could further boost system adoption and effectiveness in real-world settings.
Blockchain-based Decentralized Energy Intra-trading with Battery Storage Flexibility in a Community Microgrid System	[52]	2022	45	This paper presents a blockchain-enabled peer-to-peer (P2P) energy intra-trading system integrated with a community microgrid. It leverages smart contracts and a decentralized energy market to optimize local energy transactions among prosumers, consumers, and renewable energy source owners. The system incorporates battery energy storage and dynamic pricing strategies using cryptocurrency (Cosmos ATOM), aiming to improve energy self-sufficiency, cost efficiency, and transparency in energy exchanges.	Future research should explore the scalability of the proposed blockchain-based framework in larger microgrid networks and investigate its integration with electric vehicles and broader renewable sources. Further development of incentive mechanisms, demand-side management models, and more efficient smart contract deployment methods will enhance real-world applicability and user engagement.
Network Selection Criterion for Ubiquitous Communication Provisioning in Smart Cities for Smart Energy System	[53]	2019	45	This paper proposes a blockchain-based peer-to-peer (P2P) energy intra-trading mechanism within a community microgrid integrated with battery storage systems. The framework enables decentralized energy exchange among prosumers, consumers, and renewable energy owners using smart contracts and cryptocurrency (Cosmos ATOM). The model considers supply-demand dynamics and price incentives to optimize energy distribution and enhance local self-sufficiency while reducing grid dependency and energy costs.	Future research should explore large-scale deployment of the proposed system, integration with electric vehicles and other flexible loads, and refinement of smart contract mechanisms for greater efficiency and scalability. Enhancing the economic models, improving user participation incentives, and addressing regulatory and cybersecurity challenges will also be critical for real-world adoption.
Demand Response Program for Efficient Demand-Side Management in Smart Grid Considering Renewable Energy Sources	[54]	2022	43	This paper introduces a demand-side management (DSM) framework for smart homes using a hybrid optimization algorithm (ACTLBO) that combines ant colony optimization and teaching learning-based optimization. The proposed system incorporates renewable energy sources (RESs), battery storage systems (BSS), and user comfort metrics (thermal, visual, air quality) to minimize energy cost, carbon emissions, peak-to-average ratio (PAR), and appliance delay.	Future research should focus on real-world deployment of the proposed ACTLBO-based scheduling framework and its integration with large-scale residential energy systems. Additional work is needed to refine the balance between user comfort and optimization objectives, improve resilience to uncertain renewable supply, and ensure compatibility with evolving real-time pricing models and smart grid standards.

The analysis of the 20 most cited articles in the field of digitalization of energy grids and the transition towards sustainable models highlights the central role of technologies in the management of energy resources and the optimization of renewable energy integration.

Ali and Choi [37] illustrate how AI has become a key component for SG, improving the prediction of demand and supply, supporting the distributed management of energy resources, and enhancing cybersecurity mechanisms. In addition, recent developments in reinforcement learning applied to grid-following converters show promising results in improving system adaptability and stability. For instance, Zeng et al. (2025) [55] proposed a multi-objective controller using Easy Transfer Reinforcement Learning, which significantly reduces training effort while maintaining control performance and robustness across varying grid conditions [55]. The Internet of Energy (IoE), explored by Strielkowski et al. [45], represents an emerging paradigm that enables peer-to-peer energy exchange, the optimization of electric vehicle charging, and a more efficient distribution of renewable sources, favoring the decentralization of energy markets. On the data processing front, the study by Minh et al. [44] highlights how Edge Computing improves the operational efficiency of SG, reducing latency in monitoring systems and allowing for more flexible management of storage resources.

To address the variability of renewable sources, Meenal et al. [46] demonstrate that the use of predictive models based on machine learning and deep learning helps reduce uncertainties related to intermittent generation, optimizing network planning and management. Building on this trend, recent approaches have combined deep reinforcement learning with physics-informed models to improve the control of power converters in smart grids. For instance, Zeng et al. (2024) [56] developed a method for optimizing Input-Series Output-Parallel Dual Active Bridge (ISOP-DAB) converters, enhancing energy transfer efficiency and system stability in complex decentralized energy networks [56].

In parallel, the adoption of IoT and big data technologies in smart cities, described by Bibri and Krogstie [42], reveals the importance of integrating smart meters, smart buildings, and SGs to improve energy efficiency and reduce environmental impact. The evolution of SGs also intersects with the issue of sustainability and infrastructure security: according to Kumari et al. [38], the combination of blockchain and AI can enhance microgrid management, ensuring greater transparency in energy markets and increasing the resilience of electrical networks.

Finally, the importance of weather forecasting for improving the reliability of renewable energy networks is emphasized by Meenal et al. [46], who demonstrate how advanced predictive algorithms can mitigate the effects of the intermittency of solar and wind generation. The analysis of these studies highlights that the energy transition and the digitalization of networks are closely interconnected processes, where the adoption of solutions based on AI, IoT, Edge Computing, and blockchain represents the key to ensuring more resilient, efficient, and sustainable electrical grids. Compared to existing literature, this study stands out for its integrated approach, which jointly examines technological trajectories related to energy efficiency, digitalization, and infrastructure security. The combination of bibliometric analysis and thematic cluster interpretation makes it possible to highlight not only the areas of highest scientific intensity, but also the emerging interconnections between advanced management tools and network protection strategies. In particular, the emphasis on the relationship between AI, IoT, blockchain, and cybersecurity represents an original contribution aimed at promoting a systemic vision for the development of smart grids.

4. Discussion and Conclusions

The analysis conducted in this study has highlighted how SGs represent a key element for the energy transition, facilitating the integration of renewable sources with traditional energy systems [1,2]. The combination of advanced technologies allows for a significant improvement in the efficiency of energy production, distribution, and consumption [37,38]. The adoption of these digital solutions not only optimizes the use of resources but also enables the reduction of greenhouse gas emissions, contributing to the sustainability goals defined by the European Union and the SDGs of the United Nations [15].

The bibliometric analysis has identified two main research streams: the first, centered on energy management and SG, highlights the crucial role of digitalization in optimizing energy flows [40]; the second, focused on optimization, security, and advanced technologies, demonstrates how the integration of digital simulation tools and cybersecurity is essential to ensuring the resilience of energy infrastructures [48].

The dimension of energy efficiency has emerged as a key priority, not only from a technological standpoint but also as a strategic goal for policy and market design. The use of AI-driven forecasting models, combined with real-time IoT-based monitoring systems, enables smart grids to dynamically adapt to load variations, prevent energy waste, and reduce peak loads [44,57].

Moreover, the discussions on Artificial Intelligence (AI) and the Internet of Things (IoT) should be reinforced by recognizing their foundational role in automating energy management, improving decision-making, and creating self-healing networks. AI supports advanced forecasting, predictive maintenance, and optimization algorithms, while IoT ensures real-time communication between grid components, sensors, and user devices, forming a synergistic digital ecosystem for energy resilience [38,40,55].

Cybersecurity emerges as another cornerstone of smart grid effectiveness. The increasing reliance on connected devices and cloud platforms exposes the grid to new vulnerabilities. Therefore, the development and implementation of robust cybersecurity frameworks—capable of threat detection, anomaly resolution, and system recovery—are essential to ensure the integrity, availability, and confidentiality of energy data and services. Recent advances in AI-based cybersecurity are promising, offering adaptive responses to emerging threats and contributing to overall grid stability [45,56].

The growing interconnection between these two areas suggests that digitalization and network protection are complementary and essential aspects for a sustainable energy system. The results emerging from the literature review show that AI-based solutions are among the most studied and implemented, thanks to their ability to improve demand and supply forecasting, optimize the distributed management of resources, and strengthen network security [46].

Furthermore, blockchain is emerging as a promising option to ensure transparency and reliability in decentralized energy markets, supporting the development of peer-to-peer energy exchange models [38]. At the same time, Edge Computing offers new opportunities to reduce latency in monitoring and control operations of SG, improving the operational efficiency of the system [42]. A summary is given in Table 3.

Table 3. Summary of constructs. Our elaboration.

Construct	Items	References
Smart Grids	Digitalization of networks, smart energy management, and waste reduction.	[37]
Energy Efficiency	Optimization of demand and supply and reduction of transmission losses.	[44]
Digital Technologies	AI, IoT, Blockchain, and Edge Computing for energy management.	[38]
Cybersecurity	Protection of digital infrastructures and cybersecurity for SGs.	[56]
Renewable Energy	Integration of solar, wind, and hydropower into SGs.	[45,57]
Blockchain and Energy	Decentralized energy exchanges and peer-to-peer transactions.	[40]
Demand Forecasting	AI and machine learning models for resource management optimization.	[46]
Grid Optimization	Digital Twin, simulation models, and dynamic management of energy sources.	[41]
Energy Policies	Fit for 55, European Green Deal, SDG 7, and SDG 13.	[42]

A fundamental aspect emerging from our analysis concerns the importance of weather forecasting in managing renewable energy sources [46]. The use of advanced predictive algorithms based on machine learning and deep learning helps mitigate uncertainties related to the variability of solar and wind energy, improving grid planning and balance [48]. This approach fits into a broader dynamic energy management framework, where the system's adaptability becomes increasingly crucial.

While this study highlights the growing convergence between energy efficiency, renewable integration, and digital technologies in the evolution of SG, several challenges still remain, requiring further technological and regulatory developments. Cybersecurity of SGs represents one of the main critical issues, considering the growing risk of cyberattacks on digitalized energy infrastructures [56]. Scalability and interoperability of different digital technologies require greater standardization to ensure a smooth and effective transition [43]. Moreover, the economic sustainability of the proposed solutions must be carefully evaluated to prevent the high implementation costs from limiting the large-scale adoption of innovative technologies [41,57].

This study highlights how the digitalization of energy grids serves as a fundamental lever for the transition toward a more efficient, resilient, and sustainable energy system. The integration of AI, IoT, blockchain, and Edge Computing not only enhances resource management but also opens new perspectives for the creation of decentralized and more equitable energy models.

5. Limitations of the Study

While this study offers a comprehensive bibliometric and thematic analysis of the role of digital technologies in smart grid development and the energy transition, several limitations should be acknowledged. First, the literature review is based solely on documents indexed in the Scopus database. Although Scopus is a widely recognized and reliable academic source, this choice may have led to the exclusion of relevant studies published in other databases such as Web of Science or IEEE Xplore. Second, the selection criteria focused primarily on technical and scientific disciplines (e.g., engineering, energy, computer science), potentially underrepresenting interdisciplinary perspectives from social sciences or public policy, which are crucial for understanding the broader implications of smart grid implementation.

Moreover, the study provides a static snapshot of the current research landscape, which may not fully reflect the fast-paced evolution of technologies such as AI, IoT, and blockchain. Emerging innovations and newly published research might not yet be adequately represented. Another limitation lies in the lack of empirical validation: while the study maps conceptual trends and technological trajectories, it does not assess how these solutions are applied in real-world energy infrastructures or their actual impact on grid performance, resilience, and sustainability.

Future research should integrate empirical case studies, cross-disciplinary frameworks, and longitudinal data to enrich the understanding of how digital technologies interact with regulatory, social, and economic contexts in shaping the energy transition.

Future research should integrate empirical case studies and longitudinal data to enrich the understanding of how digital technologies interact with regulatory, social, and economic contexts in shaping the energy transition.

Author Contributions: Conceptualization, R.C. and P.C.; methodology, R.C., R.R., C.A. and P.C.; software, R.C., R.R. and C.A.; validation, R.C., R.R., C.A. and P.C.; formal analysis, R.C.; resources, R.C.; data curation, R.C.; writing—original draft preparation, R.C.; writing—review and editing, R.C., R.R., C.A. and P.C.; visualization, R.R., C.A. and P.C.; supervision, P.C.; project administration, P.C. All authors have read and agreed to the published version of the manuscript.

Funding: This research received no external funding.

Data Availability Statement: The original contributions presented in this study are included in the article. Further inquiries can be directed to the corresponding author.

Conflicts of Interest: The authors declare no conflicts of interest.

Abbreviations

AI	Artificial Intelligence
IoT	Internet of Things
SDGs	Sustainable Development Goals
SGs	Smart Grids

References

1. Jaiswal, K.K.; Chowdhury, C.R.; Yadav, D.; Verma, R.; Dutta, S.; Jaiswal, K.S.; Karuppasamy, K.S.K. Renewable and sustainable clean energy development and impact on social, economic, and environmental health. *Energy Nexus* **2022**, *7*, 100118. [CrossRef]
2. Jie, H.; Khan, I.; Alharthi, M.; Zafar, M.W.; Saeed, A. Sustainable energy policy, socio-economic development, and ecological footprint: The economic significance of natural resources, population growth, and industrial development. *Util. Policy* **2023**, *81*, 101490. [CrossRef]
3. Paramati, S.R.; Shahzad, U.; Doğan, B. The role of environmental technology for energy demand and energy efficiency: Evidence from OECD countries. *Renew. Sustain. Energy Rev.* **2022**, *153*, 111735. [CrossRef]
4. Pablo-Romero, M.P.; Sánchez-Braza, A.; Galyan, A. Renewable energy use for electricity generation in transition economies: Evolution, targets and promotion policies. *Renew. Sustain. Energy Rev.* **2021**, *138*, 110481. [CrossRef]
5. Erhueh, O.V.; Nwakile, C.; Akano, O.A.; Esiri, A.E.; Hanson, E. Digital transformation in energy asset management: Lessons for building the future of energy infrastructure. *Glob. J. Res. Sci. Technol.* **2024**, *2*, 10–37. [CrossRef]
6. Dehdarian, A.; Tucci, C.L. A complex network approach for analyzing early evolution of smart grid innovations in Europe. *Appl. Energy* **2021**, *298*, 117143. [CrossRef]
7. Hua, W.; Chen, Y.; Qadrdan, M.; Jiang, J.; Sun, H.; Wu, J. Applications of blockchain and artificial intelligence technologies for enabling prosumers in smart grids: A review. *Renew. Sustain. Energy Rev.* **2022**, *161*, 112308. [CrossRef]
8. Liargovas, P.; Papageorgiou, C. Environmental Policy: Progress, Challenges and Future Directions. In *The European Integration, Volume 2: Institutions and Policies*; Springer Nature: Cham, Switzerland, 2024; pp. 405–428.
9. Khan, K.; Su, C.W.; Rehman, A.U.; Ullah, R. Is technological innovation a driver of renewable energy? *Technol. Soc.* **2022**, *70*, 102044. [CrossRef]
10. Abbruzzese, M.; Infante, D.; Smirnova, J. The diffusion of renewable energy production in European countries: The role of incentives. *Econ. Innov. New Technol.* **2025**, *34*, 70–93. [CrossRef]
11. Deshmukh, M.K.G.; Sameeroddin, M.; Abdul, D.; Sattar, M.A. Renewable energy in the 21st century: A review. *Mater. Today Proc.* **2023**, *80*, 1756–1759. [CrossRef]
12. Hassan, Q.; Viktor, P.; Al-Musawi, T.J.; Ali, B.M.; Algburi, S.; Alzoubi, H.M.; Al-Jiboory, A.K.; Sameen, A.Z.; Jaszczur, M. The renewable energy role in the global energy Transformations. *Renew. Energy Focus* **2024**, *48*, 100545. [CrossRef]
13. Mazhar, T.; Irfan, H.M.; Haq, I.; Ullah, I.; Ashraf, M.; Shloul, T.A.; Elkamchouchi, D.H. Analysis of challenges and solutions of IoT in smart grids using AI and machine learning techniques: A review. *Electronics* **2023**, *12*, 242. [CrossRef]
14. Bianco, V.; Cascetta, F.; Nardini, S. Analysis of the carbon emissions trend in European Union. A decomposition and decoupling approach. *Sci. Total Environ.* **2024**, *909*, 168528. [CrossRef] [PubMed]
15. Hainsch, K.; Löffler, K.; Burandt, T.; Auer, H.; Del Granado, P.C.; Pisciella, P.; Zwickl-Bernhard, S. Energy transition scenarios: What policies, societal attitudes, and technology developments will realize the EU Green Deal? *Energy* **2022**, *239*, 122067. [CrossRef]
16. Walesiak, M.; Dehnel, G. Progress on SDG 7 achieved by EU countries in relation to the target year 2030: A multidimensional indicator analysis using dynamic relative taxonomy. *PLoS ONE* **2024**, *19*, e0297856. [CrossRef]
17. Skare, M.; Gavurova, B.; Rigelsky, M. Quantification of the impact of innovations in industry and infrastructure for sustainable circular economy production and consumption. *J. Innov. Knowl.* **2024**, *9*, 100456. [CrossRef]
18. Sardianou, E.; Nikou, V.; Kostakis, I. Harmonizing sustainability goals: Empirical insights into climate change mitigation and circular economy strategies in selected European countries with SDG13 framework. *Sustainability* **2024**, *16*, 296. [CrossRef]
19. Greener, S. Evaluating literature with bibliometrics. *Interact. Learn. Environ.* **2022**, *30*, 1168–1169. [CrossRef]
20. Pranckutė, R. Web of Science (WoS) and Scopus: The titans of bibliographic information in today's academic world. *Publications* **2021**, *9*, 12. [CrossRef]
21. Kraus, S.; Breier, M.; Lim, W.M.; Dabić, M.; Kumar, S.; Kanbach, D.; Mukherjee, D.; Corvello, V.; Piñeiro-Chousa, J.; Liguori, E.; et al. Literature reviews as independent studies: Guidelines for academic practice. *Rev. Manag. Sci.* **2022**, *16*, 2577–2595. [CrossRef]
22. Zhang, K.; Liang, Q.M. Recent progress of cooperation on climate mitigation: A bibliometric analysis. *J. Clean. Prod.* **2020**, *277*, 123495. [CrossRef]

23. Kumar, R.; Saxena, S.; Kumar, V.; Prabha, V.; Kumar, R.; Kukreti, A. Service innovation research: A bibliometric analysis using VOSviewer. *Compet. Rev. Int. Bus. J.* **2024**, *34*, 736–760. [CrossRef]
24. Weber, E.J.; Katz, P.P.; Waeckerle, J.F.; Callaham, M.L. Author perception of peer review: Impact of review quality and acceptance on satisfaction. *JAMA* **2002**, *287*, 2790–2793. [CrossRef]
25. Page, M.J.; Moher, D.; Bossuyt, P.M.; Boutron, I.; Hoffmann, T.C.; Mulrow, C.D.; McKenzie, J.E. PRISMA 2020 explanation and elaboration: Updated guidance and exemplars for reporting systematic reviews. *BMJ* **2021**, *372*, n160. [CrossRef]
26. Arya, S.; Kaji, A.H.; Boermeester, M.A. PRISMA reporting guidelines for meta-analyses and systematic reviews. *JAMA Surg.* **2021**, *156*, 789–790. [CrossRef]
27. Al Husaeni, D.F.; Nandiyanto, A.B.D. Bibliometric using Vosviewer with Publish or Perish (using google scholar data): From step-by-step processing for users to the practical examples in the analysis of digital learning articles in pre and post COVID-19 pandemic. *ASEAN J. Sci. Eng.* **2022**, *2*, 19–46. [CrossRef]
28. Orduña-Malea, E.; Costas, R. Link-based approach to study scientific software usage: The case of VOSviewer. *Scientometrics* **2021**, *126*, 8153–8186. [CrossRef]
29. De la Salgado-Cruz, M.P.; Salgado-Cruz, J.; García-Hernández, A.B.; Calderón-Domínguez, G.; Gómez-Viquez, H.; Oliver-Espinoza, R.; Fernández-Martínez, M.C.; Yáñez-Fernández, J. Chitosan as a coating for biocontrol in postharvest products: A bibliometric review. *Membranes* **2021**, *11*, 421. [CrossRef]
30. Wang, X.; Xu, Z.; Škare, M. A bibliometric analysis of economic research-ekonomska istrazivanja (2007–2019). *Econ. Res. Ekon. Istraživanja* **2020**, *33*, 865–886. [CrossRef]
31. Nandiyanto, A.B.D.; Biddinika, M.K.; Triawan, F. How bibliographic dataset portrays decreasing number of scientific publications from Indonesia. *Indones. J. Sci. Technol.* **2020**, *5*, 154–175. [CrossRef]
32. Eck, N.; Waltman, L. Citation-based clustering of publications using CitNetExplorer and VOSviewer. *Scientometrics* **2017**, *111*, 1053–1070. [PubMed]
33. Van Eck, N.J.; Waltman, L.; Dekker, R.; Van Den Berg, J. A comparison of two techniques for bibliometric mapping: Multidimensional scaling and VOS. *J. Am. Soc. Inf. Sci. Technol.* **2010**, *61*, 2405–2416. [CrossRef]
34. Khalil, G.M.; Crawford, C.A.G. A bibliometric analysis of US-based research on the behavioral risk factor surveillance system. *Am. J. Prev. Med.* **2015**, *48*, 50–57. [CrossRef] [PubMed]
35. Mengelkamp, E.; Gärttner, J.; Rock, K.; Kessler, S.; Orsini, L.; Weinhardt, C. Designing microgrid energy markets: A case study: The Brooklyn Microgrid. *Appl. Energy* **2018**, *210*, 870–880. [CrossRef]
36. Shi, Z.; Yao, W.; Li, Z.; Zeng, L.; Zhao, Y.; Zhang, R.; Tang, Y.; Wen, J. Artificial intelligence techniques for stability analysis and control in smart grids: Methodologies, applications, challenges and future directions. *Appl. Energy* **2020**, *278*, 115733. [CrossRef]
37. Ali, S.S.; Choi, B.J. State-of-the-art artificial intelligence techniques for distributed smart grids: A review. *Electronics* **2020**, *9*, 1030. [CrossRef]
38. Kumari, A.; Gupta, R.; Tanwar, S.; Kumar, N. Blockchain and AI amalgamation for energy cloud management: Challenges, solutions, and future directions. *J. Parallel Distrib. Comput.* **2020**, *143*, 148–166. [CrossRef]
39. Chiu, T.-C.; Shih, Y.-Y.; Pang, A.-C.; Pai, C.-W. Optimized day-ahead pricing with renewable energy demand-side management for smart grids. *IEEE Internet Things J.* **2017**, *4*, 374–383. [CrossRef]
40. Yang, Q.; Wang, H. Privacy-preserving transactive energy management for IoT-aided smart homes via blockchain. *IEEE Internet Things J.* **2021**, *8*, 11463–11475. [CrossRef]
41. Revesz, A.; Jones, P.; Dunham, C.; Davies, G.; Marques, C.; Matabuena, R.; Scott, J.; Maidment, G. Developing novel 5th generation district energy networks. *Energy* **2020**, *201*, 117389. [CrossRef]
42. Bibri, S.E.; Krogstie, J. Environmentally data-driven smart sustainable cities: Applied innovative solutions for energy efficiency, pollution reduction, and urban metabolism. *Energy Inform.* **2020**, *3*, 29. [CrossRef]
43. Tabaa, M.; Monteiro, F.; Bensag, H.; Dandache, A. Green Industrial Internet of Things from a smart industry perspective. *Energy Rep.* **2020**, *6*, 430–446. [CrossRef]
44. Minh, Q.N.; Nguyen, V.-H.; Quy, V.K.; Ngoc, L.A.; Chehri, A.; Jeon, G. Edge computing for IoT-enabled smart grid: The future of energy. *Energies* **2022**, *15*, 6140. [CrossRef]
45. Choi, C.; Esposito, C.; Wang, H.; Liu, Z.; Choi, J. Intelligent power equipment management based on distributed context-aware inference in smart cities. *IEEE Commun. Mag.* **2018**, *56*, 212–217. [CrossRef]
46. Meenal, R.; Binu, D.; Ramya, K.C.; Michael, P.A.; Vinoth Kumar, K.; Rajasekaran, E.; Sangeetha, B. Weather forecasting for renewable energy system: A review. *Arch. Comput. Methods Eng.* **2022**, *29*, 2875–2891. [CrossRef]
47. Çelik, D.; Meral, M.E.; Waseem, M. Investigation and analysis of effective approaches, opportunities, bottlenecks and future potential capabilities for digitalization of energy systems and sustainable development goals. *Electr. Power Syst. Res.* **2022**, *211*, 108251. [CrossRef]
48. Liu, Z.; Wang, D.; Wang, J.; Wang, X.; Li, H. A blockchain-enabled secure power trading mechanism for smart grid employing wireless networks. *IEEE Access* **2020**, *8*, 177745–177756. [CrossRef]

49. Strielkowski, W.; Streimikiene, D.; Fomina, A.; Semenova, E. Internet of Energy (IoE) and high-renewables electricity system market design. *Energies* **2019**, *12*, 4790. [CrossRef]
50. Geetha, B.T.; Santhosh Kumar, P.; Sathya Bama, B.; Neelakandan, S.; Dutta, C.; Vijendra Babu, D. Green energy aware and cluster-based communication for future load prediction in IoT. *Sustain. Energy Technol. Assess.* **2022**, *52*, 102244. [CrossRef]
51. Rochd, A.; Benazzouz, A.; Ait Abdelmoula, I.; Raihani, A.; Ghennioui, A.; Naimi, Z.; Ikken, B. Design and implementation of an AI-based & IoT-enabled home energy management system: A case study in Benguerir—Morocco. *Energy Rep.* **2021**, *7*, 699–719. [CrossRef]
52. Umar, A.; Kumar, D.; Ghose, T. Blockchain-based decentralized energy intra-trading with battery storage flexibility in a community microgrid system. *Appl. Energy* **2022**, *322*, 119544. [CrossRef]
53. Ahuja, K.; Khosla, A. Network selection criterion for ubiquitous communication provisioning in smart cities for smart energy system. *J. Netw. Comput. Appl.* **2019**, *127*, 82–91. [CrossRef]
54. Ali, S.; Ur Rehman, A.; Wadud, Z.; Khan, I.; Murawwat, S.; Hafeez, G.; Albogamy, F.R.; Khan, S.; Samuel, O. Demand response program for efficient demand-side management in smart grid considering renewable energy sources. *IEEE Access* **2022**, *10*, 53832–53853. [CrossRef]
55. Zeng, Y.; Jiang, S.; Konstantinou, G.; Pou, J.; Zou, G.; Zhang, X. Multi-Objective Controller Design for Grid-Following Converters with Easy Transfer Reinforcement Learning. *IEEE Trans. Power Electron.* **2025**, *40*, 6566–6577. [CrossRef]
56. Zeng, Y.; Xiao, Z.; Liu, Q.; Liang, G.; Rodriguez, E.; Zou, G.; Zhang, X.; Pou, J. Physics-informed deep transfer reinforcement learning method for the input-series output-parallel dual active bridge-based auxiliary power modules in electrical aircraft. *IEEE Trans. Transp. Electrif.* **2024**, *11*, 6629–6639. [CrossRef]
57. Bellini, F.; Campana, P.; Censi, R.; Di Renzo, M.; Tarola, A.M. Energy communities in the transition to renewable sources: Innovative models of energy self-sufficiency through organic waste. *Energies* **2024**, *17*, 3789. [CrossRef]

energies

Article

Supply of MV Island with High-Penetration of Prosumer Renewable Energy Sources

Krzysztof Dobrzynski [1], Zbigniew Lubośny [1], Jacek Klucznik [1], Paweł Bućko [1,*], Sławomir Noske [2], Mirosław Matusewicz [2], Michał Brodzicki [3], Maciej Klebba [3] and Arkadiusz Frącz [3]

1 Faculty of Electrical and Control Engineering, Gdansk University of Technology, Narutowicza 11/12, 80-233 Gdansk, Poland; krzysztof.dobrzynski@pg.edu.pl (K.D.); zbigniew.lubosny@pg.edu.pl (Z.L.); jacek.klucznik@pg.edu.pl (J.K.)

2 ENERGA-OPERATOR SA, Marynarki Polskiej 130, 80-557 Gdansk, Poland; slawomir.noske@energa-operator.pl (S.N.); miroslaw.matusewicz@energa-operator.pl (M.M.)

3 Faculty of Mechanical and Electrical Engineering, Institute of Electrical Engineering, Polish Naval Academy, Smidowicza 69, 81-127 Gdynia, Poland; m.brodzicki@amw.gdynia.pl (M.B.); m.klebba@amw.gdynia.pl (M.K.); a.fracz@amw.gdynia.pl (A.F.)

* Correspondence: pawel.bucko@pg.edu.pl

Abstract: The rapid development of prosumer renewable energy sources (RESs) observed in Poland in recent years causes problems in distribution networks such as current amplitude and voltage asymmetry increases, power and energy loss increases, and reverse power flows, and related are voltage control problems, deterioration of energy quality, etc. Moreover, in the case of planned repair/maintenance works in the network and the need to supply energy consumers in an islanded MV grid, the problem of the correct operation of such a subsystem appears. This occurs when the power production by the prosumers' energy sources at a given moment exceed the power consumption. In such a case, reverse power flows occur in MV/LV transformers, i.e., from the LV network to the MV network. This causes reverse power flow to the diesel generator, leading to its shutdown and, in extreme cases, to damage. The solution to this problem is to use a mobile system equipped with energy storage in addition to a diesel generator and an LV/MV transformer. An additional problem in the case of using a mobile system (diesel generator) to power an MV island is the islanded MV network grounding. Grid islanding changes the earth fault current and electric shock voltages. In general, MV networks in Poland operate as compensated, i.e., grounding transformers are used, the star point of which is grounded by a compensation choke. Unfortunately, in the case of powering an MV island from a mobile system, there is no real possibility of grounding the star point of the LV/MV transformer used there. This article proposes an algorithm of a diesel generator with an energy storage selection, including electric shock protection requirements verification, for the use in suppling energy via an islanded MV network.

Keywords: uninterruptible power supply of MV island; prosumer RES; energy storage

1. Introduction

Operational work carried out by a distribution system operator (DSO) in an MV network usually involves the need to disconnect a fragment of the MV network from the rest of grid. MV subsystem islanding is possible by using disconnectors and reclosers (remotely controlled disconnectors), which are located in selected locations only in the MV network. This makes it impossible to disconnect, for example, one repaired pole or single line span. This means that typically a certain fragment of the MV network is separated,

 https://doi.org/10.3390/en18102625

in which there are several to a dozen MV/LV substations supplying LV networks. In such LV networks, there may be up to several hundred LV energy consumers. This means that an MV subsystem islanding may require power supply for up to several thousand LV recipients. Currently, this is performed by independent power supply of individual MV/LV substations using diesel generators operated by a power team (operator employees). Taking into account the formal procedures accompanying the power grid work, this is a complex logistical challenge, requiring significant human resources (power teams and dispatchers) and material resources (diesel generators).

A solution to the above problem may be use of a single diesel generator supplying a separate part of the MV network via an LV/MV transformer. Due to the specific nature of MV network operation, this requires an analytical evaluation of the possibility of safe operation of such an MV island each time. This results from the fact that MV networks in Poland operate as so-called compensated, in which compensation chokes connected to grounding transformers located in 110 kV/MV substations are used.

MV subsystem islanding leads to a reduction in the value of the earth fault current and a reduction in the risk of electric shock at the point of the fault. On the other hand, supplying a part of the MV network from a diesel generator involves the need to use an LV/MV transformer, which cannot be grounded at its star point. This means that for each MV island supplied from a generator, the protection against electric shock must be analytically assessed.

An additional problem that has emerged in recent years results from the dynamic development of prosumers' RESs located in the LV network. A very large number of these sources means that during periods of high solar radiation the power generated by these sources exceeds the load, which leads to power flows from the LV to MV network and to the diesel generator. If the generator is equipped with reverse power protection, the generator is switched off, with a consequence being the of loss of power supply for electricity consumers. If the generator is not equipped with such a protection system it may be damaged, with the same effect for consumers.

This problem can be partially solved by carrying out work in months when the solar radiation is low, i.e., during winter in Northern Europe. But in practice, this significantly limits the possibility of carrying out maintenance work in the MV distribution network. A more advantageous solution is a hybrid analytical–technical system consisting of the following:

- In the technical area, such a system should consist of a diesel generator, an electrical energy storage, and an LV/MV transformer. This system is referred to in this publication as a hybrid mobile station (HMS), as shown in Figure 1. This system should be able to balance active power for at least several hours, i.e., charge and consume it depending on the power generated by prosumers' RESs.

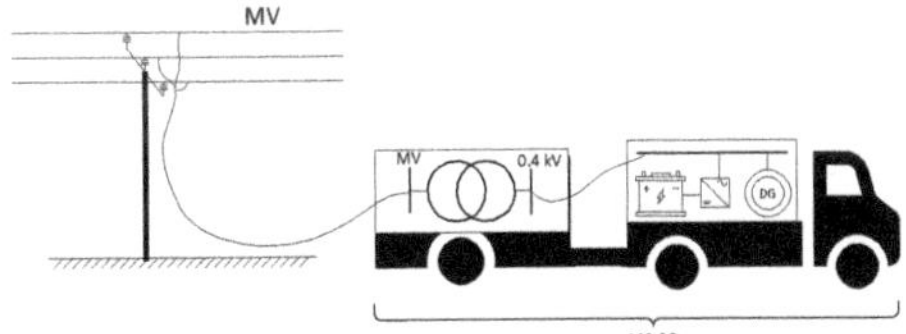

Figure 1. Hybrid mobile station (HMS).

- In the analytical area, the system should provide the network operator with information about the possibility of safe operation of a separated MV and LV network.

In order to conduct such analysis, information is required on the structure of the separated MV network, ground fault protection settings in the 110 kV/MV substation, residual current resulting from grid compensation and historical information on the load profiles of the energy consumers.

Data on consumers' load profiles are obtained from remote reading meters (individual smart meters (ISMs)), which have been gradually installed in Poland over the past few years. This results from implementation of Directives (EU) 2019/944 [1] into the Polish Energy Law, according to which 80% of consumers should have this type of meter by 2028, and all consumers by 2031. Regardless of ISMs, operators also install remote reading meters (station smart meters (SSMs)) in MV/LV substations.

If the islanded subsystem consists on the LV network only, historical measurements from IMSs should be used. And if the islanded subsystem consists of MV and LV grids (contains MV/LV substations), then historical measurements obtained from SSMs should be used. This type of case is considered in this publication.

In general, hybrid power systems (based on various technologies) are known and have been used for years. They are flexible, what makes them applicable to a wide range of applications. They are utilized in systems operating as permanent or temporal electric islands. They are used temporarily because of failures or maintenance work. In the first group are marine systems (vessels), while in the both groups are industrial, telecommunication, transportation, electric power, etc., systems.

Hybrid systems provide different operation configurations, thus proper control and management makes fulfilling the most efficient demand or control goals possible. Energy storage and diesel generator sizing is discussed in [2,3]. In a typical configuration, diesel generators operate as prime movers, being the primary source of electrical energy, while energy storage shaves the load peaks. Additionally, such a hybrid system with energy storage can limit the amount of diesel fuel used.

In the case of marine applications there are many different systems combining both mechanical and electrical drives with a hybrid power supply. Such solutions are based on the cooperation between mechanical, combustion generating sets and energy storage, which is intended for electrical supply [4–6].

In the case of land applications the discussion related to a diesel generator with RESs in an islanded grid focuses mainly on: the energy source's control (often fewer or more RESs are considered) [7–10], the optimal placement of RESs in the grid, the optimal parameters of RESs in the grid [11], the islanded grid static and dynamic features, and the RESs placement and operation costs [12–15]. Also, results of tests conducted in a real power system are presented.

Remote areas and islands isolated from the main electric power system ensure reliability and stability of power supply using diesel generation. Due to the high purchase and transportation costs of diesel fuel, isolated communities are exploring other options for efficient and reliable power supply. Renewable energy sources can reduce operating costs [12]. When, because of various reasons, it is impossible to supply loads from the hybrid system, the use of local renewable sources and a small battery control is proposed [16,17]. This way of grid control requires the system observability and controllability, including ability to control prosumer (local) RESs.

In real systems, tests are conducted to verify the ability of such subsystem islanding and resynchronization. However, stand-alone operation including voltage and frequency control is also discussed [18,19]. It is indicated that the islanded grid's low inertia and inverter features significantly impact protection, control, and grid stability, especially during large transients.

Also discussed are islanding partition and scheduling strategies which the system operator can conduct before grid maintenance to optimize islandic operation [20].

The problem under consideration is similar to microgrids operating islanded. A comprehensive review of almost 200 publications on microgrids and hybrid systems in [21] shows that island operation of microgrids is not typical and requires special attention in balancing active power to maintain the desired frequency value, which is why many authors do not address the topic of island operation of microgrids. Publications dealing with island operation of microgrids, on the other hand, assume the need for a complex central or hierarchical energy management systems which controls all of the energy sources in the grid. This paper assumes that the active power balance and frequency value will be controlled by only one source, which is the HMS. The literature focuses usually on a well-defined system, i.e., SG, PV, and energy storage, with known data and parameters. The presented paper considers the HMS sizing for a real grid, i.e., with measurements containing gaps of data and errors.

2. Materials and Methods

Evaluating the possibility of supplying power to a separate part of the MV network, the MV subsystem load profile should be determined. The assumed period of operation of the HMS and its maximum power should be considered. Determining the load profile also allows us to determine the amount of energy delivered by the diesel generator during its assumed operating time. This allows for us to also determine the necessary storage capacity. With some approximation, this can be performed using historical measurements carried out in the distribution network. In general, loads can be considered as the MV/LV distribution transformers and as the loads connected directly to the MV grid. Simultaneously, energy sources, e.g., photovoltaic farms or wind farms, connected to the MV network should be included in the considerations. Then, using the historical power/energy profiles of these devices, the HMS production/consumption profile can be determined.

In the case of the DSO under consideration, most MV/LV substations are equipped with an SSM. Measurements in the stations are performed on the LV side. The DSO has access to these measurements within the AMI (advanced metering infrastructure) system. In general, hourly values are measured in MV/LV distribution stations, including the value of active energy transmitted to the LV network (A+) and to the MV network (A−). Therefore, the "+" or "−" sign indicates the direction of current flow through the MV/LV station. In addition to the energy values, instantaneous and average values of currents and voltages are also recorded. These values can be used to determine the HMS load profile. But, the key factor here is the availability of measurements for individual MV/LV substations.

Energy measurements are sufficient to determine the capacity of the electrical energy storage in the HMS. For this case, the active and reactive energies measured in all MV/LV substations located in the MV island should be summed as follows:

$$A_{HMS} = \Sigma_t \, \Sigma_i \{k_{(i)} \cdot (A+_{(i,t)} - A-_{(i,t)})\}, \tag{1}$$

where

i — i-th MV/LV substation charged by HMS with LV/MV;
t — t-th measurement hour;
$k_{(i)}$ — factor for i-th MV/LV substation, resulting from current transformer ratio, to which a given SSM is connected.

In turn, to determine the HMS load profile, current and voltage measurements, instantaneous or average, can be used. It should be noted here that the use of instantaneous or average values carries a certain uncertainty. If average values are used, there is no information about the maximum value that occurred in an hour. The situation is similar with the instantaneous value, which is recorded at the time of measurement, i.e., every 1 h. To limit the uncertainty resulting from the measurements, the current value that is the greater of these two values can be used. Therefore, in order to determine the HMS load profile the powers of the individual phases for each MV/LV substation should be summed up as follows:

$$S_{HMS(t)} = \Sigma_i[k_{(i)} \cdot (V_{L1(i,t)} \cdot I_{L1(i,t)} + V_{L2(i,t)} \cdot I_{L2(i,t)} + V_{L3(i,t)} \cdot I_{L3(i,t)})], \quad (2)$$

where

i	—	i-th MV/LV substation charged by HMS with LV/MV;
t	—	t-th measurement hour;
V_{L1}, V_{L2}, V_{L3}	—	instantaneous or average voltage in phase L1, L2, and L3;
I_{L1}, I_{L2}, I_{L3}	—	instantaneous or average current in phase L1, L2, and L3.

It should also be noted that the determined apparent power S is on the low-voltage side, so it will be slightly higher on the medium-voltage side because it will take into account the losses incurred in the MV/LV transformer. Additionally, the HMS load profile determined in this way does not include losses incurred in the MV island transmission lines. However, neglecting these losses introduces an error in the safe direction. Since the current measurements available in the AMI system do not contain information on the flow direction (current module measurement available), the above method of determining the load profile is correct only for power flows towards the LV grid. In order to determine the flows towards the MV network, active energy measurements A+ and A− can be used. If the values both are non-zero then during the measurement hour there was a power flow in both directions. In such a case, the direction of current flow cannot be unambiguously determined. Then, it is assumed that the direction of flow is determined by the higher value. In turn, the unambiguous direction of current flow will only be for cases in which one of the values is zero. That is, if A+ > 0 and A− = 0 then there is assumed flow to the LV grid. And, if A+ = 0 and A− < 0 then there is assumed flow to the MV grid.

In general, it is not easy to precisely determine the load profile of the HMS supplying an MV island with high RES saturation. On the one hand, it is helpful to use historical measurements, which are described above. However, the currently used measurements do not provide weather information that would be helpful in assessing the share of RES generation in the power balance. And since most RESs located in the LV network are photovoltaic installations, information on cloud cover and insolation is crucial. Therefore, the HMS load profile determined on the basis of historical measurements should be verified after taking into account historical cloud cover and forecasted cloud cover for the day of the planned work.

An example of determining the HMS profile based on historical data for the MV island is presented below. The separated fragment contains 11 MV/LV distribution substations (Figure 2) for which historical measurement data are available. The analyzed fragment of the network consists only of MV overhead lines, the poles of which are marked in Figure 2 with dots.

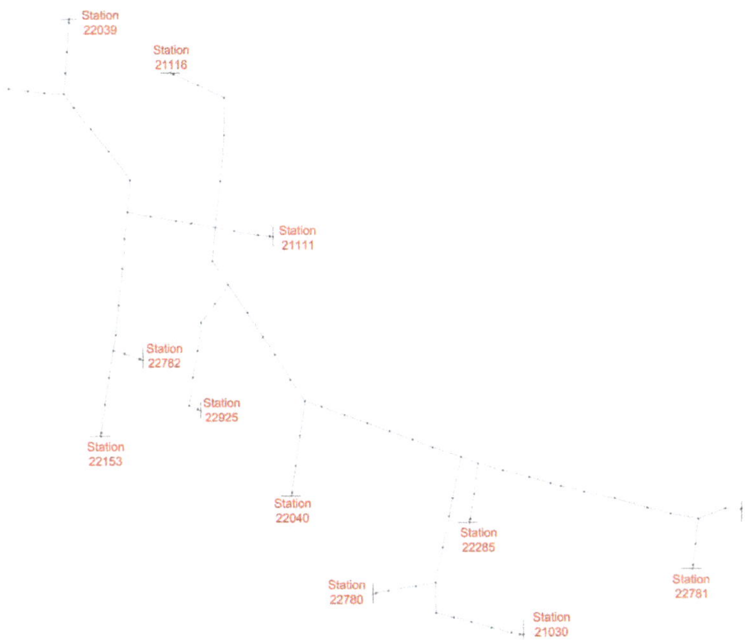

Figure 2. The considered fragment of the MV network.

3. Determination of the Load of the Mobile Power Station

As mentioned above, historical measurement data are needed to determine the HMS load profile. The period of measurement data analyzed covers the annual period from 1 October 2022 to 30 September 2023.

- It should be noted that, in general, the measurement data are not error-free or there are gaps in the measurements. For the considered network fragment some irregularities were noticed that required explanation. For example:
- In the case of distributed substations 21,030, 22,040, and 22,782, there was a multiplier of 200, which in reality turned out to be equal to 80, because 400/5 A/A current transformers are installed at these stations. Clarifying these inconsistencies required checking the DSO's paper documentation.
- In the case of distributed substations 21,030, 22,153, and 22,285, there were flows towards the MV grid during night hours. It should be noted here that each of the 11 distribution substations contain only prosumer photovoltaic installations. This was considered as a measurement error because there is no power flow into the MV grid during nighttime hours.
- There are gaps in the measurement data that concern single hours, whole days, or some number of whole days. This paper presents a simplified approach in which the gaps are filled by evenly divided energy values, calculated as the difference in first (after the gap) and end energy values measured. In the considered case the missing measurements are rare (<1%).

Despite the inaccuracies in the measurement data described above, the HMS load was assessed. As previously mentioned, the direction of power flow for the HMS was determined by comparing the active energy consumed (A+) and emitted (A−) by each of the 11 MV/LV distribution substations located in the considered area. Below, Figures 3 and 4 show load profiles for two sample MV/LV distribution substations. Out of the 11 distribution substations located in the considered area, only 1 (Figure 4) is characterized by only power transmitted towards the LV network. In the remaining cases, energy flows occur towards the MV grid, as for substation 21,116 (Figure 3).

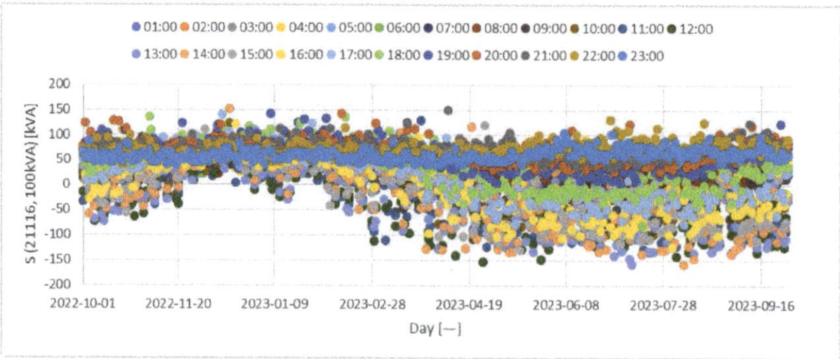

Figure 3. Load profile of MV/LV distribution substation no 21116.

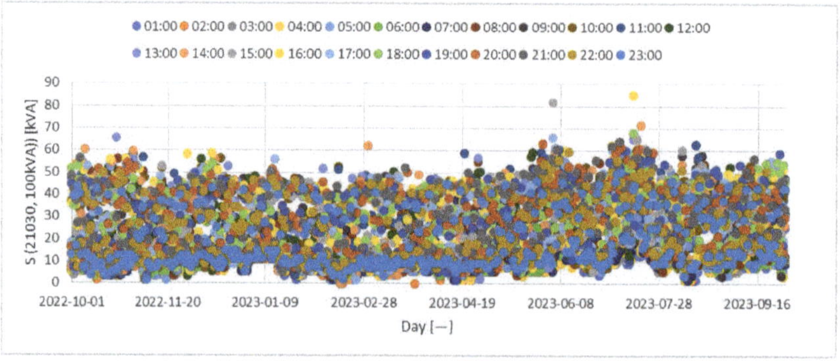

Figure 4. Load profile of MV/LV distribution substation no 21030.

As a result of summing up the load of all 11 MV/LV substations, a profile of the HMS load that would occur in the measurement year was obtained (Figure 5). Figure 5 shows the apparent power value S for each hour of the day for each day of the year. Analyzing the obtained profile, it can be seen that for most of the year the power balance has a negative value, which means that it is not possible to operate the grid with a traditional diesel generator on these days. Power flows towards the LV grid (with a few exceptions) are observed from the second half of November 2022 to the beginning of February 2023. This means that powering the analyzed part of the MV network using a classic solution (only a diesel generator) would only be possible during this period.

The data presented are historical. Therefore, while planning maintenance/repair work in an islanded grid, the weather forecast for the planned day of work should be included.

Limiting planned work to specific days of the year is a major obstacle for DSOs. In turn, the use of an HMS equipped with electricity storage significantly increases the freedom in planning these works. Nevertheless, DSOs need information on the maximum power that an HMS should have and the expected energy balance during the planned work. This is important in the context of selecting the storage capacity and its permissible charge level.

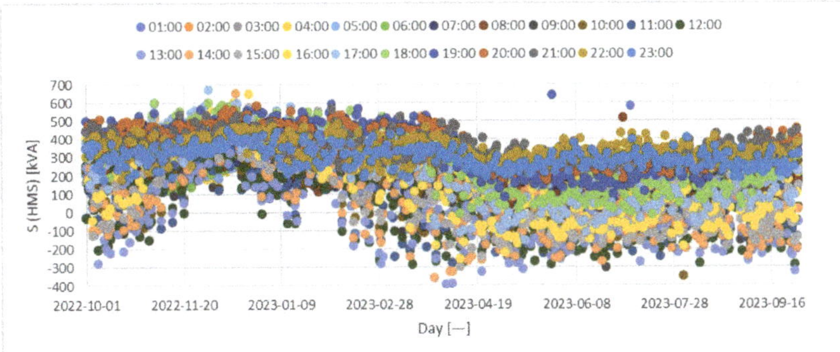

Figure 5. Load profile of a selected part of the MV grid.

Considering the analyzed fragment of the MV network, it is possible to determine the maximum power that an HMS should have and what energy it should supply or consume on a given day. To determine these values, it was assumed that the power supply to the MV island lasts no longer than 6 h. Therefore, the HMS operating time was assumed to be from 9 am to 2 pm. For such a limited time during the day, the maximum power supplied by an HMS should not exceed 652 kW (Figure 6), and the maximum power consumed should not exceed −148 kW. Taking into account the previously adopted assumption that the rated power of the HMS is 630 kVA, the day with a load of 652 kW should be excluded from the planned work. However, such a load occurs only once a year. In other cases the maximum power achieved is below 600 kW.

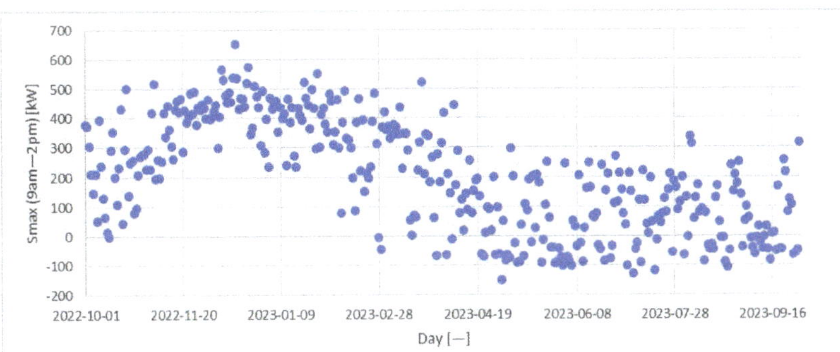

Figure 6. Maximum apparent power S_{max} load in hours from 9 am to 2 pm.

The energy required to be supplied or consumed by the HMS during 6 h of operation is greater for the supplied energy and exceeds 3.96 MWh (Figure 7). In general, this is easy to achieve because it can be provided by the diesel generator, which is present in the HMS system. In turn, the energy consumed by the HMS for the most demanding day reaches 2.12 MWh. Therefore, the DSO who wants to carry out work on any day during the measurement year should use an HMS with an energy storage with capacity equal to 2.12 MWh. On the other hand, to carry out planned work on the MV grid it needs a mobile solution that will be able to relatively quickly connect, for example, to an MV overhead line pole. In such a case, a solution built on a truck comes into play. This means that the location of the diesel generator and energy storage on such a truck will limit the storage capacity. In turn, in order not to further limit the energy storage capacity, the LV/MV transformer could be built on a trailer.

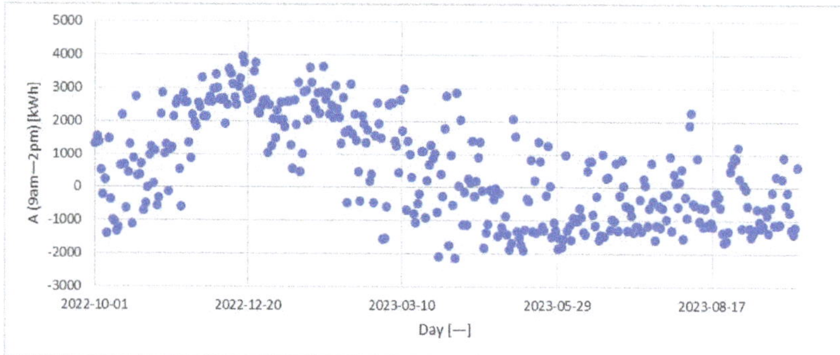

Figure 7. Total energy delivered/consumed on measurement days in hours from 9 am to 2 pm.

Analyzing the measurement data presented above, it can also be seen that using the classic solution, i.e., only with a diesel generator, the DSO can carry out work on only 155 out of 365 (42.5%) days in the measurement year (Figure 8). On the remaining days in the assumed working hours (from 9 am to 2 pm) there is at least one hour in which excess power is generated. As a result, this would result in reverse power on the diesel generator and its disconnection. Therefore, the DSO needs a solution that will ensure the balancing of active power in the MV island, including the possibility of energy collection.

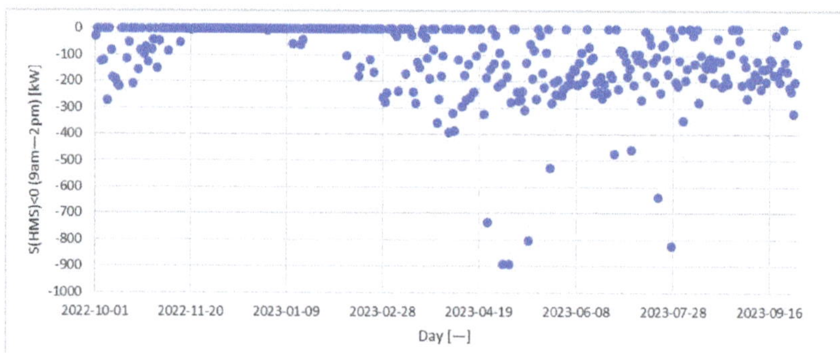

Figure 8. Maximum apparent power consumption, $S_{HMS} < 0$, in hours from 9 am to 2 pm.

4. Effect of MV Island Separation on Electric Shock Protection

The second issue related to separation of a fragment of the MV network is the risk of shock hazard assessment, i.e., the calculation of the touch voltages that may occur during an earth fault.

As mentioned in Section 1, separating a fragment of the MV network changes the method of network earthing. Prior to separation, the MV network operates as compensated, using a compensation choke installed at the 110 kV/MV substation. The MV network is often islanded without the compensation choke. Additionally, the LV/MV transformer of the HMS has no real possibility of earthing the grid star point. So, this creates a network operating with an isolated star point, which changes a ground short-circuit current I''_{k1}.

Such a short-circuit current in an islanded grid has to be evaluated (calculated). To do this for overhead transmission lines with conducting supports (poles) without a disconnector the standard EN 50341-1 [22] can be applied. This standard defines the permissible touch voltage V_D (Table 1) as a function of the earth fault duration (de facto as setting time of the earth fault protection). In the considered full MV grid (not islanded), the earth fault protection is set to $\Delta t = 1.2$ s, while the single-phase short-circuit current (residual current) is equal to $I''_{k1} = 17.5$ A.

Table 1. Largest permissible contact voltage V_{Tp} (V_D) in MV power lines according to EN 50341-1.

t_F [s]	0.05	0.1	0.2	0.5	1	2	5	10
V_{Tp} (V_D) [V]	735	633	528	204	107	90	81	80

The electric shock protection verification for MV power lines should be performed by calculation of the earth voltage V_E (Figure 9) for each MV uninsulated pole using its earth resistance R_s and comparing the voltage to the permissible touch voltage V_D, which is shown as follows:

$$V_E \leq 2\,V_D, \tag{3}$$

$$V_E = I_E \cdot R_s, \tag{4}$$

For this purpose, the earth electrode resistances of the poles in the MV island are needed. Unfortunately, often only resistances for poles with a disconnector or recloser are available (Table 2).

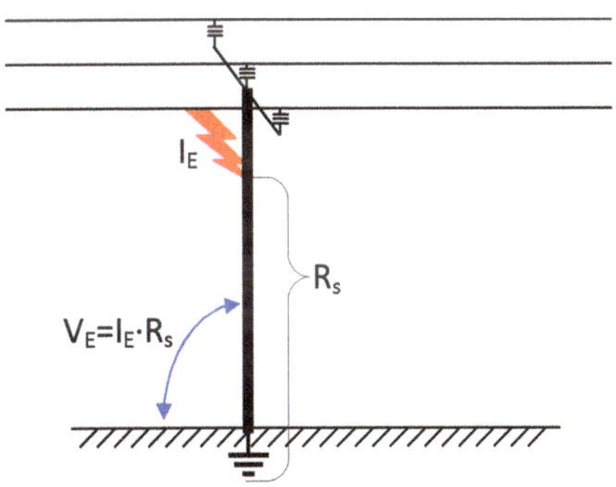

Figure 9. Earth voltage at MV support.

Table 2. Earth electrode resistances of poles with disconnectors located in the analyzed MV island. Values obtained from measurements.

Disconnector No	R_s [Ω]	V_E [V]	$2\,V_D$ for t = 1.2 s [V]	Fulfilled Shock Protection
942/II	7.44	2.3		Yes
198/II	28.2	8.7		Yes
975/II	1.56	0.5		Yes
290/II	1.2	0.4		Yes
1642/II	5.28	1.6		Yes
1274/II	33.48	10.4		Yes
420329	3.96	1.2	200	Yes
422115	12.96	4.0		Yes
964/II	20.28	6.3		Yes
1490/II	1.2	0.4		Yes
263/II	9.12	2.8		Yes
421785	69.72	21.6		Yes
961/II	3.24	1.0		Yes

Using the resistance R_s and the earth fault current I_E, the earth fault voltage V_E can be calculated. By comparing it to the permissible touch voltage V_D, the electric shock protection ability in the islanded grid can be verified.

Notice that some of the resistances are high. The highest one reaches 69.72 Ω. This means that for residual current equal to I''_{k1} = 17.5 A a ground voltage reaches a value equal to 69.72 × 17.5 = 1220 V. It leads to the conclusion that electric shock protection requirements are not fulfilled. Such a conclusion can be based only on information about the pole earthing resistance R_s. In practice, some additional electric shock protection measures can be used in the pole earthing which can lead to the fulfillment of electric shock protection requirements.

In the considered MV grid the declared time of operation of the earth fault protection is equal to 1.2 s. For this time the permissible touch voltage is equal to V_D = 100 V (Table 1). This value is significantly higher than the calculated earth voltages V_E (see Table 2). The voltages are calculated for earth current in the MV islanded grid equal to I_E = 0.31 A. The current calculation is discussed below. But, in the considered case, i.e., for each pole of an overhead line with a disconnector or recloser located in the MV island, electric shock protection is met if 1.2 s setting time of earth protection is used in the HMS.

It should be noted that the above verification of the effectiveness of electric shock protection was carried out only for poles with a disconnector. For the remaining uninsulated poles in the MV island there is no information on earthing resistances. For this reason, as well as in the situation of low reliability of measurement data, where there is no information on the application of additional protection measures it is impossible to determine the effectiveness of electric shock protection on the basis of comparing the earth voltage with the permissible voltage.

Another of the important issues related to the neutral point grounding of the MV network is the arc extinction limit current. For a 15 kV network operating with an isolated neutral point, the VDE 0228, teil 2 standard (1987) defines this value as equal to I_{limit} = 35 A. The short-circuit current should not be greater than the arc extinction limit current, which is shown as follows:

$$I''_{k1} \leq I_{limit}. \tag{5}$$

Therefore, in the considered case, while defining the MV island the calculated earth current must not exceed the value of 35 A.

Another issue to consider is the impact of the presence of cable lines on the MV island. The cable lines significantly impact the values of earth currents due to a much higher capacitance in comparison to the overhead lines. A characteristic property of MV cable lines is their construction. They contain shielding conductors. The MV networks operate with shielding conductors earthed on both sides. In such systems during short-circuit, part of the short-circuit current flows through the shielding conductor, reducing the value of the current flowing through the earth at the point of short-circuit. This reduces the earth voltage. In practice, when determining the contribution of the shielding conductors to the flow of short-circuit currents a reduction factor r is used. The factor defines what part of the short-circuit current flows through the pole earth and is shown as follows:

$$I_E \leq r \cdot I''_{k1}, \tag{6}$$

The reduction factor takes values below 1 and varies for different shielding conductor cross-sections. For example, for MV cables with polyethylene insulation, the reduction factor is equal to 0.25 for a Cu 50 mm^2 shielding conductor, 0.4 for a Cu 25 mm^2 shielding conductor, or 0.55 for a Cu 16 mm^2 shielding conductor.

Given the above, it is necessary to determine the earth current for the MV island. This current can be determined using the symmetrical components method. It results from the following series connection of impedances: zero-sequence Z_0, positive-sequence Z_1

and negative-sequence Z_2. The single-phase short-circuit current is calculated as equal to the following:

$$I''_{k1} = I_E = (c \cdot V_n \cdot \sqrt{3}) / |\underline{Z}_0 + \underline{Z}_1 + \underline{Z}_2|, \quad (7)$$

where

V_n — nominal voltage of network;
c — voltage factor (1.1 for MV) [23].

In networks operating with an inefficiently grounded neutral point, the equivalent impedances for the positive-sequence Z_1 and the negative-sequence Z_2 are many times smaller than the impedance for the zero-sequence Z_0. Therefore, the positive- and negative-sequence can be neglected, which leads to the following equation:

$$I''_{k1} = I_E = (c \cdot V_n \cdot \sqrt{3}) / |\underline{Z}_0|, \quad (8)$$

The equivalent impedance for the zero-sequence of Z_0 depends on how the neutral point of the network is earthed. For a network operating with isolated neutral points, it is equal to the following:

$$\underline{Z}_0 = 1/j\omega C_0 = 1/jB_0, \quad (9)$$

where

C_0 — transmission line zero-sequence capacitance,
B_0 — transmission line zero-sequence susceptance.

It should be noted that the value of power line capacitance (susceptance) used in the calculation of earth fault current is not the same as the capacitance (susceptance) used in the calculation of power line charging current. For short-circuit calculations, the zero-sequence of capacitance C_0 or susceptance B_0 is used, whilst for power line charging current calculations the positive-sequence of capacitance C_1 or susceptance B_1 has to be used.

The power lines used in MV networks can be overhead or cable. In the case of the MV island under consideration, there are only overhead lines, whose zero-sequence capacitance C_0 depends on the mutual arrangement of wires and their height above the earth. A comparison of the capacitance values B_0 and B_1 for an example of three arrangements of wire spacing on an overhead line support (Figure 10) is shown in Table 3. Since the value of B_0 also depends on the height of the hanging wire, values are shown for the minimum (5.5 m) and maximum (16.25 m) hanging wire used by the DSO.

By analyzing the obtained susceptances it can be concluded that the zero-sequence susceptance is 40% to 60% higher than the positive-sequence susceptance. In addition, changing the height of the hanging cable only affects the zero-sequence susceptance. This change reaches up to 29%, depending on the arrangement of the wires on the pole.

In order to analyze the impact of the arrangement of the wires on the support (pole), the hanging of the wires above earth, and the sag of the wires, the MV island was modeled in the PowerFatory 2022 software from DIgSILENT (Gomaringen, Germany). The software used gives the possibility to take into account the geometry of the arrangement of the wires, converting it to electrical parameters of sections of the transmission line. This is particularly important from the point of view of relatively accurate determination of the capacitance of individual power line sections, which depends significantly on the distance of the wires from the earth.

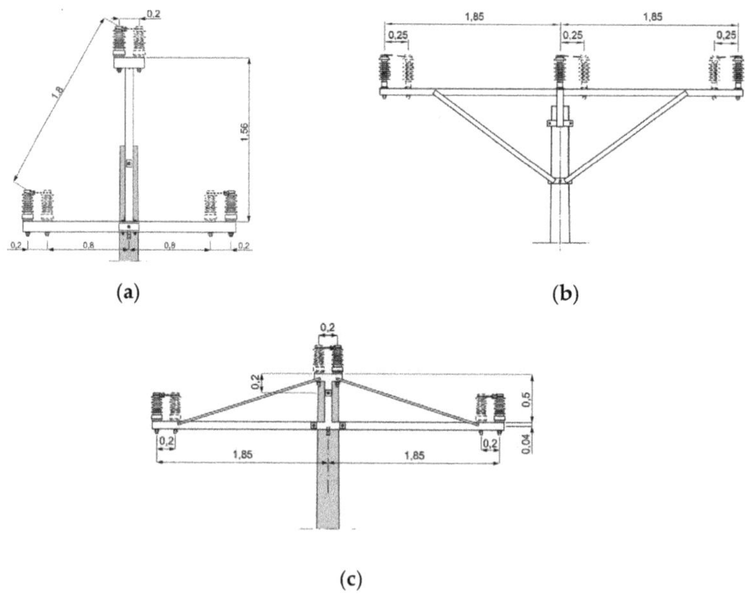

Figure 10. Examples of cable hanging arrangements on MV supports: (**a**) triangular; (**b**) flat 1; (**c**) flat 2.

Table 3. Equivalent scheme parameters of selected MV overhead lines. Type of conductor: AFL6-70.

Conductor Support	R_0	R_1	X_0	X_1	B_0	B_1	I''_{k1}
[-]	[Ω/km]	[Ω/km]	[Ω/km]	[Ω/km]	[μS/km]	[μS/km]	[A/km]
Minimum height of hanging wires above the earth (5.5 m)							
Triangular	0.58	0.43	1.55	0.38	1.54	3.01	0.044
Flat 1	0.58	0.43	1.52	0.39	1.63	2.93	0.047
Flat 2	0.58	0.43	1.52	0.39	1.47	2.91	0.042
Maximum height of hanging wires above the earth (16.25 m)							
Triangular	0.58	0.43	1.55	0.38	1.22	3.01	0.035
Flat 1	0.58	0.43	1.53	0.39	1.26	2.92	0.036
Flat 2	0.58	0.43	1.52	0.39	1.27	2.91	0.036

In general, the DSO does not have information in its IT databases about either the arrangement of the wires on the poles or the height of the wires hanging above the earth. In the case considered, this information was obtained from flyovers of the analyzed part of the network MV by reading this information from graphic files of individual spans (Figure 11). The data provided show that all MV line sections occurring in the analyzed area have a triangular arrangement of wires, and the total length of all line sections is 8.007 km.

In networks operated with an isolated neutral point it can be assumed that the earth fault current will be the same at each point of the analyzed network. This is due to the dominant value of the zero-sequence, which leads to a reasonable simplification involving the neglection of the positive- and negative-sequence in the calculation of earth fault current. Therefore, the evaluation of the effectiveness of electric shock protection can be carried out independently of the network structure, i.e., there is no need to map the network topology, as information on individual spans and cable line sections is sufficient.

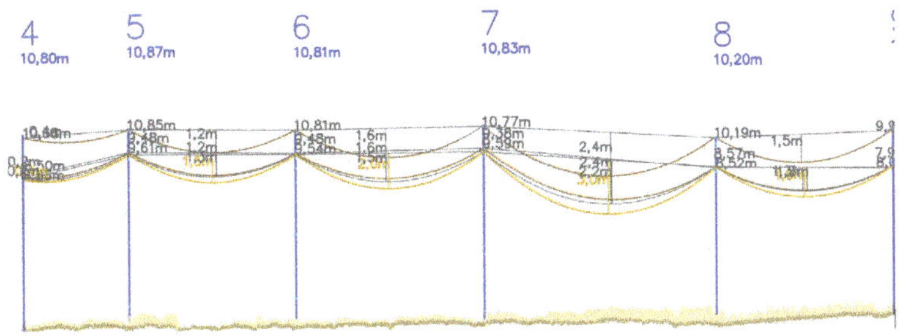

Figure 11. Profile of an example section of the overhead line.

The occurrence of the same earth current at any point of the MV island under analysis is proved by calculations carried out in the mathematical model. Figure 12 shows four different locations of short-circuit, and in each of them an earth current equal to 0.31 A is obtained. This value is relatively small but is confirmed by the kilometric parameters.

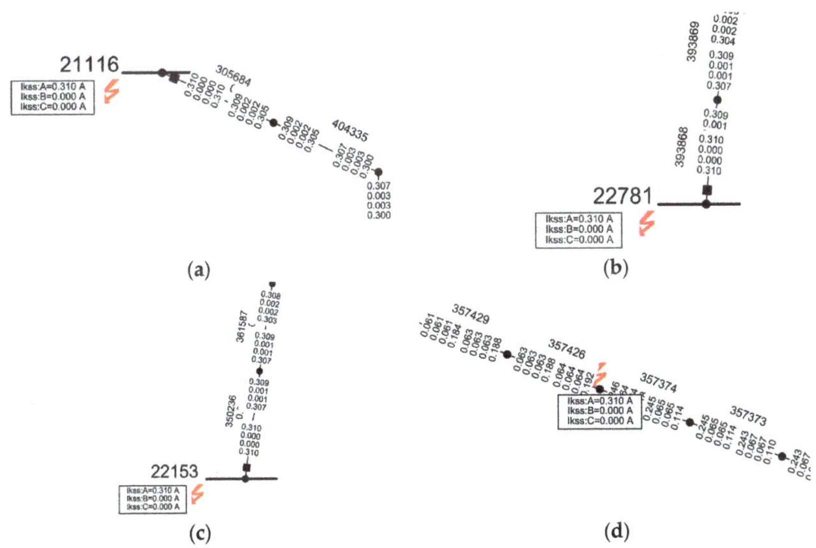

Figure 12. One-phase short-circuit at different locations of the analyzed MV island. (**a**) Short-circuit in Station 21116; (**b**) Short-circuit in Station 22781; (**c**) Short-circuit in Station 22153; (**d**) Short-circuit on the example tower.

The results obtained are also not affected by the location of the HMS connection. This is confirmed by an analysis of the various HMS connection locations shown in Figure 13a. Regardless of the connection location, the earth fault current has the same value (Figure 13b–e). So, from the point of view of electric shock protection, the HMS can be connected at any point on the MV island.

Short-circuit current is clearly influenced by the height of the hanging wires and the sag. In the general case, the lower the hanging wires, the higher the line capacitance (Table 3) and consequently the higher the earth current.

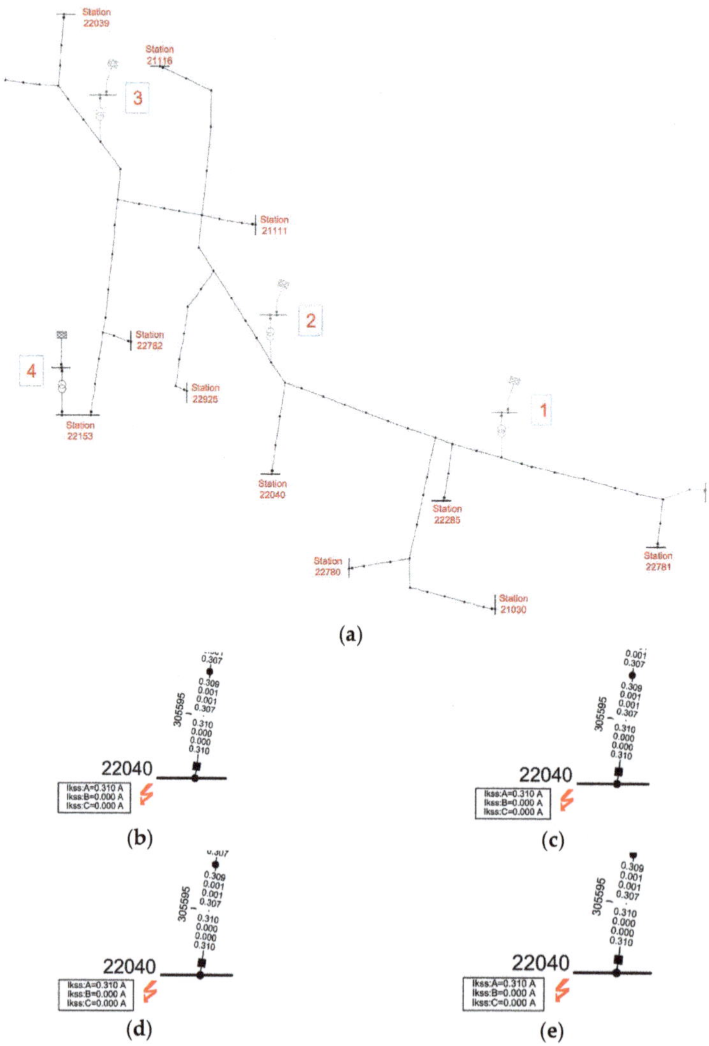

Figure 13. One-phase short-circuit on MV buses in MV/LV station. (**a**) Examples of HMS connection locations; (**b**) HMS connected in point 1; (**c**) HMS connected in point 2; (**d**) HMS connected in point 3; (**e**) HMS connected in point 4.

The effect of sag on the values of earth currents is relatively small. Using the considered network model, it can be found that the difference in earth current for the sag from the power line flyover performed at 19 °C and sag calculated at 40 °C differs by 0.6% (Figure 14). On the other hand, assuming no sag (sag equal to 0 m), the difference to the case with maximum sag does not exceed 3%. Thus, in the calculation of electric shock protection, in order to simplify the calculation, it is acceptable to assume the maximum overhang of wires.

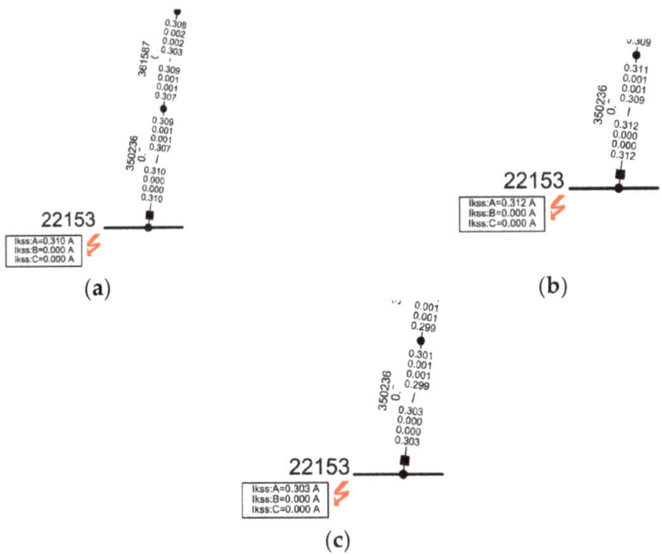

Figure 14. One-phase short-circuit on MV buses in MV/LV station: (**a**) power line sag at a temperature of 19 °C; (**b**) sag at a temperature of 40 °C; (**c**) sag equal to 0 m.

A greater influence on the values of earth currents is the hanging height of the wires. Taking into account three different heights, 5.5 m, 10 m, and 16.25 m, differences in earth currents reach up to 29%. The calculations have been performed for an exemplary AFL-6 overhead line of 35 mm^2 with a length of 1 km (Figure 15).

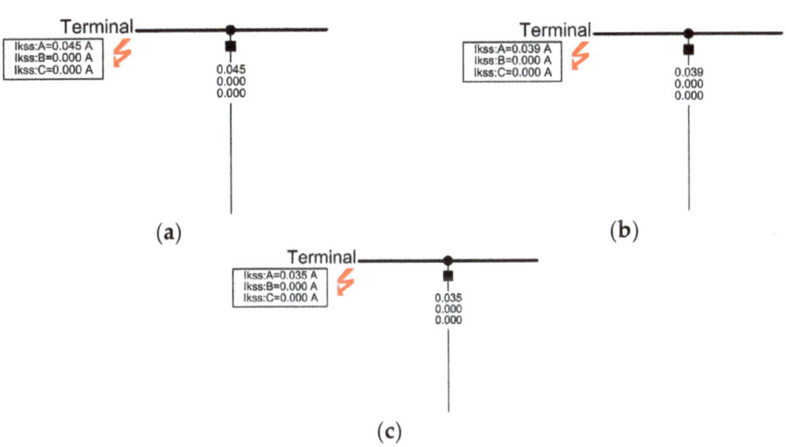

Figure 15. One-phase short-circuit current for an example AFL-6 overhead line with a cross-section of 35 mm^2 and a length of 1 km: (**a**) wires at height 5.5 m; (**b**) wires at height 10 m; (**c**) wires at height 16.25 m.

At the same time, it should be noted that calculated values of the earth short-circuit currents were not verified with measurements in the real grid. This is because the DSO has not implement the proposed method yet.

Evaluation of electric shock protection can be performed based on the earth fault protection setting for the section from which the MV island is supplied. In this case, this setting is 17.5 A, with an operating time of 1.2 s. Thus, if the MV island's earth current is less than this setting, it can be assumed that electric shock protection is satisfied. In the MV island analyzed, this condition is met because the earth fault current is equal to 0.31 A (then voltages V_D (Table 2) are very small), which is far less than the residual current (17.5 A), confirming that the electric shock protection is met.

On the other hand, if there is a case in which in the MV island the earth current is greater than the residual current, the operation time of the HMS earth fault protection should be calculated from the following relationship:

$$(V_{D(t,\ island\ MV)})/(I''_{k1(island\ MV)}) \geq (V_{D(t,\ station110\ kV/MV)})/(I_{(res.\ station\ 110kV/MV)}), \quad (10)$$

where

$V_{D(t,\ island\ MV)}$ — is the voltage resulting from the characteristics for the permissible touch voltage $V_{D(t,\ island\ MV))} = f(t_{prot\ Island\ MV})$ (Table 1) for the time of operation of the earth fault protection installed in the HMS, for which the checked inequality is satisfied;

$V_{D(t,\ station\ 110kV/MV)}$ — is the voltage resulting from the characteristics for the permissible touch voltage $V_{D(t,\ station\ 110kV/MV)} = f(t_{prot.\ station\ MV})$ (Table 1) for the operating time of the earth fault protection installed in the section where the MV island is located;

$I_{res\ station\ 110kV/MV}$ — is the minimum residual current for the section from which the MV island is supplied during normal operation;

$I''_{k1(Island\ MV)}$ — is the earth current calculated for the analyzed MV island.

5. Discussion

The current rapid development of prosumer RESs cause various problems in distribution networks. One of them is the difficulty in carrying out planned maintenance in LV and MV networks. The high saturation of RESs in LV networks may result in active power flow towards the MV network. This often makes it impossible to traditionally feed a section of the MV network from a diesel generator as this leads to generator shutdown due to reverse power. One solution is to use a hybrid mobile station consisting of a diesel generator, energy storage, and an LV/MV transformer. Such an arrangement offers the possibility of actively balancing power in a dedicated MV island. HMS capacity is determined by the load profile, which can be calculated based on historical measurements and weather forecasts.

The second problem associated with powering the MV island using the HMS is verification that electric shock protection requirements are met. The need for this assessment arises from the change in operation of the MV network from compensated to isolated. This can be calculated using the network parameters of the islanded MV network section.

Therefore, in addition to the HMS, computation software is needed to determine the HMS rating from the load profiles and to assess the electric shock protection. A simplified algorithm for such software is shown in Figure 16. The algorithm starts by indicating the day of scheduled work and checking whether historical measurements are available for the same period as the indicated day. If there are no historical measurements, then another day should be selected. Next, the electric shock protection ability of the HMS and islanded grid is considered. The earth short-circuit current $I''_{k1island}$ in the considered island is calculated. When the current $I''_{k1island}$ is higher than the residual current I_E of the full MV grid then the HMS protection settings can be set, for instance, as equal in protection relay in a 110 kV/MV substation. In another case the HMS protection settings have to be modified.

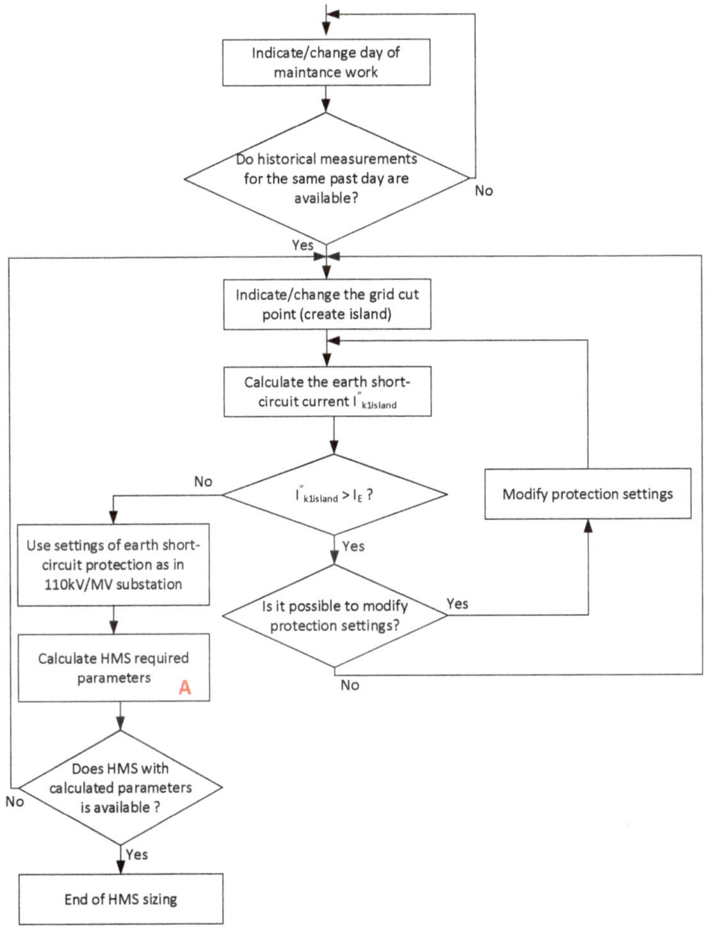

Figure 16. Simplified algorithm for verifying HMS operation capabilities. A—expansion of the algorithm.

Next, the HMS capacity sizing process, marked by letter A, can be realized. The process details are presented in Figure 17. Here, the following data are necessary: validated historical load profiles of MV/LV substations; capacity of RESs installed in considered grid now and in the past; weather forecast for day of planned work; and historical weather forecast. This allows us to determine the load profile of the HMS, the required rated power of the diesel generator, and the required energy storage capacity.

The aim of the HMS is to balance the MV island's power. The power balance variation results from variations in customer load and RES generation. These variations can be significant over short-term periods. The HMS operation basic algorithm is shown in Figure 18. Active power balancing is realized by diesel generator (DG) when the island behaves as a load, i.e., $P_{HMS} > 0$, which means power flow to grid. When energy S_{ES} stored in the energy storage is less than 20%, the DG simultaneously charges the energy storage. In the opposite case, the energy storage operates with power equal to $P_{ES} = 0$. When the island behaves as a source of energy, i.e., $P_{HMS} \leq 0$, which means power flow from grid to the HMS, the energy storage ES is responsible for power balancing, while the diesel generator DG operates with power equal to $P_{DG} = 0$.

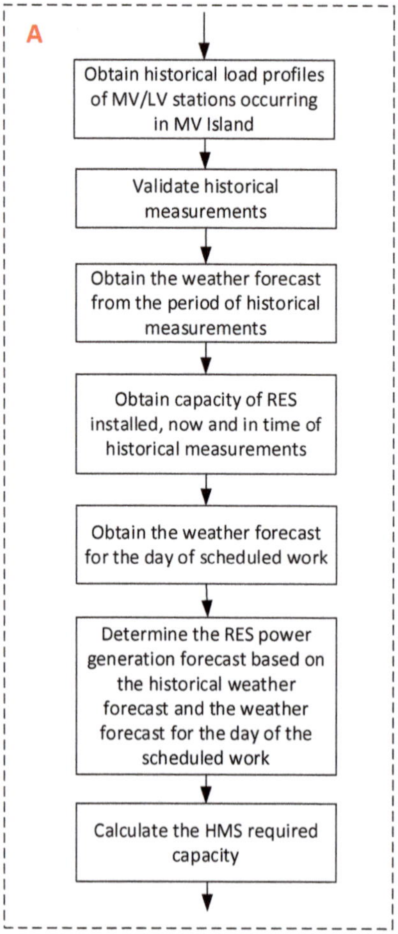

Figure 17. The HMS sizing algorithm. A—expansion of the algorithm in Figure 3.

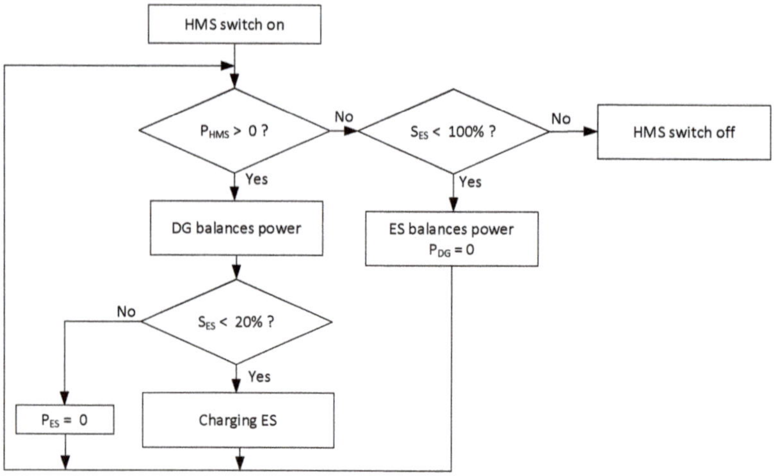

Figure 18. The HMS operation basic algorithm.

The algorithms presented above are a starting point for developing detailed algorithms. Nevertheless, such a solution would undoubtedly support distribution network operators, speeding up the planning of maintenance works while increasing power supply reliability and ensuring safety.

The solution described above assumes that the HMS consists of a diesel generator, energy storage, and an LV/MV transformer. Such a combination seems to be optimal in terms of the flexibility of its application. In this case, it is possible to supply the MV island with an uncharged energy storage and balance the power with the energy storage or diesel generator.

On the other hand, when disconnecting the transformer from the HMS it is possible to supply the LV grid with RESs. Nevertheless, another solution could be an HMS consisting only of energy storage and a transformer. But such a solution requires a more precise determination of the HMS load profile and assumed work time.

The proposed HMS is a solution for grids with an increasing capacity of RESs. The energy storage and MV/LV transformer increase the cost of such an HMS but also increase its functionality, significantly simplify logistics and saving human and material resources. For this reason, the implementation of this type of solution in practice should be expected.

Another advantage of the proposed system is the HMS mobility. This is very important for DSOs, because in practice, the HMS is connected to an MV tower which is sometimes in terrain with difficult access. Thus, a mobile solution, i.e., built on a car, is necessary. The mobility feature unfortunately limits the energy storage capacity. This leads to the state that it is not possible to use such an HMS in every case (day).

Author Contributions: Conceptualization, K.D., Z.L. and J.K.; methodology, K.D., Z.L. and J.K.; software, K.D.; validation, K.D., Z.L., J.K. and P.B.; formal analysis, K.D., Z.L. and J.K.; investigation, K.D., Z.L. and J.K.; resources, Z.L., P.B., M.B., M.K. and A.F.; data curation, K.D., S.N. and M.M.; writing—original draft preparation, K.D., Z.L. and J.K.; writing—review and editing, K.D., Z.L., J.K., P.B., S.N., M.B., M.K. and A.F.; visualization, K.D.; supervision, P.B. and S.N.; project administration, K.D. and M.M. All authors have read and agreed to the published version of the manuscript.

Funding: This research received no external funding.

Data Availability Statement: Data are contained within the article.

Conflicts of Interest: Authors Sławomir Noske and Mirosław Matusewicz were employed by the company ENERGA-OPERATOR SA. The remaining authors declare that the research was conducted in the absence of any commercial or financial relationships that could be construed as a potential conflict of interest.

Abbreviations

The following abbreviations are used in this manuscript:

DSO	Distributed System Operator
HMS	Hybrid Mobile Station
ISM	Individual Smart Meter
SSM	Station Smart Meter

References

1. *EU 2019/944*; Directive (EU) 2019/944 of the European Parliament and of the Council of 5 June 2019 on Common Rules for the Internal Market for Electricity and Amending Directive 2012/27/EU (Recast). Official Journal of the European Union L 158/125: Brussels, Belgium, 2019.
2. Xiao, J.; Bai, L.; Li, F.; Liang, H.; Wang, C. Sizing of Energy Storage and Diesel Generators in an Isolated Microgrid Using Discrete Fourier Transform (DFT). *IEEE Trans. Sustain. Energy* **2014**, *5*, 907–916. [CrossRef]

3. Luna-Rubio, R.; Trejo-Perea, M.; Vargas-Vazquez, D.; Ríos-Moreno, G.J. Optimal sizing of renewable hybrids energy systems: A review of methodologies. *Sol. Energy* **2012**, *86*, 1077–1088. [CrossRef]
4. Geertsma, R.D.; Negenborn, R.R.; Visser, K.; Hopman, J.J. Design and control of hybrid power and propulsion systems for smart ships: A review of developments. *Appl. Energy* **2017**, *194*, 30–54. [CrossRef]
5. Gutierrez, C.; Meana, D. Application of hybrid-electric power supply system in fishing vessels. *EPJ Web Conf.* **2012**, *33*, 04009. [CrossRef]
6. Sasilatha, T.; Lakshmi, D.; Vaijayanthimala, J.K.; Padmashini, R.K.; Priya, S.; Padmapriya, J.; Kavitha Kumari, K.S. Investigation of hybrid power system for marine applications. *WSEAS Trans. Power Syst.* **2022**, *17*, 207–217. [CrossRef]
7. Zhang, X.; Fu, L.; Ma, F.; Wang, Y. An Emergency Control Strategy for Isolated Power System of Three-Phase Inverter and Diesel-Engine Generator Operating in Parallel. *IEEE Access* **2018**, *6*, 66223–66234. [CrossRef]
8. Rezkallah, M.; Dubuisson, F.; Singh, S.; Singh, B.; Chandra, A.; Ibrahim, H.; Ghandour, M. Coordinated Control Strategy for Hybrid OFF-Grid System Based on Variable Speed Diesel Generator. *IEEE Trans. Ind. Appl.* **2022**, *58*, 4411–4423. [CrossRef]
9. Das, S.; Singh, B.; Chakraborty, S. Seamless Multimode Control and Operation of a DFIG-SPVA-BES-DG Based Hybrid Power Conversion System with Multiple Transition States. *IEEE Trans. Ind. Inform.* **2025**, 1–12. [CrossRef]
10. Puchalapalli, S.; Singh, B.; Das, S. Synchronizing Control of Wind Turbine Driven Doubly Fed Induction Generator System with DG in Remote Area Involving Solar PV-Battery Energy Storage. *IEEE Trans. Ind. Appl.* **2023**, *59*, 5774–5783. [CrossRef]
11. Mohammed, A.Q.; Kassim, A.A.; Hannun, R.M. Optimal Combination and Sizing of a Stand–Alone Hybrid Energy System Using a Nomadic People Optimizer. *IEEE Access* **2020**, *8*, 200518–200540. [CrossRef]
12. Semshchikov, E.; Negnevitsky, M.; Hamilton, J.; Wang, X. Cost-Efficient Strategy for High Renewable Energy Penetration in Isolated Power Systems. *IEEE Trans. Power Syst.* **2020**, *35*, 3719–3728. [CrossRef]
13. Thirugnanam, K.; Kerk, A.K.; Yuen Ch Liu, N.; Zhang, M. Energy Management for Renewable Microgrid in Reducing Diesel Generators Usage with Multiple Types of Battery. *IEEE Trans. Ind. Electron.* **2018**, *65*, 6772–6786. [CrossRef]
14. Khamis, H.A.; Mashhour, E.; Joorabian, M.; Seifossadat, S.G.; Hammadi, S.H. Operation Strategy for a Stand-Alone Hybrid Energy System Supplying a Remote Residential Home for Minimizing Total Cost and Carbon Emission Equipped with Geothermal System. *IEEE Access* **2024**, *12*, 120687–120710. [CrossRef]
15. Cecati, C.; Citro, C.; Piccolo, A.; Siano, P. Smart Operation of Wind Turbines and Diesel Generators According to Economic Criteria. *IEEE Trans. Ind. Electron.* **2011**, *58*, 4514–4525. [CrossRef]
16. Marchand, J.; Buire, J.; Debusschere, V.; Pompee, J.; Ivan, J.; Bretas, A.; Hadjsaid, N. Energy supply time in an islanded grid portion re-energized by local renewable sources. *IET Conf. Proc.* **2025**, *2024*, 315–319. [CrossRef]
17. Guo, L.; Fu, X.; Li, X.; Wang, C. Coordinated control of battery storage and diesel generators in isolated AC microgrid systems. *Proc. Chin. Soc. Elect. Eng.* **2012**, *32*, 70–78.
18. Miscio, J.; Joshi, A.; Suryanarayana, H. LVAC microgrid control for resilient operation: Pilot results. *IET Conf. Proc.* **2025**, *2024*, 232–236. [CrossRef]
19. Stanković, N.; Bertazzon, P.; Jecu, C. Experimental tests on an islanded microgrid with a diesel generator and inverter-based battery energy storage system in synchronous operation. *IET Conf. Proc.* **2025**, *2024*, 143–146. [CrossRef]
20. Sun, L.; Wang, H.; Huang, Z.; Wen, F.; Ding, M. Coordinated Islanding Partition and Scheduling Strategy for Service Restoration of Active Distribution Networks Considering Minimum Sustainable Duration. *IEEE Trans. Smart Grid* **2024**, *15*, 5539–5554. [CrossRef]
21. Ahmad, S.; Shafiullah, M.; Ahmed, C.B.; Alowaifeer, M. A Review of Microgrid Energy Management and Control Strategies. *IEEE Access* **2023**, *11*, 21729–21757. [CrossRef]
22. *EN 50341-1:2012*; Overhead Electrical Lines Exceeding AC 1 kV—Part 1: General Requirements—Common Specifications. BSI: London, UK, 2012.
23. *EN-60909-0:2016*; Short-Circuit Currents in Three-Phase A.C. Systems—Part 0: Calculation of Currents. International Electrotechnical Commission: London, UK, 2016.

energies

MDPI

Article

A Novel Technical Framework for Colombia's Distribution System Operator with Distributed Energy Resources Integration

Eduardo Gómez-Luna [1], Jonathan A. Tariacuri [1], Jorge De La Cruz [1] and Juan C. Vasquez [2,*]

[1] Grupo de Investigación en Alta Tensión—GRALTA, Escuela de Ingeniería Eléctrica y Electrónica, Universidad del Valle, Cali 760015, Colombia; eduardo.gomez@correounivalle.edu.co (E.G.-L.); jonathan.tariacuri@correounivalle.edu.co (J.A.T.); jodelacruzsaavedra@gmail.com (J.D.L.C.)

[2] Center for Research on Microgrids (CROM), AAU Energy, Aalborg University, 9220 Aalborg, Denmark

* Correspondence: juq@energy.aau.dk; Tel.: +45-9940-9724

Abstract: This paper presents a thorough examination of the technical requirements for a new Distribution System Operation (DSO) scheme in Colombia. This study contextualizes these requirements to consider local particularities by looking at national and international standards and models relevant to DSO. This study aims to align the technical requirements to the DSO technologies that offer the greatest advantages (real-time data readings to automate commercial cycle, suspension and reconnection of the service, improving reliability and quality of power supply, and environmental benefits) and the fewest implementation obstacles. Today, an electrical operator can become more proactive by integrating technologies such as advanced metering infrastructure (AMI), distributed energy resources (DER), microgrids, and advanced distribution automation (ADA). This study will provide a structured framework for the implementation of a cutting-edge DSO technology in order to assist Colombia's energy sector in becoming more dynamic and efficient with a smarter and more active electricity distribution system.

Keywords: DER; DSO; electricity distribution grids; advanced metering infrastructure; distributed energy resources; technical requirements

1. Introduction

The structure of electricity supply in Colombia is the result of a prolonged state intervention, which operated in a centralized manner until the reforms carried out in 1994 (Laws 142 and 143), which laid the foundations for its creation as we know it today. From the beginning of the 1990s until today, the distribution network operator (DNO) has not had to alter its role up to this point. However, the consumers' ability to become generators and actively alter their consumption in response to market signals [1] is changing this initial conception. Furthermore, recent events like the COVID-19 pandemic, high fossil fuel prices, and climate impacts have brought to light the costs of a centralized energy system, which is mostly dependent on fossil fuels, to the global economy [2,3]. This has made the need for an energy transition even more urgent.

Traditionally, the planning and operation of the system have adopted a top-down approach, covering the one-way flow of energy from production to end users. This approach gives end users located at the end of the chain less visibility and, therefore, limits their participation in the energy transition in favor of a more flexible and sustainable system with stable prices, deferred infrastructure investment, decarbonization, and efficient demand management to avoid congestion.

The energy transition brings with it a redefinition of energy systems, an increase in the participation of new renewable energy sources, energy efficiency, and the integration of new technologies to provide more reliable and continuous services [4,5]. As a result, the current DNO will have to take a more active role in grid management by using new technologies and real-time data for monitoring, control, and operational targets. The answer to harnessing the benefits that distributed generation brings to the system is the new operator known as the Distribution System Operator (DSO).

The transition to establish a new DSO has been discussed for more than a decade. However, no model or structural design has been fully consolidated in Colombia, which is mainly due to regulatory barriers that have failed to understand the challenges faced by the industry, as well as the lack of incentives for innovation [6,7]. Even if a DSO model that takes advantage of DERs is established somewhere in the world in the coming years, the same model will not work everywhere because a unique approach independent of local particularities does not make sense [8–10].

This paper presents the technical requirements for the new DSO in Colombia, aligned with the standards and models widely accepted in conjunction with stakeholders [11–15] for the deployment of Smart Grid technologies and functions.

This paper comprises five sections. The first section covers the fundamental concepts in relation to the DSO transition; secondly, the existing standards and models for the development of Smart Grids are presented; in the third section, the technical requirements are described, divided into three categories: physical, logical components, and communication protocols. In the fourth section, the technical requirements for the conception of the DSO in Colombia are identified and discussed. Finally, the fifth section presents the conclusions and discusses future work.

The proposed research presents discussions on Smart Grid architecture, DER, and the role of the DSO, as well as their context in the Colombian energy system. This article contributes to the research as follows:

- A comprehensive analysis of the new DSO is proposed to provide the basis for an existing operator to adapt this new operating scheme.
- New technical requirements are proposed to conceive a new DSO in Colombia, classifying them into hardware, software, and communication protocols.
- Redefinition of Smart Grid standards and models to enable the integration of new technologies (AMI, BES, ADA, and DER) and functions (DER management, real-time data readings, accurate fault location, and automatic service restoration, among others) into the grid is discussed.

2. Towards a New Distribution System Operator

As DERs become more prevalent in the electric power system, the grid is changing, with an urgent need for a more active operator role in grid management [15] and operation. To benefit from a smarter and more flexible grid, conventional operators must evolve from a passive to a more active operator (DSO).

This chapter addresses the contextualization of the new DSO in two sections. The first section starts with the understanding of the DSO, its characteristics compared to the traditional model (DNO), its new responsibilities, and concludes with the global initiatives for the adoption of the DSO.

The second section describes in detail the DER, made up of those new technologies to produce and consume energy close to the consumers, which are transforming the systems as we know them.

2.1. New Distribution System Operator (DSO)

According to a document published by OATI (Open Access Technology International, Inc.: Bloomington, MN, USA), the DSO is defined as the entity responsible for the reliable operation of the distribution system (traditional role) while providing demand-side services [16]. To achieve the energy transition in the coming years, the distribution grid will have to accommodate an increasing amount of DER, allowing the operation of electricity systems to evolve from a conventional 'rigid' system to a flexible system at lower investment costs [15]. In this new scenario, the need for a transition from DNO to DSO becomes a necessity in order to take advantage of the benefits that distributed generation offers to the system [6]. Figure 1 shows a comparison between the conventional scenario and the emerging transition scenario.

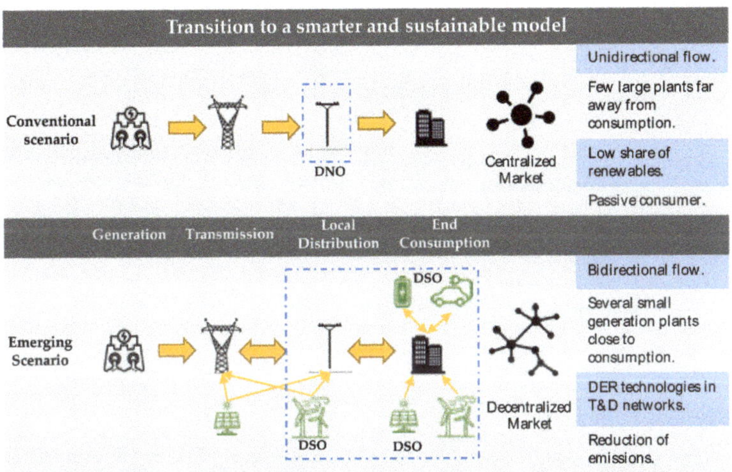

Figure 1. Shifting to a more intelligent and sustainable grid operation model.

The emergence of DERs gives the existing operator new responsibilities, including assisting the transmission system operator (TSO) with DERs, facilitating a retail market with DERs, and operating DERs to maximize grid usage and prevent congestion. However, because of bi-directional flows, which impact the protection and control devices used to manage grid variables like frequency and voltage, the large integration of DER tends to alter how systems are designed and operated [6,15].

To create a more effective, reliable, resilient, and adaptive system, planning for future networks must consider novel strategies involving currently uncommon system agents. In Table 1, some strategies that have emerged globally in international R&D projects in relation to the switch to DSO are listed.

Table 1. DSO-related initiatives worldwide.

Location	Country/State	Initiative	Remarks
United Kingdom	Scotland	NINES [17]	Provide an affordable and reliable energy system that takes advantage of renewable sources. Implement the necessary infrastructure for active demand management, generation, and storage.
	Great Britain	RIIO [18]	Regulatory framework that implements price controls on tariffs while incentivizing operators to innovate.

Table 1. *Cont.*

Location	Country/State	Initiative	Remarks
United States	New York	REV Program [19]	Adoption of DER through incentives, development of renewable sources, and microgrids. Enabling a neutral market between suppliers and consumers.
		FERC 2222 [20]	Regulation focused on enabling DERs to participate in wholesale electricity markets, thereby accelerating the energy transition.
European Union	Spain	CoordiNet [21]	Increase the efficiency and quantity of renewable energy by developing new platforms that enable interaction between different actors in the system.
		OneNet [22]	Create a replicable and scalable architecture that enables integrity in the operation of the European electricity system with a fully decentralized market.
	Sweden	CoordiNet [21]	Provide a market platform and flexible tools for DSOs to manage demand while avoiding grid infrastructure upgrades. Enable increased wind energy production through a P2P digital marketplace.
	Greece	CoordiNet [21]	Development of a local market and implementation of innovative coordination schemes that allow consumers and small actors to access the flexibility market and facilitate their active role in the management and operation of the system.

2.2. Distributed Energy Resources (DER)

DER are manageable energy resources connected in distribution networks or facilities close to the end user [15]. DER includes not only generation technologies but also storage systems and controllable loads such as electric vehicles and demand response. Figure 2 shows the classification of DERs.

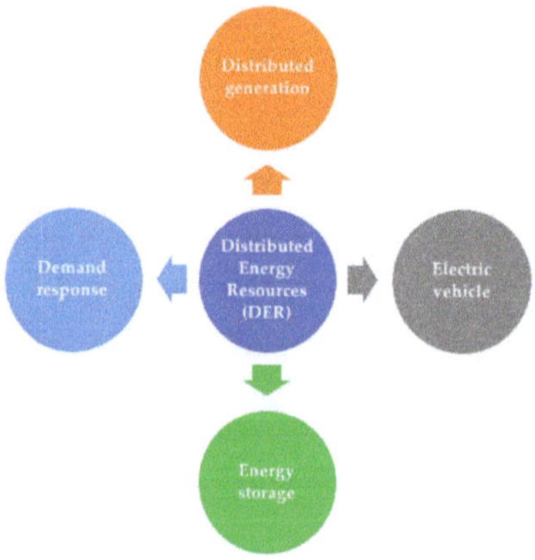

Figure 2. Classification of distributed energy resources.

The following is a brief summary of each of the categories that make up DER, as shown in Figure 2.

Distributed Generation: The production of electricity near consumption centers and connected to a local distribution system, such as solar photovoltaic installations, battery storage systems, or wind turbines, is known as distributed generation, as defined by Law 1715-2014 of the Colombian regulations. Because of its close proximity to the consumer, this type of generation minimizes technical losses. Additionally, with the right control system, it can offer supplementary services to the grid during critical times. The user benefits from savings on energy bills or from the sale of surpluses to the grid; for example, in the Colombian case, the sale of surpluses is regulated by the Resolution 174-2021 where the price is equivalent to the spot price plus a component that translates into the relief of technical losses that this type of generation offers to the system.

Energy Storage: These are systems that, because of their unpredictable and uncontrollable nature, complement the production of energy from renewable sources [15]. With energy storage systems (ESS), including battery banks, pumped hydroelectric facilities (mechanical storage), hydrogen storage, and supercapacitors, among others, conservation or storage offers the grid flexibility and backup.

Demand Response: These are consumer-oriented programs that aim to enable consumers to participate actively in the network by contributing to the flexibility of the network. Demand response is considered the most versatile DER resource because it is less expensive in money and effort to intelligently influence consumers than to deploy new infrastructure and integrate it into the electrical system [23], allowing consumers to participate in an economic incentive program to connect or disconnect loads at critical times in the grid, optimizing the use of existing distribution networks and deferring new investments [24].

Electric Vehicle: It is a type of DER designed as a mobility solution that replaces the use of fossil fuels with electricity. The electric vehicle can become an active part of the system (V2G—Vehicle to Grid or V2H—Vehicle to Home) by taking and delivering energy and offering auxiliary services to the grid, such as peak demand management and energy backup, among others [25].

3. Intelligent Grid Standards and Models

Different standards and models have emerged around the world with the purpose of contributing to the understanding and development of the transformation of the sector [11–15]. These models and standards are not isolated concepts; they are the result of synergy among local communities, academia, and governmental entities. In fact, some standards and models are interrelated with each other with the aim of complementing each other and obtaining a more comprehensive vision.

Throughout this section, the standards and models as IEEE, NIST, SGAM, IEC, and the Colombian initiative called "Colombia Inteligente" [15], illustrated in Figure 3, are described for the design of the new DSO.

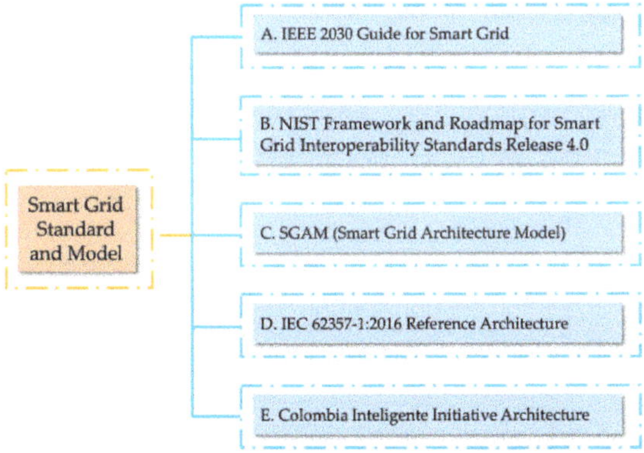

Figure 3. Standards and models to be considered by the new DSO design.

3.1. IEEE 2030 Guide for Smart Grid [11]

This guide establishes the Smart Grid Interoperability Reference Model (SGIRM), which considers the Smart Grid as a system of systems by providing a common understanding to stakeholders of the interoperability criteria from the point of view of the electricity system, communications, and information technology platforms.

SGIRM takes an approach to understanding and guiding the interoperability components of communications, power systems, and information technology platforms, becoming a key element of the Smart Grid architectural design and operation.

The IEEE 2030 guide links conceptual reference models, such as NIST and IEC, with Smart Grid applications such as AMI, plug-in electric vehicle (PEV), and others. The guide provides understanding, definitions, guidance for component design and implementation, and end-use applications for both existing and future infrastructures.

3.2. NIST Framework Release 4.0 [12]

This framework is the result of an ongoing collaborative effort involving individuals, industry organizations, academia, and government, providing invaluable feedback that helped to ensure that the work reflects a wide range of stakeholder perspectives on the issues surrounding the interoperability of the Smart Grid. Its objective is to give numerous stakeholders a high-level understanding of the system and assist smart grid stakeholders in future decision-making. Customers, markets, service providers, operations, generation including DER, transmission, and distribution are the seven domains that make up the NIST conceptual model.

3.3. SGAM (Smart Grid Architecture Model) [13]

The reference model developed by the European Smart Grid Coordination Group (SG-CG) allows not only the current state of implementations to be represented but also the evolution towards future Smart Grid scenarios.

Interoperability is a key enabler of the Smart Grid, so this feature was not an exogenous issue for the SGAM framework, which adopted an abstraction of the interoperability categories introduced by the GridWise Architecture Council (GWAC) through five layers: business goals and processes, functions, information exchange and models, communication protocols, and components.

SGAM is a three-dimensional model made up of domains (generation, transmission, distribution, DER, and customer premises), interoperability layers, and zones (process, field, station, operation, enterprise, and market). In terms of interoperability layers, SGAM provides five layers that allow the representation of entities and their relationships in the context of Smart Grid domains and information management hierarchies, taking into account interoperability aspects.

3.4. IEC 62357-1:2016 Reference Architecture [14]

The International Electrotechnical Commission (IEC) proposes a clear and comprehensive roadmap of all standards that contribute to supporting the interactions present in Smart Grids in an open and interoperable way.

The drivers of the IEC reference architecture are as follows: the need to manage increasing DERs, the need for efficient and sustainable energy, the need for secure and reliable energy to provide a resilient and cost-effective electricity system.

The IEC reference architecture is based on the domains and zones of the SGAM model to represent the information exchanges between the different elements of the Smart Grid, reflecting the possible use cases with their respective standards (IEC 57 technical committee standards and related standards).

3.5. Colombia Inteligente Initiative Architecture [15]

This is an initiative developed by a collaborative network made up of different companies and entities of the Colombian electrical sector with the purpose of promoting the integral and efficient development of intelligent systems in the infrastructure of the Colombian electrical sector.

The technological reference architecture developed by Colombia Inteligente for the integration of Smart Grids in the country is based on the reference architectures that have been developed internationally (Intelligrid, NIST, SGAM, SEPA-Gridwise, IEEE 2030, IEC 62357) and incorporates the particularities of the Colombian electrical system. The use of open architectures facilitates interoperability and streamlines implementation by meeting the requirements for consistency of equipment, scalability, interoperability, and cybersecurity.

4. Technical Requirements for the New DSO Role

This section presents the technical requirements for DSO design identified in the standards and Smart Grid models studied, based on a system-level approach (Figure 4) through the hardware (logical components), software (logical components), and communication protocols that make possible the interface between the physical and logical parts.

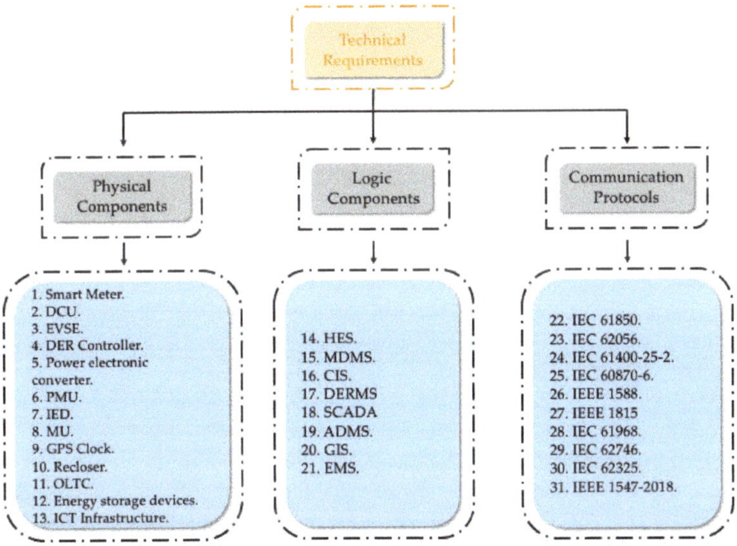

Figure 4. Grouping of DSO technical requirements.

4.1. Physical Components

These components comprise metering devices, sensors, and devices capable of supporting the bi-directional flows and interactions existing in the Smart Grid. Within this category, we can find components such as:

- Smart meter;
- Data concentrator unit (DCU);
- Electric vehicle supply equipment (EVSE);
- DER controller;
- Power electronic converter;
- Phasor measurement unit (PMU);
- Intelligent electronic device (IED);
- Merging unit (MU);
- GPS clock;
- Recloser;

- On-load tap changer (OLTC);
- Energy storage devices;
- Information communication technology (ICT) infrastructures.

4.2. Logic Components

These comprise the set of instructions, data, or programs designed to execute specific tasks. With the digital revolution, most organizations today have incorporated software solutions into their processes to increase productivity and facilitate the use and access to information.

The following subsections describe the logical components identified in the standards and models of Smart Grids in order to know and identify the key points of each one of them at the time of their implementation.

4.2.1. Head End System (HES)

Head end systems, or HES, manage connectivity and schedule data collection from the metering infrastructure, including metering and communication devices. Additionally, they allow secure access to meters for configuration, updates, and specific requests [26].

4.2.2. Meter Data Management System (MDMS)

It is the central system of the smart metering infrastructure, which has long-term storage capacity for all data collected from the meters by the HES for further processing. The MDM is the connection point for other systems to reach the smart meters, i.e., it is a gateway to and from the HES [27].

4.2.3. Customer Information System (CIS)

A Customer Information System (CIS) contains a database of customer information. In addition to being used to store and record customer information, the CIS serves as a tool to support the organization in making decisions about customers in order to increase its competitiveness [28].

4.2.4. Distributed Energy Resources Management System (DERMS)

The DERMS is a system responsible for monitoring the DERs, which, in turn, exchange signals with the SCADA system for the operation of the distribution system. Sensors installed in the electrical network, such as PMUs, are responsible for taking information directly from the network and sending this information about the behavior of the DER to the DERMS [15].

In addition to enabling optimal management of medium and large-scale DERs, DERMS can manage the DERs connected behind the meter through a specialized set of tools, allowing it to overcome the problems caused by the high penetration of these resources and maximize the benefits delivered to the system [29].

4.2.5. Supervisory Control and Data Acquisition (SCADA)

The SCADA system allows the supervision, control, and data acquisition of the processes of generation, transmission, and distribution of energy in an electrical network.

Since the management of an electrical system is not a trivial task, the need arises to implement SCADA systems that assist operators due to the complexity of the processes involved in a power system [30]. Some of the capabilities that a SCADA system must provide are as follows:

- Ensure uninterrupted power supply in a safe environment;
- Provide real-time status of the electrical system;
- Monitor and control systems located in remote areas.

4.2.6. Advanced Distribution Management Systems (ADMS)

This system is the evolution of the distribution management system (DMS), which enables relatively advanced analysis and optimization programs and functions. An ADMS includes functions that automate outage recovery and optimize the performance of the distribution network. The following are some of the main components of an ADMS [31].

- Power distribution system monitoring.
- Decision-making and support tools.
- Power distribution system control.
- Power distribution system operational planning.
- Power distribution system engineering studies.
- Retail power market.
- Training simulator system.
- Quality control (development) system.

4.2.7. Geographic Information System (GIS)

Computer system with the ability to capture, store, analyze, and display information from assets that are spatially distributed [32]. This system is especially useful for companies in the electrical sector where the equipment that makes up the power system is generally distributed tens and hundreds of kilometers away, allowing information to be obtained quickly through catalogs or key identifier search engines, which are assigned by the company.

4.2.8. Energy Management System (EMS)

Energy management system (EMS) is a system of computer-aided tools used to perform functions such as estimation, supervision, measurement, and control of both generation and demand for a reliable and safe operation of an electrical network [33]. The operating scheme of an EMS consists of four key factors: generation, demand, storage, and control to enable the massive penetration of renewable energy.

4.3. *Communication Protocols*

Communication protocols comprise the set of formal rules that describe how to transmit and exchange data over a network. For utilities, due to the dynamics of their processes, it is essential that these exchanges take place in real time. This requires the development of an IT architecture that allows fast and secure data recovery.

Among the actions required for DSO deployment is the development of standards (both at the physical layer and information and communications technology levels) to improve coordination across aggregators, DSOs, TSOs, and consumers.

The main existing communication protocols for information exchange are described below.

4.3.1. IEC 61850: Communication Networks and Systems for Power Utility Automation

IEC 61850 is recognized as one of the pillars for the realization of Smart Grid goals for interoperability and device management [14]. They cover everything from basic connectivity to semantic understanding, even addressing aspects of the business context. IEC 61850 is the state of the art for communication protocols in power system automation [13]. IEC 61850 allows the DSO to deal with issues such as the integration of DER within the electrical distribution network.

4.3.2. IEC 62056: Electricity Metering Data Exchange (DLMS/COSEM)

It is a set of standards for the exchange of electrical measurement data from the International Electrotechnical Commission (IEC), mainly adopted by European countries [34]. The IEC 62056 standards are the international standard version of the DLMS/COSEM specification.

4.3.3. IEC 61400-25-2: Communications for Monitoring and Control of Wind Power Plants

The standard establishes the basis for communication between wind farms and SCADA systems, providing guidelines for establishing the wind farm information model, information exchange methods, and mapping to a standardized communications model [35].

4.3.4. IEC 60870-6: Telecontrol in Electrical Engineering and Power System Automation Applications

The IEC 60870-6 set of standards establishes guidelines for the exchange of critical time information using wide area network (WAN) or local area network (LAN) infrastructures, thus enabling the exchange of data between control centers. The standard is composed of three main parts: TASE.2/ICCP protocol and services (Inter-Control Center Protocol), functional profiling to provide services to the end system and object model.

4.3.5. IEEE 1588: Precision Time Protocol (PTP)

A standard designed to provide accurate clock synchronization in packet-based network systems. Clock synchronization can be achieved even in heterogeneous systems that include clocks with different precision and resolution [36].

4.3.6. IEEE 1815: Standard for Electric Power Systems Communications-Distributed Network Protocol (DNP3)

It is a communication protocol designed for utilities to facilitate remote communication between computer systems and equipment geographically scattered throughout the electrical system (remote station) [37].

4.3.7. IEC 61968: Application Integration at Electric Utilities System Interfaces for Distribution Management

This standard aims to facilitate the integration between applications of different distributed software application systems. IEC 61980 allows the exchange of information between EMS, DMS, and other related external computer systems [38].

4.3.8. IEC 62746: Systems Interface Between Customer Energy Management System and the Power Management System

Defines the interfaces and communication protocols that cover the chain between a Smart Grid and the Smart Home/Building/Industry area. It specifies the communication of application-level services that can incentivize the response of distributed energy resources located on the customer side, known as OpenADR [39].

4.3.9. IEC 62325: Common Information Model (CIM) for Energy Market Communication

It establishes a universal exchange mechanism for the electricity market information model, involving power generation models, physical network models, and power user models. In addition, IEC 62325 defines a message exchange mechanism that allows different applications or systems to access public data and exchange information without relying on internal subscriptions [40].

4.3.10. IEEE 1547-2018: Common Information Model (CIM) for Energy Market Communication

This standard establishes criteria and requirements for the interconnection of DER with the electric power system (EPS). The specifications and technical requirements contained in the standard are universally necessary for the interconnection and interoperability of DER [41].

5. Energy System Context and Definition of the Technical Requirements for the Design of a New DSO in Colombia

Research on DSO agrees that it must be designed for each site according to local particularities: policy, regulatory framework, distribution network conditions, among others [8]. Therefore, the following section addresses the context of the Colombian energy system, focusing on aspects such as regulation, the large integration of IT/smart technologies for grid management, and the current situation of the electricity system in the distribution operation.

5.1. The Context of the Colombian Energy System

For the electricity sector with respect to energy objectives, similar visions can be seen in the proposals of the Unidad de Planeación Minero Energética (UPME) [42] and initiatives such as Colombia Inteligente [15]. Figure 5 shows the main lines of Colombia's strategic energy objectives:

Figure 5. Colombia's strategic energy objectives.

Additionally, in Colombia, there are two areas of application of Smart Grid technologies: the National Interconnected System, or SIN for its acronym in Spanish, and the Non-Interconnected Zones, or ZNI [43]. These areas have different characteristics, which, although they may share similar functional aspects, require an adequate analysis and approach according to the particularities of each one of them.

As a result of the above, the present paper will focus on the SIN area where large-scale benefits could be achieved and later extended to the ZNI through experience and lessons learned.

As for the required architecture, it will vary depending on the Smart Grid functionalities to be deployed; therefore, it is recommended to work initially with those functionalities that contribute to the achievement of Colombia's energy objectives and that, at the moment, have few barriers to their development. In [44], a map is shown that relates Smart Grid functionalities with Colombia's energy objectives and the barriers to their implementation.

Those functionalities with greater benefits and closer to their implementation are mainly translated into functionalities related to AMI.

5.1.1. Policy and Regulatory Context

This section presents an analysis based on the study presented by UPME [43], identifying the opportunities and barriers within Colombia's regulatory framework for deploying Smart Grid technologies, such as AMI, ADA, and distributed energy resources–microgrids (DER-MG), of particular interest to achieve Colombia's strategic energy objectives.

The analysis includes aspects related to AGPE and distributed generation (CREG 174 of 2021), integration of non-conventional solar photovoltaic and wind non-conventional sources (CREG 148 of 2021), integration of energy communities (CREG 101 072 of 2025), requirements for operation and management of AMI (NTC 6079-2021), and sectoral policy ICT.

Table 2 below summarizes the results of the analysis with the opportunities and barriers identified within Colombia's regulatory framework.

Table 2. Opportunities and policy/regulatory barriers to the deployment of Smart Grid technologies in Colombia.

Technology	Policy/Regulation	Opportunities	Barriers
AMI	CREG 038-2014 [45]	Free agreement between the parties to define the ownership of the measurement system.	Stakeholders do not consider the positive externalities of smart metering deployment; instead, they focus on how and by whom the implementation costs will be recognized.
	CREG 015-2018 [46]	Methodology for the calculation of hourly charges that discriminates between different periods according to the system load level.	This feature is only available to users with smart meters.
	NTC 6079-2021 [47]	Colombian technical standard that establishes minimum requirements for the management and operation of AMI systems.	No barriers were identified in relation to the technology studied.
	Sectoral policy ICT	Protection of the privacy of information of constitutional origin. Consumption data are collected and processed with the owner's authorization. Implementation of information security systems by the independent system operator (ISO). National policy on cybersecurity since 2011.	A Colombian Electricity Sector Computer Security Incident Response Team (CSIRT) has not been set up.
ADA	Ley 1715-2014 [48]	Promotes the implementation of non-conventional renewable energy sources, rational use of SIN energy, and DR projects.	No barriers were identified in relation to the technology studied.
	Executive Order 2492/2014 [49]	Design of mechanisms to encourage the efficient use of infrastructure and reduce service provision costs.	No barriers were identified in relation to the technology studied.
	CREG 015-2018 [46]	It encourages infrastructure modernization through incentives and compensation schemes that improve service quality.	Some features are only available to users with smart meters.
	CREG 101 019-2022 [50]	Voluntary disconnectable demand scheme as a safety ring for the reliability charge.	This feature is only available to users with smart meters.

Table 2. *Cont.*

Technology	Policy/Regulation	Opportunities	Barriers
ADA	CREG 011-2015 [51]	DR scheme in the wholesale market under critical conditions.	This feature is only available to users with smart meters. This scheme only applies during periods of shortage.
DER-MG	Ley 1715-2014 [48]	Procedures for the connection, operation, backup, and commercialization of distributed generation. Delivery of surpluses to the grid from self-generation.	Delivery of surpluses to the grid is conditioned on the implementation of smart meters.
	CREG 174-2021 [52]	Operational and commercial aspects to enable the integration of self-generation and distributed generation.	Integration is dependent on the implementation of smart meters.
	CREG 148-2021 [53]	Connection and operation of solar photovoltaic and wind power plants with a capacity of 5 MW or more in the local distribution system.	Prior consultation with ethnic communities may be required before licences to develop projects are granted.
	CREG 101 072-2025 [54]	Operational and commercial aspects to enable the integration of energy communities.	No barriers were identified in relation to the technology studied.

5.1.2. Technological Context

With regard to Smart Grid technologies, Colombia is in an initial implementation stage where most of the functionalities deployed are inherent to the technology and do not require advanced data and information processing, a scenario caused mainly by regulatory and economic barriers [55].

Deploying new Smart Grid functionalities requires incorporating new technologies and capabilities into current network operators' processes. AMI is a catalyst technology that provides access to these new functionalities [15]. However, AMI has not yet been deployed on a significant scale in Colombia, with coverage percentages that are not significant in relation to the total number of customers served [15,55], due to the high investment required in the initial stages of deployment.

On the other hand, the infrastructure of Colombia's distribution system offers significant technical challenges for the deployment of new technologies by operators due to its predominantly radial topology in overhead arrangement [56], which, in some cases, may be obsolete, particularly in rural areas where reliable communication infrastructure is lacking to support the real-time data flow generated by Smart Grids.

Despite the current limitations, a significant number of pilot projects focusing on AMI, DER, and ADA have been successfully deployed in Colombian cities such as Cali, Bogotá, Medellín, and Barranquilla [55], allowing technologies to be validated in different scenarios before their implementation on a large scale.

5.1.3. Current System Status and Distribution Operation

The Colombian energy matrix is mostly clean, with a good participation of conventional renewable energy (water resources), but there is still no significant participation of non-conventional resources (see Figure 6), causing the country to be exposed to high volatility in short-term energy prices (stock market) [57,58], as these are subject to weather conditions.

In addition, distribution and commercialization in Colombia are carried out by 29 operators, which are located in different regions of the country and provide an electricity service to nearly 13 million users of the SIN. In terms of distribution networks, Colombia has over 200,000 km of lines divided into over 5000 circuits, with an average of almost 100 transformers per circuit [56], which means that considerable investments are needed to achieve the adoption of Smart Grid technologies.

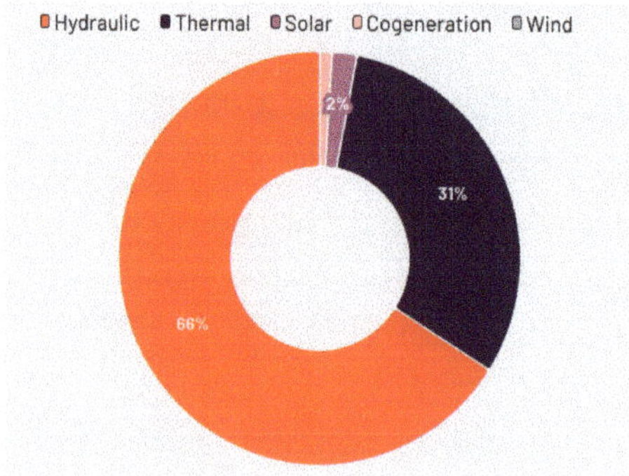

Figure 6. Effective capacity by type of generation in Colombia in 2023.

Regarding the infrastructure for distribution automation applications, such as remote reclosers and phasor measurement units (PMU) for network monitoring, clearing, and reconfiguration functionalities, it is observed that most operators are still in an early stage of deployment [55], which has repercussions in the area of service continuity and other technical issues [43].

In the area of continuity of energy supply, measured through the SAIDI (System Average Interruption Duration Index) indicator, characterized by measuring the time of interruptions perceived by a user, respectively, Figure 7 shows the SAIDI indicator evolution in Colombia between 2019 and 2023.

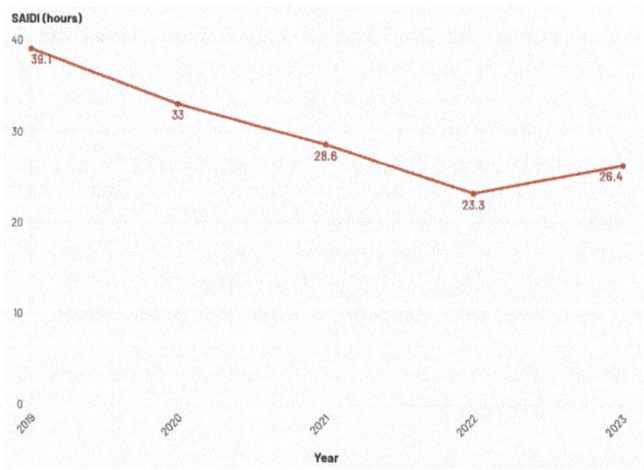

Figure 7. The evolution of the average annual duration of SAIDI outages in Colombia.

Although the efforts made by the operators have led to improvements in the quality of service over the last few years, Colombia still has much room for improvement compared to countries such as Brazil and Chile, with an average interruption of 10.4 and 13.6 h/year, respectively, given that Colombia's average is 30 h/year.

Thus, the distribution domain is identified as the domain of the electricity sector chain where the greatest opportunities for improvement are identified. Therefore, the design of a new DSO would offer mechanisms to obtain potential results in this domain of the electricity sector.

5.2. Definition of Technical Requirements for the New DSO in Colombia

This section will define the technical requirements necessary for the design of a new DSO in Colombia in view of the integration of DERs, based on the reference architecture proposed by Colombia Inteligente [59], which has a clear vision of the system to be achieved and is also aligned with the strategic objectives of the energy sector, and on those technologies that currently have fewer barriers to their deployment and greater benefits according to the analysis carried out in the previous section (AMI, ADA, and DER-Microgrids). Similarly, as mentioned above, this study will focus on the National Interconnected System so that later, through experience and lessons learned, it can be extended to the ZNIs.

Figure 8 illustrates, in a general way, the obtaining of the technical requirements for the DSO by grouping them into three categories: physical components, logic components, and communication protocols.

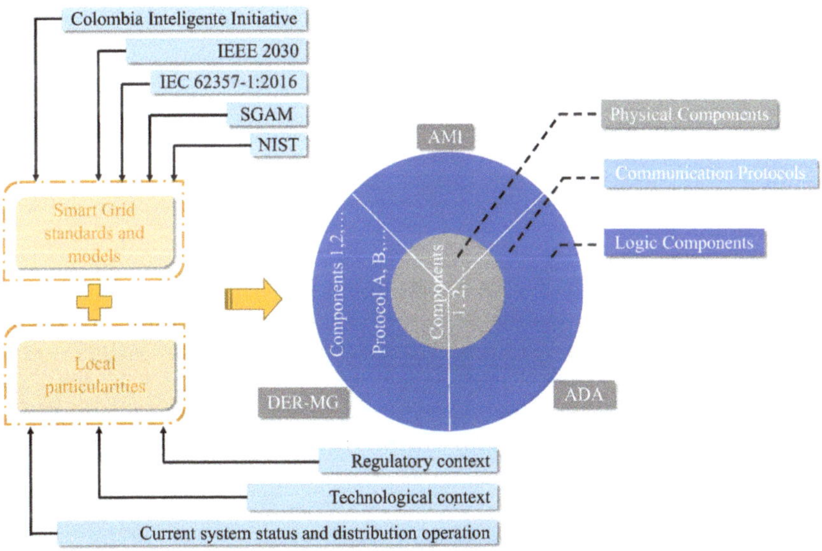

Figure 8. Technical requirements proposed for the new DSO.

Below, Table 3 summarizes the technical requirements for the conception of the DSO in Colombia in view of the integration of the DER, based on the inputs obtained throughout the development of this work, aligned with those technologies that have fewer barriers and greater benefits in terms of their deployment.

Table 3. Summary of technical requirements for a new DSO in Colombia.

Technology	Physical Components	Logic Components	Communication Protocols
AMI (Advanced Metering Infrastructure)	Smart meter	Meter data management system	IEC 62056-2/62056-5-3
	Data concentrator unit		
	Head end system/measurement data collector	Web access portals	IEC 61968-9/61968-100
	ICT infrastructure		ANSI C12.19/C12.21/C12.22

Table 3. *Cont.*

Technology	Physical Components	Logic Components	Communication Protocols
ADA (Advanced Distribution Automation)	Recloser	Advanced distribution management systems	IEC 61850
	Field equipment (circuit breaker, current transformer/CT, transformer, compensator)		IEC 60870
	Intelligent electronic device	Supervisory control and data acquisition/human–machine interface	IEC 62746
	GPS clock		IEEE 1588
	Merging unit	Geographic Information System	IEEE 1815
	Phasor measurement unit		
	Remote terminal unit		
	On-load tap changer	Customer Information System	IEEE C37.239 2010
	ICT infrastructure		
DER—Microgrid (Distributed Energy Resources–Microgrid)	Controllable loads	Energy management system	IEC 61850
	Energy storage devices		IEC 62746 (Open ADR)
	DER controller	Distributed energy resources management system	IEEE 1588
			IEEE 1815
	Power electronic converter		IEC 61400
	Electric vehicle supply equipment		IEC 60870
	Distributed generation	Web access portals	IEC 61724
	ICT infrastructure		IEC 61970

6. Conclusions

The transition from DNO to DSO is the answer to the challenge faced by current operators to take a more active role in network management and to take advantage of the benefits that distributed generation offers to the system. This transition comes with a paradigm shift in terms of system planning and operation, giving greater visibility to the end consumer who can become an active player in the system (prosumer), contributing to a more flexible grid, thanks to its ability to generate and store energy, and to respond to market signals.

The technical requirements for DSO in Colombia were defined according to local particularities, based on national and international Smart Grid standards and models, and those technologies that have fewer barriers and greater benefits in terms of deployment. From the analysis developed in this paper on the current Colombian regulatory framework, it is identified that technologies such as AMI (advanced metering infrastructure), ADA (advanced distribution automation), and DER-Microgrids (distributed energy resources and microgrids) are technologies of particular national interest.

Most conventional grid operators are struggling to adapt to the ongoing evolution of the energy sector due to the limitations of their traditional infrastructure. Planning future networks requires innovative approaches that incorporate new and diverse system actors, aiming to build a more efficient, reliable, resilient, and flexible power system.

To accelerate this transition in Colombia, several key aspects must be addressed to evolve the traditional role of DNO to a more dynamic and proactive DSO. First, the regulatory framework must be clearly defined to establish the roles and responsibilities of DSOs. Second, promote the massification of AMI in the territory; initial investments by operators are high, and the regulation does not clearly define how these investments will be recovered. Third, data collection and exchange processes should be standardized to ensure interoperability and transparency. Finally, operators require a robust ICT infrastructure capable of supporting the complex real-time information flows of Smart Grids.

Future work will further refine the requirements for the conception of the DSO in Colombia by incorporating additional aspects such as cybersecurity, the common information model (CIM), monitoring and forecasting systems for variable generation sources, and advanced digital technologies—including artificial intelligence, the Internet of Things, digital twins, and blockchain—as well as the development of a pilot with an operator in Colombia that will allow us to put into practice the feasibility an applicability of the proposed requirements.

Author Contributions: Conceptualization, investigation, and writing—original draft preparation, J.A.T.; validation, formal analysis, writing—review and editing, E.G.-L., J.A.T., J.C.V. and J.D.L.C. All authors have read and agreed to the published version of the manuscript.

Funding: This research received no external funding.

Data Availability Statement: The original contributions presented in the study are included in the article. Further inquiries can be directed to the corresponding author.

Acknowledgments: The authors of this paper would like to thank the GRALTA research group at Universidad del Valle, Colombia, and the Center for Research on Microgrids (CROM) for the facilities and support provided for the completion of this paper.

Conflicts of Interest: The authors declare no conflicts of interest.

Nomenclature

ADA	Advanced Distribution Automation
ADMS	Advanced Distribution Management System
AMI	Advanced Metering Infrastructure
BES	Battery Energy Storage
CIM	Common Information Model
CIS	Customer Information System
COSEM	Companion Specification for Energy Metering
DCU	Data Concentrator Unit
DER	Distributed Energy Resources
DMLS	Device Language Message Specification
DMS	Distribution Management System
DNO	Distribution Network Operator
DOE	U.S. Department of Energy
DRMS	Demand Response Management Systems
DSO	Distribution System Operator
EMS	Energy Management System
EVSE	Electric Vehicle Supply Equipment
GIS	Geographic Information System
GPS	Global Positioning System
IED	Intelligent Electronic Device
MDMS	Meter Data Management System
MU	Merging Unit
OLTC	On-Load Tap Changer
OMS	Outage Management System
PMU	Phasor Measurement Unit
PTP	Precision Time Protocol
SCADA	Supervisory Control and Data Acquisition
SGAM	Smart Grid Architecture Model
SGIRM	Smart Grid Interoperability Reference Model

References

1. MIT Energy Initiative. *Utility of the Future*; Massachusetts Institute of Technology: Cambridge, MA, USA, 2016.
2. International Renewable Energy Agency (IRENA). *World Energy Transitions Outlook 2022*; International Renewable Energy Agency: Abu Dhabi, United Arab Emirates, 2022.
3. Cerón, J.A.; Gómez-Luna, E.; Vasquez, J.C. Driving the Energy Transition in Colombia for Off-Grid Regions: Microgrids and Non-Conventional Renewable Energy Sources. *Energies* **2025**, *18*, 1010. [CrossRef]
4. Lopez Alzate, Y.; Gómez-Luna, E.; Vasquez, J.C. Innovative Microgrid Services and Applications in Electric Grids: Enhancing Energy Management and Grid Integration. *Energies* **2024**, *17*, 5567. [CrossRef]
5. Escobar-Orozco, L.F.; Gómez-Luna, E.; Marlés-Sáenz, E. Identification and Analysis of Technical Impacts in the Electric Power System Due to the Integration of Microgrids. *Energies* **2023**, *16*, 6412. [CrossRef]
6. International Renewable Energy Agency. *Innovation Landscape Brief: Future Role of Distribution System Operators*; International Renewable Energy Agency: Abu Dhabi, United Arab Emirates, 2019.
7. Vlerick Energy Centre. *Outlook on the European DSO Landscape 2020*; Vlerick Energy Centre: Brussels, Belgium, 2020.
8. Specht, M.; Rohjans, S.; Trefke, J.; Uslar, M.; Gonzalez Vazquez, J.M. International Smart Grid Roadmaps and Their Assessment. *EAI Endorsed Trans. Energy Web* **2013**, *1*, e2. Available online: https://eudl.eu/doi/10.4108/trans.ew.2013.01-06.e2 (accessed on 22 April 2025).
9. Nyangon, J. Distributed Energy Generation Systems Based on Renewable Energy and Natural Gas Blending: New Business Models for Economic Incentives, Electricity Market Design and Regulatory Innovation. Ph.D. Thesis, University of Delaware, Newark, DE, USA, 2017. [CrossRef]
10. Apostolopoulou, D.; Bahramirad, S.; Khodaei, A. The Interface of Power: Moving Toward Distribution System Operators. *IEEE Power Energy Mag.* **2016**, *14*, 46–51. [CrossRef]
11. *IEEE Std 2030-2011*; IEEE Guide for Smart Grid Interoperability of Energy Technology and Information Technology Operation with the Electric Power System (EPS), End-Use Applications, and Loads. IEEE: Piscataway, NJ, USA, 2011.
12. Gopstein, A.; Nguyen, C.; O'Fallon, C.; Hastings, N.; Wollman, D. *NIST Framework and Roadmap for Smart Grid Interoperability Standards, Release 4.0*; NIST: Gaithersburg, MD, USA, 2021. [CrossRef]
13. Smart Grid Coordination Group. Smart Grid Reference Architecture, CEN-CENELEC-ETSI. 2012. Available online: https://www.cencenelec.eu/media/CEN-CENELEC/AreasOfWork/CEN-CENELEC_Topics/Smart%20Grids%20and%20Meters/Smart%20Grids/reference_architecture_smartgrids.pdf (accessed on 24 February 2024).
14. *IEC TR 62357-1*; Power Systems Management and Associated Information Exchange—Part 1: Reference Architecture. The International Electrotechnical Commission (IEC): Geneva, Switzerland, 2016.
15. Colombia Inteligente. Recursos Energéticos Distribuidos. In *Acciones Para Su Integración*; Colombia Inteligente: Medellín, Colombia, 2020.
16. Rahimi, F.; Mokhtari, S. *A New Distribution System Operator Construct*; Open Access Technology International, Inc.: Bloomington, MN, USA, 2014; pp. 1–20.
17. Scottish & Soutern Electricity Networks. NINES. Available online: https://ssen-innovation.co.uk/nines/ (accessed on 24 February 2024).
18. OFGEM. Network Price Controls 2021-2028 (RIIO-2). Available online: https://www.ofgem.gov.uk/energy-policy-and-regulation/policy-and-regulatory-programmes/network-price-controls-2021-2028-riio-2 (accessed on 29 February 2024).
19. NYISO. Reforming the Energy Vision (REV)-DSIP/DSP. Available online: https://www.nyiso.com/documents (accessed on 24 February 2024).
20. The Federal Energy Regulatory Commission (FERC). FERC Order No. 2222 Explainer: Facilitating Participation in Electricity Markets by Distributed Energy Resources. Available online: https://www.ferc.gov/ferc-order-no-2222-explainer-facilitating-participation-electricity-markets-distributed-energy (accessed on 28 February 2024).
21. European Distribution System Operators (E.DSO). CoordiNet. Available online: https://www.edsoforsmartgrids.eu/eu-projects/coordinet/ (accessed on 22 April 2025).
22. OneNet. OneNet-Demos. Available online: https://www.onenet-project.eu/demos/ (accessed on 28 February 2024).
23. Palensky, P.; Dietrich, D. Demand Side Management: Demand Response, Intelligent Energy Systems, and Smart Loads. *IEEE Trans. Ind. Inform.* **2011**, *7*, 381–388. [CrossRef]
24. Shen, J.; Jiang, C.; Li, B. Controllable Load Management Approaches in Smart Grids. *Energies* **2015**, *8*, 11187–11202. [CrossRef]
25. Stefano de Souza Pelegrino, L.; Heldwein, M.L.; Waltrich, G. Integrated System for Power Flow Control between Electric Vehicle, Utility Grid and Residence. *IET Power Electron.* **2020**, *13*, 953–960. [CrossRef]
26. Xia, W.; Hou, E.; Mao, X.; Wang, L. Hyperscale AMI System Design and Construction Practice. In Proceedings of the 11th IET International Conference on Advances in Power System Control, Operation and Management (APSCOM 2018), Hong Kong, China, 11–15 November 2018; pp. 1–5. [CrossRef]

27. *IEC TR 62357-2:2019*; Power Systems Management and Associated Information Exchange—Part 2: Use Cases and Role Model. The International Electrotechnical Commission (IEC): Geneva, Switzerland, 2019.
28. Norfazlina, G.; Akma, A.S.S.; Adrina, S.N.; Noorizan, M.M. Customer Information System Satisfaction and Task Productivity: The Moderating Effect of Training. *Procedia Econ. Financ.* **2016**, *37*, 7–12. [CrossRef]
29. Strezoski, L.; Padullaparti, H.; Ding, F.; Baggu, M. Integration of Utility Distributed Energy Resource Management System and Aggregators for Evolving Distribution System Operators. *J. Mod. Power Syst. Clean Energy* **2022**, *10*, 277–285. [CrossRef]
30. Ordean, M.; Chiorean, D.; Rogoz, I.; Lehene, C.; Stoian, I.; Stancel, E. SCADA Systems—Support for the Maintenance Management of Hydro Power Plants. In Proceedings of the 2006 IEEE International Conference on Automation, Quality and Testing, Robotics, Cluj-Napoca, Romania, 25–28 May 2006; Volume 1, pp. 238–242. [CrossRef]
31. Amiri, S.S.; Rahmani, M.; McDonald, J.D. An Updated Review on Distribution Management Systems within a Smart Grid Structure. In Proceedings of the 2021 11th Smart Grid Conference (SGC), Tabriz, Iran, 7–9 December 2021; pp. 1–5. [CrossRef]
32. Abdulrahman, I.; Radman, G. Power System Spatial Analysis and Visualization Using Geographic Information System (GIS). *Spat. Inf. Res.* **2020**, *28*, 101–112. [CrossRef]
33. Ahmad, S.; Shafiullah, M.; Ahmed, C.B.; Alowaifeer, M. A Review of Microgrid Energy Management and Control Strategies. *IEEE Access* **2023**, *11*, 21729–21757. [CrossRef]
34. Otani, T. A Primary Evaluation for Applicability of IEC 62056 to a Next-Generation Power Grid. In Proceedings of the 2010 First IEEE International Conference on Smart Grid Communications, Gaithersburg, MD, USA, 4–6 October 2010; pp. 67–72. [CrossRef]
35. Lee, J.-H.; Seo, M.-J.; Kim, G.-S.; Lee, H.-H. IEC 61400-25 Interface Using MMS and Web Service for Remote Supervisory Control at Wind Power Plants. In Proceedings of the 2008 International Conference on Control, Automation and Systems, Seoul, Republic of Korea, 14–17 October 2008; pp. 2719–2723. [CrossRef]
36. *IEEE Std 1588-2019 (Revision ofIEEE Std 1588-2008)*; IEEE Standard for a Precision Clock Synchronization Protocol for Networked Measurement and Control Systems. IEEE: Piscataway, NJ, USA, 2020; pp. 1–499. [CrossRef]
37. *IEEE Std 1815-2012 (Revision of IEEE Std 1815-2010)*; IEEE Standard for Electric Power Systems Communications-Distributed Network Protocol (DNP3). IEEE: Piscataway, NJ, USA, 2012; pp. 1–821. [CrossRef]
38. Lv, G.; Zhao, J.; Su, J.; Zhang, D. Research on IEC 61968 Standard Oriented Function Framework of Adapter in Smart Distribution Grid. In Proceedings of the 2013 Fourth International Conference on Digital Manufacturing & Automation, Shinan, China, 29–30 June 2013; pp. 1061–1065. [CrossRef]
39. Keko, H.; Keserica, H.; Sučić, S.; Miranda, V. Technical Backbone for the Democratization of Flexibility: Standards-Based Demand Response Infrastructure. In Proceedings of the 2019 16th International Conference on the European Energy Market (EEM), Ljubljana, Slovenia, 18–20 September 2019; pp. 1–5. [CrossRef]
40. Shao, P.; Ye, F.; Guo, Y.; Liu, Y.; Zhang, J.; Wang, G. Power Market Unified Model Technology Architecture Based on IEC 62325. In Proceedings of the 2019 IEEE 3rd Information Technology, Networking, Electronic and Automation Control Conference (ITNEC), Chengdu, China, 15–17 March 2019; pp. 1474–1477. [CrossRef]
41. *IEEE Std 1547-2018 (Revision of IEEE Std 1547-2003)*; IEEE Standard for Interconnection and Interoperability of Distributed Energy Resources with Associated Electric Power Systems Interfaces. IEEE: Piscataway, NJ, USA, 2018; pp. 1–138. [CrossRef]
42. UPME. *Plan Energético Nacional Colombia: Ideario Energético 2050*; Unidad Planeación Min. Energética, Repub: Colomb, Sri Lanka, 2015; Volume 184.
43. La Unidad de Planeación Minero Energética—UPME. *Smart Grids Colombia Visión 2030—Parte I*; UPME: Bogotá, Colombia, 2016.
44. Gómez, J.R. Smart Grids Colombia, Visión 2030 Hoja de Ruta. In Proceedings of the XIII ASOCODIS Jornada de Distribución de Energía Eléctrica, Bogotá, Colombia. 2016. Available online: https://www.asocodis.org.co/docs/XIII-Jornada/Sesion%20II/BID_SmartGrid_Colombia_2016.pdf (accessed on 22 April 2025).
45. *Resolución CREG 038 de 2014*; Modificación Código de Medida Contenido en el Anexo General del Código de Redes; Comisión de Regulación de Energía y Gas: Bogotá, Colombia, 2014. Available online: https://gestornormativo.creg.gov.co/gestor/entorno/docs/resolucion_creg_0038_2014.htm (accessed on 22 April 2025).
46. *Resolución CREG 015 de 2018*; Metodología para la Remuneración de la Actividad de Distribución de Energía Eléctrica en el Sistema Interconectado Nacional; Comisión de Regulación de Energía y Gas: Bogotá, Colombia, 2018. Available online: https://gestornormativo.creg.gov.co/gestor/entorno/docs/resolucion_creg_0015_2018.htm (accessed on 22 April 2025).
47. *NTC 6079 de 2021*; Requisitos para Sistemas de Infraestructura de Medición Avanzada (AMI) en Redes de Distribución de Energía Eléctrica. Instituto Colombiano de Normas Técnicas y Certificación: Bogotá, Colombia, 2021.
48. *Law 1715 de 2014*; Regulación para la Integración de las Energías Renovables No Convencionales al SISTEMA Energético Nacional; Congreso de Colombia: Bogotá, Colombia, 2014. Available online: https://www.funcionpublica.gov.co/eva/gestornormativo/norma.php?i=57353 (accessed on 22 April 2025).
49. *Decreto 2492 de 2014*; Adopción de Disposiciones en Materia de Implementación de Mecanismos de Respuesta de la Demanda; Presidencia de la República de Colombia: Bogotá, Colombia, 2014. Available online: https://www.funcionpublica.gov.co/eva/gestornormativo/norma.php?i=60174 (accessed on 22 April 2025).

50. *Resolución CREG 101 019 de 2022*; Regulación del Anillo de Seguridad del Cargo por Confiabilidad Denominado Demanda Desconectable Voluntaria; Comisión de Regulación de Energía y Gas: Bogotá, Colombia, 2022. Available online: https://gestornormativo.creg.gov.co/gestor/entorno/docs/resolucion_creg_101-19_2022.htm (accessed on 22 April 2025).
51. *Resolución CREG 11 de 2015*; Regulación Programa de Respuesta de la Demanda para el Mercado Diario en Condición Crítica; Comisión de Regulación de Energía y Gas: Bogotá, Colombia, 2015. Available online: https://gestornormativo.creg.gov.co/gestor/entorno/docs/resolucion_creg_0011_2015.htm (accessed on 22 April 2025).
52. *Resolución CREG 174 de 2021*; Regulación Actividades de Autogeneración a Pequeña Escala y de Generación Distribuida en el Sistema Interconectado Nacional; Comisión de Regulación de Energía y Gas: Bogotá, Colombia, 2021. Available online: https://gestornormativo.creg.gov.co/gestor/entorno/docs/resolucion_creg_0174_2021.htm (accessed on 22 April 2025).
53. *Resolución CREG 148 de 2021*; Regulación para Permitir Conexión y Operación de Plantas Solares Fotovoltaicas y Eólicas en el SDL Con Capacidad Efectiva Neta o Potencia Máxima Declarada Igual o Mayor a 5 MW y se Dictan Otras Disposiciones; Comisión de Regulación de Energía y Gas: Bogotá, Colombia, 2021. Available online: https://gestornormativo.creg.gov.co/gestor/entorno/docs/resolucion_creg_0148_2021.htm (accessed on 22 April 2025).
54. *Resolución CREG 101 072 de 2025*; Regulación para la Integración de las Comunidades Energéticas al Sistema Energético Nacional y se Dictan otras Disposiciones; Comisión de Regulación de Energía y Gas: Bogotá, Colombia, 2025. Available online: https://gestornormativo.creg.gov.co/gestor/entorno/docs/originales/Resoluci%C3%B3n_CREG_101_072_2025/Resoluci%C3%B3n_CREG_101_072_2025.pdf (accessed on 22 April 2025).
55. Carbon Trust. *Reporte Final del Proyecto: Apoyo al Despliegue de Tecnologías de Redes Inteligentes en Colombia*; Carbon Trust: London, UK, 2020. Available online: https://www1.upme.gov.co/DemandayEficiencia/Documents/Reporte_Final_Despliegue_Redes_Inteligentes.pdf (accessed on 22 April 2025).
56. La Unidad de Planeación Minero Energética—UPME. *Smart Grids Colombia Visión 2030—Parte II*; UPME: Bogotá, Colombia, 2016.
57. Quiroga Rubio, L. Con un Riesgo de Racionamiento de Energía Latente, Gobierno Estudia Medidas Adicionales. *El País*. 16 March 2016. Available online: https://www.eltiempo.com/economia/sectores/con-un-riesgo-de-racionamiento-de-energia-sobre-la-mesa-gobierno-estudia-nuevas-medidas-3334105 (accessed on 26 February 2024).
58. Reynoso, L. Colombia Se Enfrenta Al Riesgo de Quedarse a Oscuras Ante La Falta de Energía. *El País*, 17 April 2024.
59. Colombia Inteligente. *Guía Arquitectura Tecnológica Sistema Interconectado Nacional*; Colombia Inteligente: Medellín, Colombia, 2020.

energies

MDPI

Article

Flexible Demand Side Management in Smart Cities: Integrating Diverse User Profiles and Multiple Objectives

Nuno Souza e Silva [1,2,*] and Paulo Ferrão [3,*]

1 Instituto Superior Técnico, University of Lisbon, 1049-001 Lisbon, Portugal
2 Portugal and R&D Nester, 2685-038 Sacavém, Portugal
3 Center for Innovation, Technology and Policy Research (IN+/LARSyS), Instituto Superior Técnico, University of Lisbon, 1049-001 Lisbon, Portugal
* Correspondence: nuno.s.silva@tecnico.ulisboa.pt (N.S.e.S.); ferrao@tecnico.ulisboa.pt (P.F.)

Abstract

Demand Side Management (DSM) plays a crucial role in modern energy systems, enabling more efficient use of energy resources and contributing to the sustainability of the power grid. This study examines DSM strategies within a multi-environment context encompassing residential, commercial, and industrial sectors, with a focus on diverse appliance types that exhibit distinct operational characteristics and user preferences. Initially, a single-objective optimization approach using Genetic Algorithms (GAs) is employed to minimize the total energy cost under a real Time-of-Use (ToU) pricing scheme. This heuristic method allows for the effective scheduling of appliance operations while factoring in their unique characteristics such as power consumption, usage duration, and user-defined operational flexibility. This study extends the optimization problem to a multi-objective framework that incorporates the minimization of CO_2 emissions under a real annual energy mix while also accounting for user discomfort. The Non-dominated Sorting Genetic Algorithm II (NSGA-II) is utilized for this purpose, providing a Pareto-optimal set of solutions that balances these competing objectives. The inclusion of multiple objectives ensures a comprehensive assessment of DSM strategies, aiming to reduce environmental impact and enhance user satisfaction. Additionally, this study monitors the Peak-to-Average Ratio (PAR) to evaluate the impact of DSM strategies on load balancing and grid stability. It also analyzes the impact of considering different periods of the year with the associated ToU hourly schedule and CO_2 emissions hourly profile. A key innovation of this research is the integration of detailed, category-specific metrics that enable the disaggregation of costs, emissions, and user discomfort across residential, commercial, and industrial appliances. This granularity enables stakeholders to implement tailored strategies that align with specific operational goals and regulatory compliance. Also, the emphasis on a user discomfort indicator allows us to explore the flexibility available in such DSM mechanisms. The results demonstrate the effectiveness of the proposed multi-objective optimization approach in achieving significant cost savings that may reach 20% for industrial applications, while the order of magnitude of the trade-offs involved in terms of emissions reduction, improvement in discomfort, and PAR reduction is quantified for different frameworks. The outcomes not only underscore the efficacy of applying advanced optimization frameworks to real-world problems but also point to pathways for future research in smart energy management. This comprehensive analysis highlights the potential of advanced DSM techniques to enhance the sustainability and resilience of energy systems while also offering valuable policy implications.

Keywords: demand side management; energy management; load-shifting; multi-objective optimization; genetic algorithm; smart electric grids; smart city

 https://doi.org/10.3390/en18154107

1. Introduction

In recent years, the growing emphasis on energy efficiency and the rising importance of renewable energy sources has led to an increased focus on the optimization of electrical power consumption [1,2]. This is particularly pertinent in the context of Demand Side Management (DSM), which plays a critical role in modern energy systems by aligning consumer energy usage with supply conditions. DSM not only facilitates significant cost reductions and enhances the reliability of power systems but also contributes to the overarching goal of reducing carbon footprints. As the penetration of renewable energy sources increases, the flexibility that DSM offers becomes crucial in managing the variability associated with these sources.

The advancement of smart grid technologies has paved the way for a new era of possibilities for DSM [3]. Smart grids integrate various digital information and communication technologies into traditional power networks, enabling real-time monitoring and management of energy flows. This integration helps in optimizing the production and distribution of electricity and revolutionizes the way that energy is used and saved [4,5]. Smart cities can leverage smart grid developments to manage the load profiles of their diverse and complex mix of residential, commercial, and industrial consumers [5,6].

Very often the management of load in a smart city is performed with the aim of addressing one particular objective function, such as minimization of cost for its consumers, minimization of CO_2 emissions, or minimization of Peak-to-Average Ratio (PAR) load in the grid (aiming at improving grid resilience). Recent approaches increasingly try to address a combination of those and other objective functions [7–11], mindful of multiple stakeholders and technical and societal concerns [12]. In particular, most early studies did not explicitly consider the impact of the load shifting on consumers, especially regarding discomfort or business implications. Consumer discomfort is mostly considered from a thermal point of view or appliance availability, often considering single environment settings (e.g., residential or industrial), and usually in relation to particular optimization objectives [13–16]. As a result, the benefits proposed by these studies may not be sufficiently realistic [17–19].

Heuristic-based approaches emerged as powerful tools in optimization due to the increasing complexity and scale of real-world problems, which often render traditional exact methods impractical or infeasible [20]. Traditional optimization techniques, such as linear programming or gradient descent, rely on precise mathematical formulations and assumptions. These methods can become computationally expensive and inefficient when dealing with large-scale, high-dimensional, or non-linear problems with numerous local optima. Heuristic approaches, inspired by natural processes such as evolution, swarm behavior, and annealing, provide flexible, adaptive, and robust strategies for exploring complex search spaces. They are also frequently used when objective functions or/and constraints of the problem are non-linear. They do not guarantee an optimal solution but can find good-enough solutions within reasonable time frames, making them particularly suitable for problems where obtaining an exact solution is less critical than finding a feasible and satisfactory one. This adaptability and efficiency in handling diverse and intricate problems have driven the widespread adoption and development of heuristic-based optimization methods.

This paper delves into the application of heuristics optimization such as a GA and NSGA-II within a DSM framework, categorizing solutions across residential, commercial, and industrial sectors, with a large number of appliances and distinct constraints. Reasons for using a GA include its widespread use in the energy domain, being simple and robust in this type of problem, and easily handling the existing constraints. The Non-dominated Sorting Genetic Algorithm II (NSGA-II) is a potent tool in this domain, addressing the multi-objective optimization of scheduling electrical appliances to minimize cost, reduce CO_2 emissions, and alleviate user discomfort simultaneously. NSGA-II preserves trade-off by finding a Pareto front of optimal solutions, eliminating the need to aggregate objectives into a single one and thereby avoiding arbitrary weighting. This study highlights the algorithm's efficacy in navigating the complex trade-offs between competing objectives and underscores the practical implications of such optimizations in real-world settings. Furthermore, it allows for the comparison of results against a single-objective approach where only one objective function (cost) is minimized at the expense of other objective functions (CO_2 emissions, user discomfort, Peak-to-Average Ratio). It is, however, more computationally intensive and requires post-analysis to select adequate solutions from the Pareto front.

In the context of DSM, it is also important to consider the hourly profile of pricing schemes and CO_2 emissions when addressing objective functions aimed at minimizing costs or emissions. Indeed, most pricing schemes vary depending on the year period (e.g., summer vs. winter), and hourly CO_2 emissions depend on the generation mix offered to the load, which also varies seasonally.

The focus of this paper is on the intersection of the following aspects:

- Smart city context with multiple user profiles (with analysis of differentiated impact of DSM in the different profiles);
- The consideration of multiple objective optimization (cost reduction, emissions reduction, discomfort reduction) given the increasing interest and importance of optimizing solutions, taking into consideration multiple aspects, and comparison with single objective optimization, for the evaluation of Demand Side Management;
- Analysis of the impact of seasonality and annual generation mixes on the optimal outcomes of Demand Side Management;
- The use of real data (electricity prices, Time-of-Use profiles, emissions annual profile, device characteristics);
- Extraction of insights for decision-makers with different preferences, including policy, such as the design of Time-of-Use profiles.

The focus is on demonstrating the usefulness of heuristics optimization, rather than advancing existing algorithms (such as the GA and NSGA-II). In this context, analyzing the performance of the energy system across different sectors and periods provides insights into specific challenges and opportunities, offering a comprehensive overview of the strategic application of DSM to enhance energy efficiency and inform potential policy measures.

2. Demand Side Management Techniques

Demand Side Management (DSM) encompasses a range of techniques aimed at encouraging consumers to modify their level and pattern of electricity usage [16]. The primary goal of DSM is to optimize energy consumption and improve the overall efficiency of the electrical grid. The main techniques of DSM can be broadly classified into the following categories of impact:

1. Energy Efficiency: This involves measures that lead to a reduction in the energy used by specific end-use devices and systems, without affecting the services provided. These measures include high-efficiency appliances, improved insulation, and more efficient heating and cooling systems. Energy efficiency programs are cost-effective over time, reducing energy consumption and utility bills.
2. Load Shifting: Also frequently known as demand response, this technique encourages consumers to increase or decrease their electricity use during specific periods where energy demand is high or when renewable energy availability is high. Load shifting helps in managing the load curve, reducing the need for peak power plants, and

enhancing grid reliability. Technologies such as smart thermostats and appliances capable of responding to signals from energy providers to delay or advance their operation time are central to this strategy.

3. Peak Load Reduction: This strategy involves programs designed to cut down energy use during peak demand times, such as on hot summer days. Techniques include offering incentives for reduced consumption and temporary increases in electricity prices during peak periods (Time-of-Use pricing).
4. Load Filling: Load filling promotes increased energy usage during periods of low demand to maintain consistent electricity production levels. This technique is particularly relevant in systems with high levels of renewable energy generation, where excess energy can be used more effectively.
5. Energy Conservation: Broad programs aimed at promoting a culture of energy savings among consumers through behavioral change. Education and information dissemination play key roles in these programs, encouraging sustained energy-saving habits.
6. Integrated DSM: This approach combines several DSM strategies to achieve optimal energy savings. For instance, integrating real-time energy monitoring tools with consumer incentives and education can lead to more significant energy savings and operational efficiencies.

DSM techniques are instrumental in stabilizing the grid and making energy use more sustainable and cost-effective. The implementation of DSM can vary significantly depending on regional regulatory policies, the maturity of the local energy market, and consumer engagement levels. In the present study, load shifting was considered.

3. Problem Formulation and Algorithm Description

3.1. Problem Formulation

The problem under study addresses the optimization of electrical appliance scheduling within residential, commercial, and industrial sectors. This section provides a detailed formulation of the variables, objective functions considered throughout this work, and constraints involved, for the implementation of the GA and NSGA-II.

A. Variables:

$g = (a, s, d)$: Represents a gene in the chromosome, where

a : Appliance identifier.

s : Start time of the appliance's operation within a 24 h period.

d : Duration of the appliance's operation.

A chromosome is composed of sequences of genes; thus, it has as many genes as appliances in the considered problem.

B. Objective Functions:

1. Cost (C) Minimization:

$$C = \sum_{h=0}^{23} \left(\sum_{g \in G} P_g \times I_{g,h} \right) \times E_h \tag{1}$$

where P_g is the power consumption of the appliance in gene g, $I_{g,h}$ indicates if the appliance in gene g operates during hour h, G is the set of all genes, and E_h is the electricity price at hour h.

2. Discomfort (D) Minimization:

$$D = \sum_{g \in G} \left| s_g - s_{g,pref} \right| \times \delta_g, \tag{2}$$

where s_g is the effective start time (hour) for the appliance in gene g, $s_{g,pref}$ is the preferred start time (hour) for the appliance in gene g, and δ_g is the hourly discomfort penalty associated with the appliance in gene g.

3. CO_2 Emission (E) Minimization:

$$E = \sum_{h=0}^{23} \left(\sum_{g \in G} P_g \times I_{g,h} \right) \times X_h, \quad (3)$$

where X_h represents the CO_2 emission factor at hour h.

C. Constraints:

i. Power Constraint: Ensures that the power consumed at any hour does not exceed the maximum allowable limit:

$$\forall h \in \{0, \ldots, 23\}, \sum_{g \in G} P_g \times I_{g,h} \leq L_{max}, \quad (4)$$

where L_{max} the peak power limit.

From the implementation point of view, this restriction can be included in the cost function with the following additional term:

$$\sum_{h=0}^{23} max(0, L_h - L_{max})^2, \quad (5)$$

where L_h is the total load at hour h.

ii. Operational Constraints: Each appliance must adhere to its operational limits regarding minimum and maximum usage durations and daily usage frequency.

3.2. Algorithm Description

A. Genetic Algorithm (GA)

A Genetic Algorithm, developed in Python, version 3.12.2, is used to search for optimal solutions for the single-objective problem of minimizing the cost function. It is a heuristic optimization technique inspired by the principles of natural selection and genetics. It is used to find approximate solutions to complex optimization problems by iteratively improving a population of candidate solutions. The main components of a GA include selection and tournament, crossover (recombination), mutation, and elitism. These components are used to explore the search space and exploit the best solutions found during the search process.

The population size determines the number of individuals (candidate solutions) in the population. A larger population size increases the diversity of solutions but also requires more computational resources.

The crossover rate is the probability with which crossover (recombination) occurs between pairs of individuals to create offspring. A higher crossover rate promotes the exchange of genetic material, leading to greater exploration of the search space. A uniform crossover mechanism allows genes (appliance schedules) from parent chromosomes to mix, promoting variability in the offspring.

The mutation rate refers to the probability of introducing random mutations into individuals within a population. A higher mutation rate increases diversity and helps prevent premature convergence in local optima but can also disrupt good solutions.

The number of generations is the number of iterations the algorithm will run. More generations allow the algorithm to refine solutions further but require more computational time.

The selection mechanism is the method used to select individuals for reproduction based on their fitness. Common mechanisms include tournament selection and roulette wheel selection. They influence the algorithm's convergence rate and diversity. In our case we applied a tournament selection process.

Elitism aims at ensuring that the best solutions are carried over to the next generation. It preserves high-quality solutions and helps maintain the best-found solutions throughout the evolutionary process.

In our study the following parameters were used in the Genetic Algorithm (GA), based on multiple tests using different combinations of parameters:

- Population Size: 400;
- Number of Generations: 100;
- Crossover Probability: 0.9 (random single point);
- Mutation Probability: 0.1;
- Tournament size: 3;
- Elite Size: 10.

The stopping criteria for the algorithm was the total number of generations (200), thus limiting the maximum amount of time that the algorithm would run.

B. Non-Dominated Sorting Genetic Algorithm II (NSGA-II)

The Non-dominated Sorting Genetic Algorithm II (NSGA-II) [21] is an advanced multi-objective optimization algorithm that extends the principles of GAs to handle multiple objectives. NSGA-II aims to find a set of Pareto-optimal solutions, where no single solution is strictly better than another across all objectives. It uses mechanisms like non-dominated sorting, crowding distance, and elitism to ensure diversity and convergence to the Pareto front.

The selection mechanism selects individuals for reproduction based on their rank and crowding distance, thus balancing convergence and diversity by preferring non-dominated individuals and maintaining spread.

The crowding distance provides a measure of the density of solutions surrounding a particular solution in the objective space. This ensures diversity by preserving a spread of solutions across the Pareto front.

The non-dominated sorting mechanism classifies individuals into different fronts based on Pareto dominance. This helps identify and rank solutions based on their dominance, facilitating multi-objective optimization.

In our study the following parameters were used in the NSGA-II algorithm, also developed in Python, based on multiple tests using different combinations of parameters:

- Population Size: 400;
- Number of Generations: 100;
- Crossover Probability: 0.9 (random single point);
- Mutation Probability: 0.1.

The stopping criteria considered were the same as for the GA.

4. Scenario and Model Specification

4.1. Scenario Details

The scenario was designed to reflect realistic household, commercial, and industrial environments in the context of a smart grid for a smart city where energy consumption optimization is crucial. This study evaluates a mixed residential, commercial, and industrial setting with the following characteristics:

- Residential Sector: Includes common appliances like refrigerators, washing machines, dryers, and ovens. Each appliance has varying power requirements, preferred operational times, and flexibility in operation, reflecting typical household energy consumption patterns.

- Commercial Sector: Focuses on energy-intensive appliances like water dispensers, ovens, and air conditioning units commonly found in office buildings and small businesses. These appliances typically have less flexibility in operation times but contribute significantly to the energy footprint of commercial establishments.
- Industrial Sector: Comprises heavy machinery such as water heaters, welding machines, and induction motors. These units are high-power consumers with strict operational schedules to maintain industrial productivity.

This study uses hourly electricity pricing and CO_2 emissions factors that vary throughout the day, reflecting the dynamic nature of energy markets and environmental impacts associated with electricity generation. Furthermore, different periods of the year are considered to account for distinct daily hourly profiles for energy prices and for CO_2 emissions. The optimization aims to balance cost, discomfort, and emissions, providing a comprehensive strategy for DSM in mixed-use environments.

4.2. Appliance Characterization

Appliances across the three sectors are characterized by their power requirements, operational timings, flexibility, and user-defined preferences. Realistic data was collected from the literature [22–25] to reflect the characteristics of a smart city. Each appliance type is detailed in the tables below, including their rated power (in W), minimum and maximum usage durations (in hours), maximum daily usage, preferred start time (between hour 1 and hour 23), start time flexibility (in number of hours), discomfort penalty (an integer index), and quantity. The "start time" is the moment of the day when it is expected to start using such an appliance. The "start time flexibility" represents the number of hours that such a starting moment is allowed to be anticipated or delayed. For example, a dish washer or a washing machine could in principle operate any time of the day; therefore, a 24 h "start time flexibility" is assigned. A toaster or a TV would in principle be used at a certain time of the day, hence a value of 1. Appliances that are non-controllable or are not allowed to have starting time flexibility have a starting time flexibility value of 0 (zero). As indicated in the study of [26], commercial loads offer more limited flexibility than residential due to the prioritization of occupant comfort and safety, and maintaining business operations, over energy savings, and the flexibility of industrial electricity demand is recognized as being considerably lower, due to technical constraints and greater immediate negative costs for stoppages or reductions, than for other sectors. For these reasons, the starting time flexibility is higher in most residential appliances than in commercial ones and higher in commercial appliances than in industrial ones [27]. This is also in line with empirical appliance utilization scheduling tables [25,28], constraints in appliance usage [26,29], and starting times, end times, and usage duration of appliances [25,30,31]. Accordingly, the discomfort penalty was considered to be between 1 and 2 for Residential appliances, 5 for Commercial appliances, and 10 for Industrial appliances in order to reflect a lower ability to move schedules in commercial and industrial environments. As mentioned in Section 1 above, most of the discomfort quantification approaches in the literature address it from either the thermal or waiting times points of view [32–34], which, although allowing for a detailed and complex analysis of some appliances' behavior, do not capture the complete range of discomfort reasons in the residential, commercial, and industrial settings in a uniform manner (e.g., rescheduling of staff, ability to anticipate use, operational restrictions or implications). The discomfort penalty allows for encompassing several factors, in a uniform way and simplified manner, at the cost of more detail. Together with the "start time flexibility", it can be used as a proxy for the concept of "flexibility". By modifying the discomfort penalty values and flexibility ranges, the present framework can be adjusted for different cities' characteristics, according to the type of industry present, the type of existing commercial and service buildings, and the characterization of the residential framework. This allows us to study the flexibility associated with different city settings. The fine tuning of the discomfort penalty values used and the consequent impact on metrics of the characteristics of different cities deserves dedicated consideration and is out of the scope of the present paper and is planned to be studied in follow-up research.

We have considered a total of 4621 appliances, with 3704 residential appliances across 17 appliance types, 808 commercial appliances across 9 appliance types, and 109 industrial appliances across 9 appliance types. These correspond to an amount of nominal power of the same order of magnitude for the different categories (2.88 MW for residential, 2.38 MW for commercial, and 3.64 MW for industrial). Tables 1–3 below list the appliances, their quantities, and characteristics. Figure 1 depicts the load profile of the commercial setting without DSM activated.

Table 1. Residential appliances.

Appliance	Power (W)	Min Duration (h)	Max Duration (h)	Preferred Start Time (h)	Start Time Flexibility (h)	Discomfort Penalty	Quantity
Dryer	1200	1	1	21	24	1	189
Dish washer	700	1	2	21	24	1	288
Washing machine	500	1	2	21	24	1	268
Oven	1300	1	1	19	3	2	279
Iron	1000	1	1	18	10	1	340
Vacuum cleaner	400	1	1	11	10	1	158
Fan	200	2	5	13	5	1	288
Kettle	2000	1	1	17	2	1	406
Toaster	900	1	1	8	1	1	48
Rice cooker	850	1	1	19	3	2	59
Hair dryer	1500	1	1	8	1	1	58
Blender	300	1	1	17	10	1	66
Frying pan	1100	1	1	19	3	2	101
Coffee maker	800	1	1	8	1	2	56
Tv	300	2	4	20	1	1	300
Lights	200	3	6	19	1	2	400
Continuous loads	400	24	24	0	0	2	400

Table 2. Commercial appliances.

Appliance	Power (W)	Min Duration (h)	Max Duration (h)	Preferred Start Time (h)	Start Time Flexibility (h)	Discomfort Penalty	Quantity
Water dispenser am	2500	5	7	11	3	5	78
Water dispenser pm	2500	5	7	16	3	5	78
Dryer	3500	4	6	11	3	5	117
Kettle	3000	2	3	16	3	5	123
Oven	5000	1	2	13	2	5	77
Coffee maker	2000	2	3	13	2	5	99
AC/fan	3000	2	3	14	3	5	93
AC	3500	3	3	14	3	5	56
Lights	1750	24	24	0	0	5	87

Table 3. Industrial appliances.

Appliance	Power (W)	Min Duration (h)	Max Duration (h)	Preferred Start Time (h)	Start Time Flexibility (h)	Discomfort Penalty	Quantity
water heater	12,500	3	5	10	3	10	39
welding machine	25,000	5	5	9	1	10	35
fan/AC	30,000	5	6	11	2	10	16
arc furnace	50,000	6	7	9	2	10	8
induction motor	100,000	6	6	9	2	10	5
dc motor am	150,000	3	4	9	2	10	3
dc motor pm	150,000	3	4	15	2	10	3

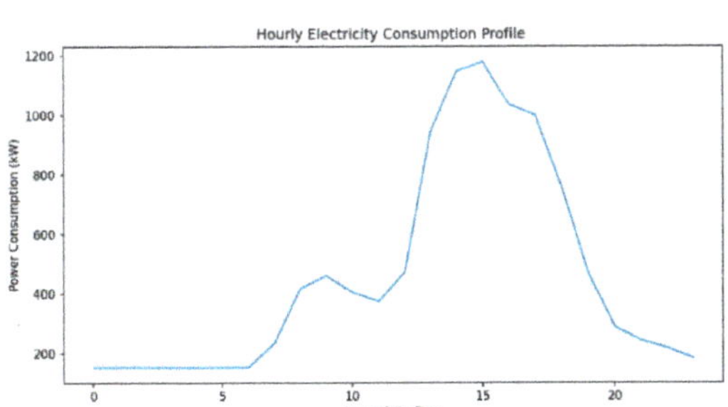

Figure 1. Load profile without DSM, commercial category.

4.3. Pricing Schemes

The scenario adopted employs a Time-of-Use electricity pricing considering real-world conditions. The ToU scheme varies throughout the day, with higher prices during peak demand periods and lower prices during off-peak times, encouraging load shifting. Real values from existing commercial offers in Portugal were considered [35]. Since, in the country, ToU presents different periods in summer and in winter, two sets of pricing schemes were considered where the ToU blocks are distributed through different hours of the day, as per the national regulator [36]. Although these data are specific to Portugal, they reflect a common practice in Europe, therefore making the results realistic to wider geographies.

The Time-of-Use Pricing values are the following (EUR/kWh):

- Peak Hours: 0.30;
- Mid-Peak Hours: 0.18;
- Off-Peak Hours: 0.15;
- Super Off-Peak Hours: 0.11.

Table 4 presents the daily pricing profile.

Table 4. Daily pricing profile (EUR/kWh).

Hour		**0**	**1**	**2**	**3**	**4**	**5**	**6**	**7**	**8**	**9**	**10**	**11**
Price	Summer	0.15	0.15	0.11	0.11	0.11	0.11	0.15	0.15	0.18	0.18	0.18	0.3
	Winter	0.15	0.15	0.11	0.11	0.11	0.11	0.15	0.15	0.18	0.3	0.3	0.18
Hour		**12**	**13**	**14**	**15**	**16**	**17**	**18**	**19**	**20**	**21**	**22**	**23**
Price	Summer	0.3	0.18	0.18	0.18	0.18	0.18	0.18	0.18	0.3	0.18	0.15	0.15
	Winter	0.18	0.18	0.18	0.18	0.18	0.18	0.3	0.3	0.3	0.18	0.15	0.15

4.4. CO_2 Emissions

CO_2 emission factors vary on an hourly basis, reflecting the changing environmental impact of electricity generation throughout the day [37]. These factors are essential for quantifying the carbon footprint associated with appliance operation schedules. In this study, real data from Portugal's 2023 generation mix was used, based on publicly available information from the national system operator (REN) [38]. The data, originally provided at 15 min intervals, was aggregated to an hourly resolution to enable analysis of the hourly emission profile. Emission factors specific to each generation source were considered for the Portuguese context [39,40]. With this data, an hourly profile of carbon intensity in consumption was generated. Portugal has a high penetration of renewable energy in its electricity mix, with more than 61% of demand met by renewables in 2023. As the generation mix fluctuates both daily and seasonally, CO_2 emissions vary accordingly. Therefore, it is essential to consider multiple days when evaluating the impact of Demand Side Management (DSM) on CO_2 emissions under varying generation conditions. In this study, five distinct days were analyzed based on their specific characteristics, as detailed in Tables 5 and 6. The percentages shown represent the share of renewable energy sources (RESs) in the total energy supply for each day.

Table 5. Periods analyzed.

Period	**Date**	**Characterization**	
0	6 July 2023	Summer sunny	41% RES
1	30 January 2023	Winter sunny	62% RES
2	10 January 2023	Winter wet/natural gas	70% RES

Table 5. *Cont.*

Period	Date	Characterization	
3	10 September 2023	Low RES (annual minimum)	21% RES
4	11 November 2023	High RES (annual maximum)	85% RES

Table 6. CO_2 emissions daily profiles (hourly CO_2 emission factors—g CO_2 eq/kWh).

Hour	Period				
	0	1	2	3	4
0	221.6	67.9	77.6	162.9	29.8
1	231	68.5	74.5	179.5	31.3
2	233.2	63.8	73	189.7	33.9
3	234.3	75.3	79.9	195.1	34.9
4	240.5	90	84	196.9	33.9
5	210.8	103.6	84.2	196	32
6	185.9	94.9	87.7	178.6	28.8
7	205.8	89.7	96.7	148	29.2
8	211.6	100.1	108.8	126.6	33.3
9	199.9	128	115	119.7	42.1
10	193.5	148.9	108.1	111.1	36.3
11	194.2	162.3	108.9	108.5	37
12	191.1	170.8	117.4	109.9	40.7
13	180.8	177.9	124	113.8	41
14	177.5	192	119	114.7	38.7
15	176.9	196.3	118.7	113.3	35.4
16	186.5	207.3	112.1	129.4	35.4
17	192.5	179.4	106.4	136.3	34.4
18	187.4	162.7	107.2	147.1	29.7
19	183.2	157.6	112.7	155.3	30
20	170.45	149.2	90.4	168	34.3
21	168.8	143.4	81.4	193	36.4
22	190.1	152.8	78.8	203	42.1
23	206.3	167.7	62.3	215.2	32.8

5. Results and Discussion

Results are presented for a single-objective approach and a multi-objective approach. In the single-objective approach, the objective function is the minimization of cost, and the resulting CO_2 emission, discomfort, and PAR value are also computed and analyzed. In the multi-objective approach, the simultaneous objective functions are minimization of cost, minimization of CO_2 emissions, and minimization of user discomfort. The resulting PAR value is also computed and analyzed.

A. Single-objective with Genetic Algorithm

The convergence of the GA throughout the 200 generations considered for an exemplifying case (Period 0, Industrial category) is depicted in Figure 2. The results obtained

(Tables 7–11) are discussed in the subsequent paragraphs and show that the DSM strategy achieved cost reductions in all cases analyzed, across different categories of loads (residential, commercial, and industrial) or annual periods (summer or winter, with distinct hourly price profiles). In each table, shaded columns indicate the objective function being considered. Cost reductions are calculated as a percentage difference between the cost with the DSM strategy and the cost without the DSM strategy.

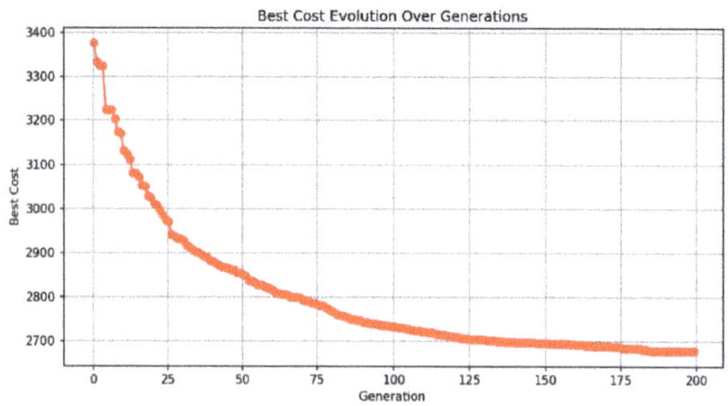

Figure 2. Convergence of GA, Period 0, Industrial category (units: "Best Cost" is in EUR/day).

Table 7. Period 0 results with GA.

Period 0	**6 July 2023**	**Summer Sunny**		**41% RES**
-	**Cost Reduction (with DSM vs. w/o DSM)**	**CO_2 Reduction (with DSM vs. w/o DSM)**	**Discomfort Index (with DSM/w/o DSM)**	**PAR Reduction (with DSM vs. w/o DSM)**
Residential	8.3 %	−1.0 %	3.9/0	56.2 %
Commercial	10.7 %	3.1 %	8.5/0	21.0 %
Industrial	32.5 %	4.3 %	37.8/0	7.8 %

Table 8. Period 1 results with GA.

Period 1	**30 January 2023**	**Winter Sunny**		**62% RES**
	Cost Reduction (with DSM vs. w/o DSM)	**CO_2 Reduction (with DSM vs. w/o DSM)**	**Discomfort Index (with DSM/w/o DSM)**	**PAR Reduction (with DSM vs. w/o DSM)**
Residential	6.2 %	2.7 %	4.1/0	54.3 %
Commercial	5.4 %	5.5 %	7.4/0	19 %
Industrial	23.3 %	0.5 %	32.6/0	12.3 %

Table 9. Period 2 results with GA.

Period 2	**10 January 2023**	**Winter Wet/Natural Gas**		**70% RES**
-	**Cost Reduction (with DSM vs. w/o DSM)**	**CO_2 Reduction (with DSM vs. w/o DSM)**	**Discomfort Index (with DSM/w/o DSM)**	**PAR Reduction (with DSM vs. w/o DSM)**
Residential	5.9 %	0.7 %	4.0/0	54.0 %
Commercial	4.4 %	3.3 %	7.0/0	14.9 %
Industrial	25.6 %	7.8 %	30.8/0	9.6 %

Table 10. Period 3 results with GA.

Period 3	10 September 2023	Low RES (Annual Minimum)		21% RES
-	Cost Reduction (with DSM vs. w/o DSM)	CO_2 Reduction (with DSM vs. w/o DSM)	Discomfort Index (with DSM/w/o DSM)	PAR Reduction (with DSM vs. w/o DSM)
Residential	8.6 %	1.1 %	4.0/0	55.1 %
Commercial	10.8 %	0.3 %	8.7/0	21.3 %
Industrial	31.9 %	−15.7 %	37.2/0	8.2 %

Table 11. Period 4 results with GA.

Period 4	11 November 2023	High RES (Annual Maximum)		85% RES
-	Cost Reduction (with DSM vs. w/o DSM)	CO_2 Reduction (with DSM vs. w/o DSM)	Discomfort Index (with DSM/w/o DSM)	PAR Reduction (with DSM vs. w/o DSM)
Residential	5.6 %	0.2 %	4.0/0	55.1 %
Commercial	5.9 %	4.8 %	7.1/0	13.4 %
Industrial	26.0 %	11.0 %	31.1/0	12.8 %

B. Multi-objective with Non-dominated Sorting Genetic Algorithm-II.

1. Result for different categories

The Industrial category consistently shows the highest cost reduction, with values ranging from 23.3% to 32.5% (depending on the period of the year considered). Although this category includes fewer appliances to manage, their higher power consumption enables significant cost reductions even with small hourly load shifts. Since the strategy and algorithm are only aiming at cost reduction, this reduction is achieved at the cost of CO_2 emissions varying between an increase of 15.7% (in the period with the lowest penetration of RESs) and a reduction of 11%. The observation that CO_2 emissions can increase as a result of a price incentive offers valuable insight for policymakers, highlighting the need to carefully design measures that effectively stimulate emissions reductions. Also, the Discomfort Indexes (calculated as the average of the Discomfort (D), as calculated in Formula (2) in Section 3.1, across the different appliances) present values varying between 30.8 and 37.8 (absolute values), rather than zero (0), which characterize the scenario without DSM. Since each hour of deviation from the preferred operation time has a discomfort value of 10 for Industrial appliances, this corresponds to an average deviation of 3–4 h. Finally, the reduction in the PAR values (which reflect the effectiveness of the load shifting mechanism) is in the 7.8–12.8% range. These are the lowest among the different categories. This may be explained by the long operation hours of the appliances and lower flexibility when compared to other categories.

Most frequently the Commercial category presents the second highest cost reduction, with values between 4.4 and 10.8%. CO_2 reductions obtained are between 0.3 and 5.5%. Discomfort Indexes increase to values between 7 and 8.7. Given that each hour of deviation from the preferred operation time has a discomfort value of 5 for Commercial appliances, this corresponds to an average deviation of 1–2 h. Finally, the reduction in the PAR values is in the range 13.4–21.3%.

Finally, the Residential category presents the lowest cost reduction with values between 5.6 and 8.6%. Again, since the strategy and algorithm are only aiming at cost reduction, this reduction is achieved at the cost of CO_2 emissions, varying between an increase of 1% (representing a price incentive producing an increase in CO_2 emissions) and a reduction of 2.7%. Discomfort Indexes increase to values between 3.9 and 4.1. Given that each hour of deviation from the preferred operation time has a discomfort value of most frequently 1 and in some cases 2 for Residential appliances, this corresponds to an average deviation of 3–4 h. Finally, the reduction in the PAR values is in the order of 54–56%, representing the largest percentual load shifting effect. This is the highest range within the categories and can be attributed to the fact that, given the low number of hours of operation of each appliance, it is easier to avoid simultaneity of operation and consequently peak synchronization.

These values of cost reduction, CO_2 reduction, Discomfort Indexes, and PAR reduction for each category, considering cost as the single objective function, will be used as a basis for comparison with the results obtained when considering multiple-objective functions.

2. Results for different periods

Different periods have different hourly cost profiles (due to winter vs. summer ToU structure) and different hourly CO_2 profiles (due to electricity generation mix).

Summer periods (Periods 0 and 3) exhibit higher cost reductions compared to winter periods (Periods 1, 2, and 4), regardless of the consumer category (Residential, Commercial, or Industrial). This trend may be attributed to the Time-of-Use (ToU) pricing blocks defined for each season, which enable load shifts resulting from DSM to align with lower-priced time slots, while remaining within the allowed flexibility constraints for each appliance.

The periods with higher consumption of electricity produced from RESs frequently present higher CO_2 emissions reductions than periods with lower consumption of electricity produced from RESs. This is evident in the Industry and Commercial categories and can be explained by the fact that shifts to lower-cost ToU blocks are more likely to coincide with periods of high renewable energy generation and, thus, lower CO_2 emissions. This result is not evident for the Residential category, which may be attributed to a much higher number of appliances, lower load per appliance, and higher flexibility in load shifting, resulting in a widespread distribution of loads throughout the day.

An example of the convergence process of the NSGA-II through the 200 generations (Period 0, Industry category) is depicted in Figure 3. The results obtained are presented in Tables 12–16 and show that the DSM strategy also achieved cost reductions in all cases analyzed, both across categories of loads (Residential, Commercial, And Industrial) and across different annual periods (summer or winter, with distinct hourly price profiles). However, the cost reductions are lower than those achieved with the single-objective algorithm (GA) for all categories and all periods. This is a consequence (and trade-off) of the improvements achieved in objective functions of minimization of CO_2 emissions and minimization of Discomfort Index, as discussed below and represented in Tables 17–19 where comparison between results from cost-only optimization (GA) and from multi-objective optimization (NSGA-II), for the three categories, is presented.

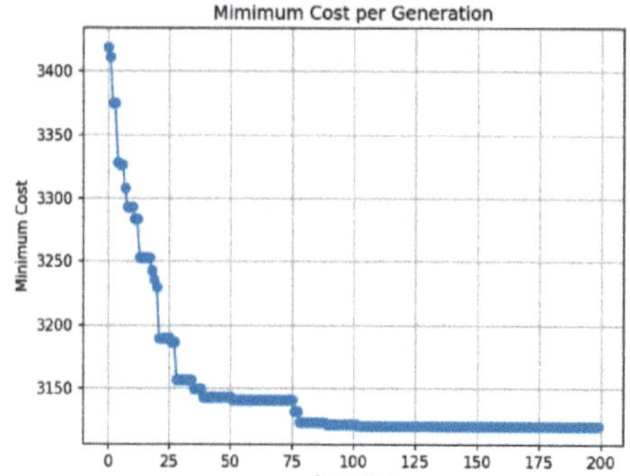

Figure 3. Convergence of NSGA-II, Period 0, Industrial category (units: "Minimum Cost" is in EUR/day).

Table 12. Period 0 results with NSGA-II.

Period 0	6 July 2023	Summer Sunny		41% RES
-	Cost Reduction (with DSM vs. w/o DSM)	CO_2 Reduction (with DSM vs. w/o DSM)	Discomfort Index (with DSM)	PAR Reduction (with DSM vs. w/o DSM)
Residential	6.5–7.6 %	0.3–0.5 %	3.2–3.4	51.7–54.4 %
Commercial	7.0–8.0 %	4.1–4.5 %	5.6–6.0	19.5–21.2 %
Industrial	10.3–19.9 %	4.2–7.0 %	13.1–24.4	20.2–49.6%

Table 13. Period 1 results with NSGA-II.

Period 1	30 January 2023	Winter Sunny		62% RES
-	Cost Reduction (with DSM vs. w/o DSM)	CO_2 Reduction (with DSM vs. w/o DSM)	Discomfort Index (with DSM)	PAR Reduction (with DSM vs. w/o DSM)
Residential	2.5–4.1 %	2.5–3.6 %	3.4–3.7	51.9–55.4 %
Commercial	2.3–3.3 %	7.9–9.2 %	5.9–6.6	26.3–30.6 %
Industrial	11.6–17.3 %	8.0–21.4 %	12.3–24.4	22.1–50.3 %

Table 14. Period 2 results with NSGA-II.

Period 2	10 January 2023	Winter Wet/Natural Gas		70% RES
-	Cost Reduction (with DSM vs. w/o DSM)	CO_2 Reduction (with DSM vs. w/o DSM)	Discomfort Index (with DSM)	PAR Reduction (with DSM vs. w/o DSM)
Residential	3.4–4.3 %	0.6–1.1 %	3.4–3.6	52.2–53.7 %
Commercial	3.6–4.3 %	6.2–6.7 %	5.9–6.4	25.0–28.1 %
Industrial	13.2–18.7 %	6.4–12.3 %	12.9–23.4	21.7–46.7%

Table 15. Period 3 results with NSGA-II.

Period 3	10 September 2023	Low RES (Annual Minimum)		21% RES
-	Cost Reduction (with DSM vs. w/o DSM)	CO_2 Reduction (with DSM vs. w/o DSM)	Discomfort Index (with DSM)	PAR Reduction (with DSM vs. w/o DSM)
Residential	5.7–7.0 %	1.6–2.4 %	3.4–3.8	56.6–58.8 %
Commercial	5.2–7.1 %	3.2–4.0 %	5.7–6.2	19.6–22.5 %
Industrial	8.8–19.5 %	−1.6–6.6 %	12.3–24.1	14.6–45.4 %

Table 16. Period 4 results with NSGA-II.

Period 4	11 November 2023	Hi RES (Annual Maximum)		85% RES
-	Cost Reduction (with DSM vs. w/o DSM)	CO_2 Reduction (with DSM vs. w/o DSM)	Discomfort Index (with DSM)	PAR Reduction (with DSM vs. w/o DSM)
Residential	2.9–4.2 %	0.4–1.0 %	3.4–3.7	52.9–57.8 %
Commercial	3.8–4.4 %	6.0–6.5 %	5.6–5.9	20.2–24.1 %
Industrial	10.7–17.8 %	8.2–12.4 %	12.8–20.1	19.8–45.9 %

Table 17. Comparison between results from cost-only optimization and results from multi-objective optimization, for Residential category (values presented are in the form: "result from GA optimization" → "lower value in the Pareto front from NSGA-II optimization"–"higher value in the Pareto front from NSGA-II optimization" "mid-point from Pareto front", with this last element in italic).

Residential												
Cost Reduction (%)			CO_2 Reduction (%)			Discomfort Index (#)			PAR Reduction (%)			Period
GA	NSGA-II	*NSGA-II Midpoint*	GA	NSGA-II	*NSGA-II Midpoint*	GA	NSGA-II	*NSGA-II Midpoint*	GA	NSGA-II	*NSGA-II Midpoint*	
8.3%	→ 6.5% – 7.6%	*7.1%*	−1.0%	→ 0.3% – 0.5%	*0.4%*	3.9	→ 3.2 – 3.4	*3.3*	56.2%	→ 51.7% – 54.4%	*53.1%*	0 (41% RES)
6.2%	→ 2.5% – 4.1%	*3.3%*	2.7%	→ 2.5% – 3.6%	*3.1%*	4.1	→ 3.4 – 3.7	*3.6*	54.3%	→ 51.9% – 55.4%	*53.7%*	1 (62% RES)
5.9%	→ 3.4% – 4.3%	*3.9%*	0.7%	→ 0.6% – 1.1%	*0.9%*	4	→ 3.4 – 3.6	*3.5*	54.0%	→ 52.2% – 53.7%	*53.0%*	2 (70% RES)
8.6%	→ 5.7% – 7.0%	*6.4%*	1.1%	→ 1.6% – 2.4%	*2.0%*	4	→ 3.4 – 3.8	*3.6*	55.1%	→ 56.6% – 58.8%	*57.7%*	3 (21% RES)
5.6%	→ 2.9% – 4.2%	*3.6%*	0.2%	→ 0.4% – 1.0%	*0.7%*	4	→ 3.4 – 3.7	*3.6*	55.1%	→ 52.9% – 57.8%	*55.4%*	4 (85% RES)

Table 18. Comparison between results from cost-only optimization and results from multi-objective optimization, for Commercial category (values presented are in the form: "result from GA optimization" → "lower value in the Pareto front from NSGA-II optimization"-"higher value in the Pareto front from NSGA-II optimization" "mid-point from Pareto front").

Commercial												
Cost Reduction (%)			CO_2 Reduction (%)			Discomfort Index (#)			PAR Reduction (%)			Period
GA	NSGA-II	*NSGA-II Midpoint*	GA	NSGA-II	*NSGA-II Midpoint*	GA	NSGA-II	*NSGA-II Midpoint*	GA	NSGA-II	*NSGA-II Midpoint*	
10.7%	→ 7.0% – 8.0%	*7.5%*	3.1%	→ 4.1% – 4.5%	*4.3%*	8.5	→ 5.6 – 6.0	*5.8*	21.0%	→ 19.5% – 21.2%	*20.4%*	0 (41% RES)
5.4%	→ 2.3% – 3.3%	*2.8%*	5.5%	→ 7.9% – 9.2%	*8.6%*	7.4	→ 5.9 – 6.6	*6.3*	19.0%	→ 26.3% – 30.6%	*28.5%*	1 (62% RES)
4.4%	→ 3.6% – 4.3%	*4.0%*	3.3%	→ 6.2% – 6.7%	*6.5%*	7	→ 5.9 – 6.4	*6.2*	14.9%	→ 25.0% – 28.1%	*26.6%*	2 (70% RES)
10.8%	→ 5.2% – 7.1%	*6.2%*	0.3%	→ 3.2% – 4.0%	*3.6%*	8.7	→ 5.7 – 6.2	*6.0*	21.3%	→ 19.6% – 22.5%	*21.1%*	3 (21% RES)
5.9%	→ 3.8% – 4.4%	*4.1%*	4.8%	→ 6.0% – 6.5%	*6.3%*	7.1	→ 5.6 – 5.9	*5.8*	13.4%	→ 20.2% – 24.1%	*22.2%*	4 (85% RES)

Table 19. Comparison between results from cost-only optimization and results from multi-objective optimization, for Industrial category (values presented are in the form: "result from GA optimization" → "lower value in the Pareto front from NSGA-II optimization"-"higher value in the Pareto front from NSGA-II optimization" "mid-point from Pareto front").

Industry												
Cost Reduction (%)			CO_2 Reduction (%)			Discomfort Index (#)			PAR Reduction (%)			Period
GA	NSGA-II	*NSGA-II Midpoint*	GA	NSGA-II	*NSGA-II Midpoint*	GA	NSGA-II	*NSGA-II Midpoint*	GA	NSGA-II	*NSGA-II Midpoint*	
32.5%	→ 10.3% – 19.9%	*15.1%*	4.3%	→ 4.2% – 7.0%	*5.6%*	37.8	→ 13.1 – 24.4	*18.8*	7.8%	→ 20.2% – 49.6%	*34.9%*	0 (41% RES)
23.3%	→ 11.6% – 17.3%	*14.5%*	0.5%	→ 8.0% – 21.4%	*14.7%*	32.6	→ 12.3 – 24.4	*18.4*	12.3%	→ 22.1% – 50.3%	*36.2%*	1 (62% RES)
25.6%	→ 13.2% – 18.7%	*16.0%*	7.8%	→ 6.4% – 12.3%	*9.4%*	30.8	→ 12.9 – 23.4	*18.2*	9.6%	→ 21.7% – 46.7%	*34.2%*	2 (70% RES)
31.9%	→ 8.8% – 19.5%	*14.2%*	−15.7%	→ −1.6% – 6.6%	*2.5%*	37.2	→ 12.3 – 24.4	*18.4*	8.2%	→ 14.6% – 45.4%	*30.0%*	3 (21% RES)
26.0%	→ 10.7% – 17.8%	*14.3%*	11.0%	→ 8.2% – 12.4%	*10.3%*	31.1	→ 12.8 – 20.1	*16.5*	12.8%	→ 19.8% – 45.9%	*32.9%*	4 (85% RES)

The results presented in Tables 12–16 show the range of values associated with the three-dimensional Pareto front from each of the objective metrics (cost, CO_2 emissions, and Discomfort Index). Given that there is often a conflict between the objective functions, the solution with the lower end value in one of the metrics does not coincide with the solution with the lower end value in other metrics. This enables decision-makers to identify, within the Pareto front, the solution(s) that best align with their preferred trade-off among the three metrics. Figure 4 depicts a typical example and the trade-off just described presented in a two-dimensional view where the third objective (CO_2 emissions, in the figure) is represented with a color code. As can be seen in Figure 4, a decision-maker prioritizing lower cost would select an operation point towards the left side of the graphic. This region also corresponds to higher levels of discomfort (upper left) and fewer dark blue points—indicating areas with higher CO_2 emissions. Alternatively, a decision-maker who prioritizes maintaining appliance usage patterns with minimal impact on operational activities, such as in an industrial setting, would opt for an operating point toward the lower part of the graph, corresponding to a lower Discomfort Index. This region is associated with higher costs (lower right side of the graph) but still offers a range of options in terms of CO_2 emissions, as indicated by the presence of both light green and dark blue points.

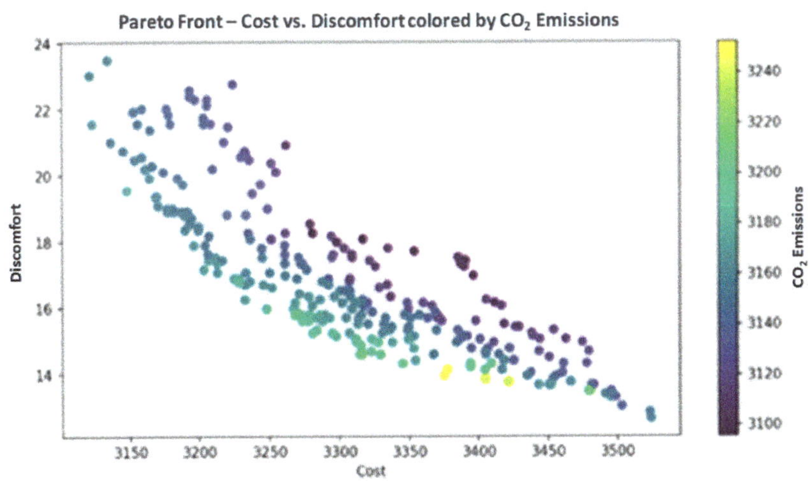

Figure 4. Population and Pareto front, Period 0, Industrial category (units: "Cost" is in EUR/day; "CO_2 emissions" is in tonCO_2 eq/day; "Discomfort" is in Discomfort Index units as defined in Section 5 above—average of discomfort D across appliances).

Figure 5 depicts the same population set in a 3-D representation from different rotation angles to have a visual idea of the 3-D Pareto front. The same effect regarding the mentioned trade-offs can be observed. It is particularly evident in the case between lower discomfort and higher cost (and vice versa). The implications on CO_2 emissions depend more on the energy mix throughout the day and the seasons (as visible in Tables 12–16), because load shifting, while pursuing lower cost or lower discomfort, can result simultaneously in lower emissions if the dislocation of load is carried out towards hours when more RESs are used. This insight can offer valuable guidance to authorities and other decision-makers in shaping policies, particularly those aimed at decarbonization.

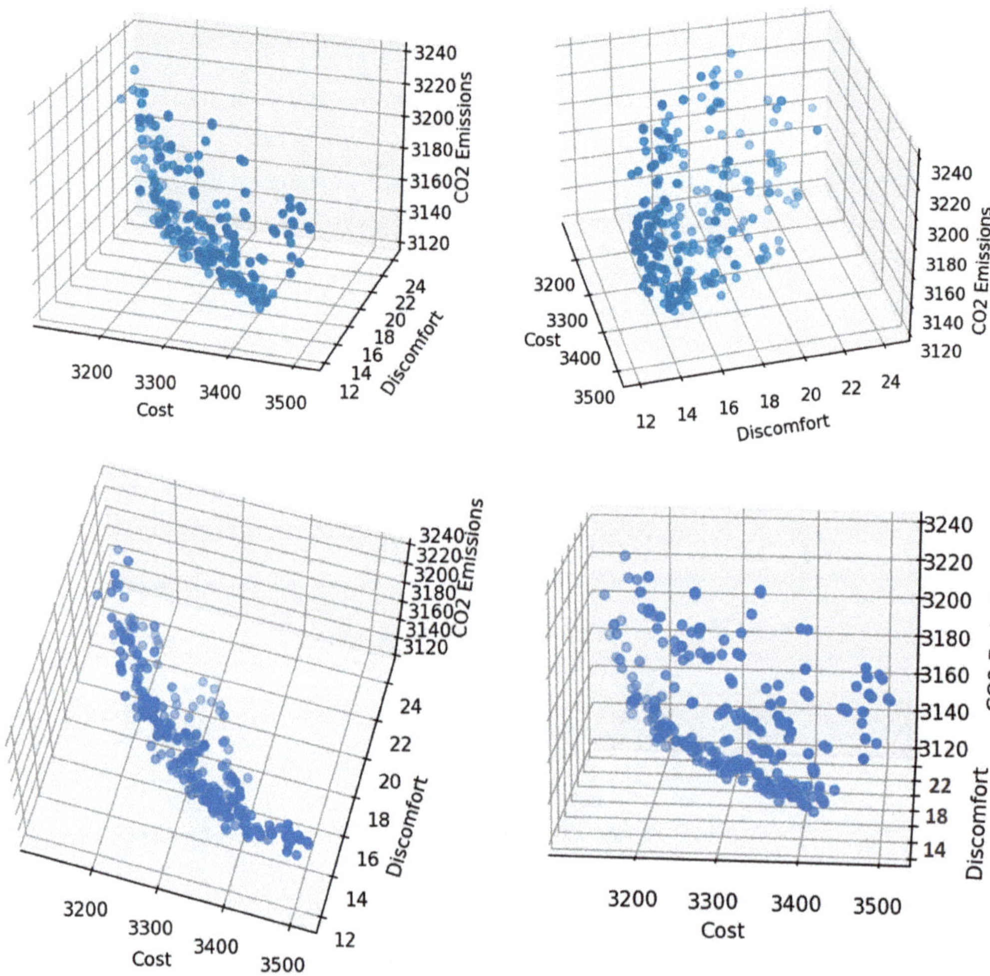

Figure 5. Population and Pareto front, Period 0, Industrial category (3-D view) (units: "Cost" is in EUR/day; "CO_2 emissions" is in tonCO_2 eq/day; "Discomfort" is in Discomfort Index units as defined in Section 5 above—average of discomfort D across appliances).

3. Results for different categories

The same patterns between the different categories as seen with the single-objective algorithm are observed.

The results for cost reduction for the Industrial category are again consistently the highest, with values between 8.8% (Table 15) and 19.9% (Table 12) (depending on the period of the year considered). This compares to the reduction of 23.3% to 32.5% obtained above when only considering cost reduction as the objective function, as can be seen in Table 19, thus showing a more modest cost reduction benefit as mentioned above in the graphical analysis. The results of CO_2 emissions improved to between an increase of 1.6% (Table 15, period with 21% RES) and a reduction of 21.4% (Table 13, period with 62% RES). This compares with values that ranged from an increase of 15.7% to a reduction of 11%, when considering only cost as the objective function, as presented in Table 19, thus showing the improvement in emissions when also including it in the objective functions. Also, the Discomfort Indexes improved to values that vary between 12.3 and 24.4 (absolute values) (Tables 12 and 13, respectively). Since each hour of deviation from the preferred operation time has a discomfort value of 10 for Industrial appliances, this corresponds to an average deviation of 1–2.5 h. This compares to Discomfort Indexes between 30.8 and 37.8 (corresponding to average deviations between 3 and 4 h), when considering only cost as the objective function (see Table 19), thus showing the improvement in convenience (less discomfort) when also including it in the objective functions. Finally, the reduction in PAR values is in the 14.6–50.3% range (Tables 13 and 15, respectively). This compares with PAR values between 7.8% and 12.8% that were obtained when considering only cost as the objective function (see Table 19). This shows that, when cost reduction is the sole objective, load shifting tends to concentrate load in periods when costs are lower, thus resulting in a lower reduction in PAR values (only 7.8–12.8%). When multiple objectives are considered (adding CO_2 emissions reduction and Discomfort Index reduction), loads are shifted into a wider range of daily periods and, consequently, the reduction in PAR values is higher (14.6–50.3%). This is particularly noticeable in this Industry category due to the high value of load that each appliance represents.

Regarding the Commercial category, it consistently shows the second highest cost reduction, with values between 2.3% (in Table 13) and 8.0% (in Table 12). This compares to the reduction of 4.4% to 10.8% obtained above when only considering cost reduction as the objective function, as can be seen in Table 18, thus evidencing the reduced benefit in cost reductions. On the other hand, the CO_2 reductions improved to values between 4.1 and 9.2%. This compares with more modest values that ranged between 0.3% and 5.5%, when considering only cost as the objective function, as presented in Table 18. Also, Discomfort Indexes have reduced and are between 5.6 and 6.6. Given that each hour of deviation from the preferred operation time has a discomfort value of 5 for Commercial appliances, this corresponds to an average deviation of just above 1 h. This compares to Discomfort Indexes between 7.0 and 8.7, when considering only cost as the objective function (see Table 18), thus showing the improvement in convenience (less discomfort) when also including it in the objective functions. Finally, the reduction in PAR values also improved and is in the range 19.5–30.6% (instead of reductions between 13.4% and 21.3% obtained when only cost reduction was the objective function).

Finally, the Residential category again shows the lowest cost reductions, now ranging between 2.5% and 7.6%. This compares to the reductions between 5.6% to 8.6% obtained above when only considering cost reduction as the objective function, as can be seen in Table 17, thus evidencing a slightly reduced benefit in cost reductions. However, CO_2 emission reductions improved, with values between 0.3% and 3.6%. This compares with an increase of 1.0% and a reduction of 2.7%, when considering only cost as the objective function, as presented in Table 17. The Discomfort Index also showed slight improvement, falling within the range of 3.2 to 3.8. Given that each hour of deviation from the preferred operation time typically corresponds to a discomfort value of 1 (and occasionally 2 for certain residential appliances), this indicates an average deviation of approximately 3 h. This compares with values that ranged between 3.9 and 4.1, when considering only cost as the objective function, as presented in Table 17. Lastly, the reduction in Peak-to-Average Ratio (PAR) values ranged between 51.7% and 58.8%. This compares with PAR values between 54.0% and 56.2% that were obtained when considering only cost as the objective function (see Table 17).

Analysis of Tables 17–19 further allows us to extract insights on the trade-offs at play that we will discuss now. For that purpose, Tables 20–22 present the improvement/decline of each of the metrics, as a result of considering additional objectives besides cost reduction, by calculating the ration between (i) the results obtained with the multi-objective optimization (for simplification, the mid-point value of the Pareto front segment presented in Tables 17–19 is used) and (ii) the results obtained with the single-objective cost reduction optimization. For example, it can be seen that, for the Industrial category, Period 0 with 41% RES, the cost reduction under multi-objective optimization is 46% (0.46) of what is achieved under single-objective optimization. At the same time, a 30% (1.30) improvement is obtained in terms of emissions reduction, a 50% (0.50) improvement in discomfort is achieved, and an almost five-fold (447%) improvement is achieved in terms of Peak-to-Average Ratio (PAR) reduction is obtained. This helps to understand the order of magnitude of the trade-offs involved. As another example, we can consider the Commercial category and Period 2 with 70% RES. In this case, we can see that the cost reduction under multi-objective optimization is 90% (0.9) of what is achieved under single-objective optimization (thus, eventually considered a small difference). At the same time, a two-fold (195%) improvement is obtained in terms of emissions reduction, discomfort levels reduce to 88% (0.88) of what they were with cost-only optimization, and an almost two-fold (175%) improvement is achieved in terms of Peak-to-Average Ratio (PAR) reduction. This analysis allows us to have a proxy of a "substitution rate" between different objectives or different metrics.

Different categories naturally present different "substitution rates". From Table 22 we can see that, for the Industrial category, the consideration of multi-objectives has a consequence of having cost reductions that are approximately halved, with, however, PAR reductions that are three to four times higher, discomfort levels that are also halved, and some emission reductions (although with very heterogeneous values across different periods).

Table 20. Residential category—ratios between the results obtained with the multi-objective optimization (NSGA-II) and the results obtained with the single-objective cost reduction optimization (GA), from Table 17.

Period	Cost Reduction Ratio	CO_2 Reduction Ratio	Discomfort Index Ratio	PAR Reduction Ratio
0 (41% RES)	0.85	n/a	0.85	0.94
1 (62% RES)	0.53	1.13	0.87	0.99
2 (70% RES)	0.65	1.21	0.88	0.98
3 (21% RES)	0.74	1.82	0.90	1.05
4 (85% RES)	0.63	3.50	0.89	1.00

Table 21. Commercial category—ratios between the results obtained with the multi-objective optimization (NSGA-II) and the results obtained with the single-objective cost reduction optimization (GA), from Table 18.

Period	Cost Reduction Ratio	CO_2 Reduction Ratio	Discomfort Index Ratio	PAR Reduction Ratio
0 (41% RES)	0.70	1.39	0.68	0.97
1 (62% RES)	0.52	1.55	0.84	1.50
2 (70% RES)	0.90	1.95	0.88	1.78
3 (21% RES)	0.57	12.00	0.68	0.99
4 (85% RES)	0.69	1.30	0.81	1.65

Table 22. Industrial category—ratios between the results obtained with the multi-objective optimization (NSGA-II) and the results obtained with the single-objective cost reduction optimization (GA), from Table 19.

Period	Cost Reduction Ratio	CO_2 Reduction Ratio	Discomfort Index Ratio	PAR Reduction Ratio
0 (41% RES)	0.46	1.30	0.50	4.47
1 (62% RES)	0.62	29.40	0.56	2.94
2 (70% RES)	0.62	1.20	0.59	3.56
3 (21% RES)	0.44	n/a	0.49	3.66
4 (85% RES)	0.55	0.94	0.53	2.57

From Table 20 we can see that, for the Residential category, the consideration of multi-objectives has a consequence of having cost reductions that are approximately two-thirds of what were with cost-only optimization, with PAR reductions that are almost the same, discomfort levels that are almost 85–90% of what they were with cost-only optimization, and some emission reductions (again, with very heterogeneous values across different periods).

This analysis offers insight into the worthiness of the trade-offs for different categories and in different periods, seasons, or RES penetration. Also, it provides insights on policy effectiveness under those categories, periods, and RES penetration levels.

4. Results for different periods

The same patterns between the different periods as seen with the single-objective algorithm are observed. Once again, summer periods (periods 0 and 3) present higher cost reductions than winter periods (periods 1, 2, and 4), independently of the category (Residential, Commercial, or Industrial). Also, for the Industrial and Commercial categories, the periods with lower consumption of electricity produced from RES present lower CO_2 emissions reductions than periods with higher consumption of electricity produced from RES. Again, this pattern is not observed for the Residential category, as mentioned above.

6. Conclusions

Demand Side Management is one of the promising strategies to address several of the challenges posed by the current and future smart grids in the context of an increasing penetration of renewable energy sources in the electricity system. This paper evaluates the impact of load shifting as one of the Demand Side Management approaches. In particular, the context of a smart grid in a smart city was considered where residential, commercial, and industrial contexts co-exist, with a large set of controllable and non-controllable devices in each area. The problem was initially mathematically formulated as a single-objective cost minimization problem, with restrictions on the flexibility of time shifting of each appliance type, and a heuristic-based evolutionary algorithm (Genetic Algorithm) was developed to solve the problem. Implications on CO_2 emissions, discomfort levels, and the Peak-to-Average Ratio in the load hourly profile were analyzed. The problem was subsequently formulated as a multi-objective problem aiming at simultaneously minimizing cost, CO_2, and discomfort while still monitoring the PAR. To assess the impact of the price daily profile and CO_2 emissions daily profile on the considered metric and objective functions, real data was used for five specific days throughout the year, providing some sensitivity analysis to the parameters. The results show that, although the single-objective approach achieves higher reductions in terms of cost, the multi-objective approach still obtains noticeable cost reductions while providing an additional reduction in terms of

CO_2 emissions, discomfort level of consumers (residential, commercial, and industrial), and PAR values. It should be noted that, in addition to the PAR, the absolute peak value also decreased across all categories, Residential, Commercial, and Industrial, and in all periods considered. The novel Discomfort Index incorporated into a multi-objective, multi-environment optimization approach allows us to obtain more realistic results regarding the feasible improvements obtained with DSM. Also, the simultaneous consideration of CO_2 emissions sheds light on policy measures in terms of pricing period definition (winter and summer periods), the number of ToU blocks and periods allowed, and potential different approaches for different periods of the year given the corresponding CO_2 intensity of the electricity consumed. Besides public policies, the results can also guide private companies to identify strategies to attract customers via reduced discomfort or other metrics.

The work performed offers potential for future developments. One is the approach regarding the appliance's characterization in terms of the preferred time of operation, the allowed flexibility for each appliance, and the discomfort penalties associated with the allowed flexibility. Although the values used are qualitatively sound, more detailed information regarding appliance use, including elasticity concepts, should be used in future developments in order to have more quantitatively precise results. The other is the use of the same prices for Residential, Commercial, and Industrial users. That is, however, frequently not the case. The obtained results, nevertheless, provide a qualitative indication of the order of magnitude of the metrics analyzed. Finally, this study considers that the load profiles based on the appliances considered are the same in summer and in winter. However, usage of energy services such as lighting or heating and cooling differ across seasons. The impact of this fact in the results is, however, mitigated by the fact that we are performing a comparative study and not analyzing absolute values.

The work performed also identified future areas to explore. In particular, it would be important to evaluate results under different pricing schemes and understand the gains with DSM when the variation in prices throughout the day is lower (with less ToU blocks), higher (such as with dynamic pricing), inclining blocks or real-time pricing. Besides the daily analysis, it would also be important to perform a longer cycle analysis, such as an annual analysis. Also, to ensure robustness of emissions-based DSM scheduling, comparisons can be drawn with the current literature that considers CO_2-aware and uncertainty-aware planning frameworks [41]. It could provide insights not only in terms of the DSM performance but also in terms of heuristics-based optimization techniques in addressing such problems. The detailed appliance characterization or the inclusion of electric vehicles [42] would also provide richer future studies. In particular, methods for improving the accuracy of load forecasting under DSM can be used [43]. From the optimization technique point of view, it could be appropriate to explore coding within the GA (e.g., binary encoding) that could impact processing speed or even other (multi-objective) heuristics approaches (e.g., associated with particle swarm optimization, or with ant colony optimization). Additionally, other optimization techniques could be considered [44–46]. Finally, it should be highlighted that, although outside of the scope of the current study, such large-scale DSM programs should consider the fairness and distribution of stakeholders' benefits, as identified and discussed in [47,48].

DSM continues to be an effective and promising mechanism for achieving challenging objectives in the context of the energy transition. It is key to understand the flexibility available on the users' side, taking into consideration users' preferences and comfort levels. The increasing digitalization and automation introduced by smart grids in smart cities favor their deployment and the ability to capture their benefits.

Author Contributions: Conceptualization, N.S.e.S.; methodology, N.S.e.S.; software, N.S.e.S.; validation, P.F.; formal analysis, N.S.e.S. and P.F.; investigation, N.S.e.S.; resources, N.S.e.S.; data curation, N.S.e.S.; writing—original draft preparation, N.S.e.S.; writing—review and editing, P.F.; visualization, N.S.e.S.; supervision, P.F.; project administration, P.F. All authors have read and agreed to the published version of the manuscript.

Funding: This research received no external funding.

Data Availability Statement: The code presented in this study is available upon request from the corresponding author.

Conflicts of Interest: The authors declare no conflicts of interest.

References

1. Mimi, S.; Ben Maissa, Y.; Tamtaoui, A. Optimization Approaches for Demand-Side Management in the Smart Grid: A Systematic Mapping Study. *Smart Cities* **2023**, *6*, 1630–1662. [CrossRef]
2. Leitão, J.; Gil, P.; Ribeiro, B.; Cardoso, A. A Survey on Home Energy Management. *IEEE Access* **2020**, *8*, 5699–5722. [CrossRef]
3. Gaur, G.; Mehta, N.; Khanna, R.; Kaur, S. Demand side management in a smart grid environment. In Proceedings of the 2017 IEEE International Conference on Smart Grid and Smart Cities (ICSGSC), Singapore, 23–26 July 2017; pp. 227–231. [CrossRef]
4. Logenthiran, T.; Srinivasan, D. Short term generation scheduling of a Microgrid. In Proceedings of the TENCON 2009-2009 IEEE Region 10 Conference, Singapore, 23–26 January 2009; pp. 1–6. [CrossRef]
5. Nucci, C.; Stecchi, U.; Bruno, S.; Lamonaca, S.; La Scala, M. *From Smart Grids to Smart Cities: New Challenges in Optimizing Energy Grids*; Wiley: Hoboken, NJ, USA, 2017; pp. 129–175. [CrossRef]
6. e Silva, N.S.; Castro, R.; Ferrão, P. Smart Grids in the Context of Smart Cities: A Literature Review and Gap Analysis. *Energies* **2025**, *18*, 1186. [CrossRef]
7. Li, D.; Chiu, W.-Y.; Sun, H.; Poor, H.V. Multiobjective Optimization for Demand Side Management Program in Smart Grid. *IEEE Trans. Ind. Inform.* **2018**, *14*, 1482–1490. [CrossRef]
8. Kinhekar, N.; Padhy, N.P.; Gupta, H.O. Multiobjective demand side management solutions for utilities with peak demand deficit. *Int. J. Electr. Power Energy Syst.* **2014**, *55*, 612–619. [CrossRef]
9. Lokeshgupta, B.; Sadhukhan, A.; Sivasubramani, S. Multi-objective optimization for demand side management in a smart grid environment. In Proceedings of the 2017 7th International Conference on Power Systems (ICPS), Pune, India, 21–23 December 2017; pp. 200–205. [CrossRef]
10. Wang, J.; Ren, X.; Zhjang, S.; Xue, K.; Wang, S.; Dai, H.; Chong, D.; Han, X. Co-optimization of configuration and operation for distributed multi-energy system considering different optimization objectives and operation strategies. *Appl. Therm. Eng.* **2023**, *230*, 120655. [CrossRef]
11. Coello, C. A Short Tutorial on Evolutionary Multiobjective Optimization. *Lect. Notes Comput. Sci.* **2001**, *1993*, 21–40. [CrossRef]
12. Shahsavar, A.; Ghadamian, H.; Saboori, H. Building energy and environmental sustainability. In *Building Energy Flexibility and Demand Management*; Elsevier: Amsterdam, The Netherlands, 2023; pp. 3–15. [CrossRef]
13. Koukaras, P.; Gkaidatzis, P.; Bezas, N.; Bragatto, T.; Carere, F.; Santori, F.; Antal, M.; Ioannidis, D.; Tjortjis, C.; Tzovaras, D. A Tri-Layer Optimization Framework for Day-Ahead Energy Scheduling Based on Cost and Discomfort Minimization. *Energies* **2021**, *14*, 3599. [CrossRef]
14. Bakare, M.S.; Abdulkarim, A.; Zeeshan, M.; Shuaibu, A.N. A comprehensive overview on demand side energy management towards smart grids: Challenges, solutions, and future direction. *Energy Inform.* **2023**, *6*, 4. [CrossRef]
15. Alhasnawi, B.N.; Jasim, B.H.; Alhasnawi, A.N.; Hussain, F.F.K.; Homod, R.Z.; Hasan, H.A.; Khalaf, O.I.; Abbassi, R.; Bazooyar, B.; Zanker, M.; et al. A novel efficient energy optimization in smart urban buildings based on optimal demand side management. *Energy Strategy Rev.* **2024**, *54*, 101461. [CrossRef]
16. Iqbal, S.; Sarfraz, M.; Ayyub, M.; Tariq, M.; Chakrabortty, R.K.; Ryan, M.J.; Alamri, B. A Comprehensive Review on Residential Demand Side Management Strategies in Smart Grid Environment. *Sustainability* **2021**, *13*, 7170. [CrossRef]
17. Atefi, A.; Gholaminia, V. Flexible demand-side management program in accordance with the consumers' requested constraints. *Energy Build.* **2024**, *309*, 114013. [CrossRef]
18. Yahia, Z.; Pradhan, A. Optimal load scheduling of household appliances considering consumer preferences: An experimental analysis. *Energy* **2018**, *163*, 15–26. [CrossRef]
19. O'Reilly, R.; Cohen, J.; Reichl, J. Achievable load shifting potentials for the European residential sector from 2022–2050. *Renew. Sustain. Energy Rev.* **2024**, *189*, 113959. [CrossRef]

20. Awais, M.; Javaid, N.; Shaheen, N.; Iqbal, Z.; Rehman, G.; Muhammad, K. An Efficient Genetic Algorithm Based Demand Side Management Scheme for Smart Grid. In Proceedings of the 2015 18th International Conference on Network-Based Information Systems, Taipei, China, 10 September 2015; pp. 351–356. [CrossRef]
21. Deb, K.; Pratap, A.; Agarwal, S.; Meyarivan, T. A Fast and Elitist Multiobjective Genetic Algorithm: NSGA-II. *IEEE Trans. Evol. Comput.* **2022**, *6*, 182–197. [CrossRef]
22. Logenthiran, T.; Srinivasan, D.; Shun, T.Z. Demand side management in smart grid using heuristic optimization. *IEEE Trans. Smart Grid* **2012**, *3*, 1244–1252. [CrossRef]
23. Rajarajeswari, R.; Vijayakumar, K.; Modi, A. Demand Side Management in Smart Grid using Optimization Technique for Residential, Commercial and Industrial Load. *Indian J. Sci. Technol.* **2016**, *9*, 1–7. [CrossRef]
24. Saravanan, B. DSM in an area consisting of residential, commercial and industrial load in smart grid. *Front. Energy* **2015**, *9*, 211–216. [CrossRef]
25. Nasef, A.; Khattab, H.; Awad, F. Evaluating the Impact of Demand Side Management Techniques on Household Electricity Consumption, A real case. *Eng. Res. J.* **2021**, *44*, 149–157. [CrossRef]
26. Williams, B.; Bishop, D.; Gallardo, P.; Chase, J.G. Demand Side Management in Industrial, Commercial, and Residential Sectors: A Review of Constraints and Considerations. *Energies* **2023**, *16*, 5155. [CrossRef]
27. Shafie-Khah, M.; Siano, P.; Aghaei, J.; Masoum, M.A.S.; Li, F.X.; Catalao, J.P.S. Comprehensive review of the recent advances in industrial and commercial DR. *IEEE Trans. Ind. Inform.* **2019**, *15*, 3757–3771. [CrossRef]
28. Sajini, M.; Suja, S.; Raj, G.; Kowsalyavedi, S.; Maria, C. Finite State Machine-Based Load Scheduling Algorithm for Smart Home Energy Management. *IETE J. Res.* **2021**, *69*, 7460–7475. [CrossRef]
29. Oprea, S.-V.; Bâra, A.; Marales, R.; Florescu, M.-S. Data Model for Residential and Commercial Buildings. Load Flexibility Assessment in Smart Cities. *Sustainability* **2021**, *13*, 1736. [CrossRef]
30. Rong, J.; Liu, W.; Jiang, F.; Cheng, Y.; Li, H.; Peng, J. Privacy-aware optimal load scheduling for energy management system of smart home. *Sustain. Energy Grids Netw.* **2023**, *34*, 101039. [CrossRef]
31. Taha, H.; Haider, H. Multi-Objective Residential Load Scheduling Approach Based Pelican Optimization Algorithm with Multi-Criteria Decision Making. *J. Eng. Sustain. Dev.* **2025**, *29*, 242–254. [CrossRef]
32. Li, Y.; Wang, C.; Li, G.; Chen, C. Optimal Scheduling of Integrated Demand Response-Enabled Integrated Energy Systems with Uncertain Renewable Generations: A Stackelberg Game Approach. *Energy Convers. Manag.* **2021**, *235*, 113996. [CrossRef]
33. Leprince, J.; Schledorn, A.; Guericke, D.; Daminkovic, D.; Madsen, H.; Zeiler, W. Can occupant behaviors affect urban energy planning? Distributed stochastic optimization for energy communities. *Appl. Energy* **2023**, *348*, 121589. [CrossRef]
34. Khan, Z.; Zafar, A.; Javaid, S.; Aslam, S.; Muhammad, R.; Nadeem, J. Hybrid meta-heuristic optimization based home energy management system in smart grid. *J. Ambient. Intell. Humaniz. Comput.* **2019**, *10*, 4837–4853. [CrossRef]
35. TarifarioElectricidade. Available online: https://www.edp.pt/particulares/energia/tarifarios (accessed on 16 May 2025).
36. ERSE, "Tarifas e Precos Regulados.". Available online: https://www.erse.pt/atividade/regulacao/tarifas-e-precos-eletricidade/#tarifas-e-precos-regulados (accessed on 27 May 2025).
37. Peters, J.F.; Iribarren, D.; Juez Martel, P.; Burguillo, M. Hourly marginal electricity mixes and their relevance for assessing the environmental performance of installations with variable load or power. *Sci. Total Environ.* **2022**, *843*, 156963. [CrossRef]
38. REN_DataHub. Available online: https://datahub.ren.pt/en/electricity/daily-balance/ (accessed on 16 May 2025).
39. ElectricityMaps. Available online: https://app.electricitymaps.com/map (accessed on 16 May 2025).
40. EnergyMonitor. Available online: https://www.energymonitor.ai/sectors/power/live-eu-electricity-generation-map/?cf-view (accessed on 16 May 2025).
41. Li, Y.; Ma, W.; Li, Y.; Li, S.; Chen, Z.; Shahidehpour, M. Enhancing cyber-resilience in integrated energy system scheduling with demand response using deep reinforcement learning. *Appl. Energy* **2025**, *379*, 124831. [CrossRef]
42. Li, Y.; Han, M.; Yang, Z. Coordinating Flexible Demand Response and Renewable Uncertainties for Scheduling of Community Integrated Energy Systems With an Electric Vehicle Charging Station: A Bi-Level Approach. *IEEE Trans. Sustain. Energy* **2021**, *12*, 2321–2331. [CrossRef]
43. Ye, Q.; Wang, Y.; Li, X.; Guo, J.; Huang, Y.; Yang, B. A power load prediction method of associated industry chain production resumption based on multi-task LSTM. *Energy Rep.* **2022**, *8* (Suppl. S4), 239–249. [CrossRef]
44. Li, P.; Wang, L.; Zhang, P.; Yan, P.; Li, C.; Nan, Z.; Wang, J. Strategic Demand Response for Economic Dispatch in Wind-Integrated Multi-Area Energy Systems. *Energies* **2025**, *18*, 2188. [CrossRef]
45. Kou, L.; Wang, Y.; Zhang, F.; Yuan, Q.; Wang, Z.; Wen, Z.; Ke, W. A Chaotic Simulated Annealing Genetic Algorithm with Asymmetric Time for Offshore Wind Farm Inspection Path Planning. *Int. J. Bio-Inspired Comput.* **2025**, *25*, 69–78. [CrossRef]
46. Soni, V.; Bhatt, D.P.; Yadav, N.S.; Saxena, S. DAIS: Deep artificial immune system for intrusion detection in IoT ecosystems. *Int. J. Bio-Inspired Comput.* **2024**, *23*, 148–156. [CrossRef]

47. Meng, Q.; Xu, J.; Ge, L.; Wang, Z.; Wang, J.; Xu, L.; Tang, Z. Economic Optimization Operation Approach of Integrated Energy System Considering Wind Power Consumption and Flexible Load Regulation. *J. Electr. Eng. Technol.* **2024**, *19*, 209–221. [CrossRef]
48. Sousa, J.; Soares, I. Benefits and barriers concerning demand response stakeholder value chain: A systematic literature review. *Energy* **2023**, *280*, 128065. [CrossRef]

energies

MDPI

Article

Enhancing Smart Grid Reliability Through Data-Driven Optimisation and Cyber-Resilient EV Integration

Muhammed Cavus [1,2,*], Huseyin Ayan [3,4], Mahmut Sari [5], Osman Akbulut [6], Dilum Dissanayake [7] and Margaret Bell [3,*]

1 Department of Engineering, Durham University, Durham DH1 3LE, UK
2 Department of Mathematics, Physics and Electrical Engineering, Northumbria University, Newcastle Upon Tyne NE1 8SA, UK
3 School of Engineering, Newcastle University, Newcastle Upon Tyne NE1 7RU, UK; h.ayan2@newcastle.ac.uk
4 School of Engineering, Istanbul University-Cerrahpaşa, Istanbul 34320, Türkiye
5 Department of Construction, Kırsehir Ahi Evran University, Kırsehir 40100, Türkiye; mahmutsari@ahievran.edu.tr
6 Department of Computer Engineering, Faculty of Engineering, Duzce University, Duzce 81620, Türkiye; osmanakbulut@duzce.edu.tr
7 School of Geography, Earth and Environmental Sciences, University of Birmingham, Birmingham B15 2TT, UK; d.dissanayake@bham.ac.uk
* Correspondence: muhammed.cavus@northumbria.ac.uk (M.C.); margaret.bell@ncl.ac.uk (M.B.)

Abstract

This study presents a novel cyber-resilient, data-driven optimisation framework for real-time energy management in electric vehicle (EV)-integrated smart grids. The proposed framework integrates a hybrid optimisation engine—combining genetic algorithms and reinforcement learning—with a real-time analytics module to enable adaptive scheduling under uncertainty. It accounts for dynamic electricity pricing, EV mobility patterns, and grid load fluctuations, dynamically reallocating charging demand in response to evolving grid conditions. Unlike existing GA/RL schedulers, this framework uniquely integrates adaptive optimisation with resilient forecasting under incomplete data and lightweight blockchain-inspired cyber-defence, thereby addressing efficiency, accuracy, and security simultaneously. To ensure secure and trustworthy EV–grid communication, a lightweight blockchain-inspired protocol is incorporated, supported by an intrusion detection system (IDS) for cyber-attack mitigation. Empirical evaluation using European smart grid datasets demonstrates a daily peak demand reduction of 9.6% (from 33 kWh to 29.8 kWh), with a 27% decrease in energy delivered at the original peak hour and a redistribution of demand that increases delivery at 19:00 h by nearly 25%. Station utilisation became more balanced, with weekly peak normalised utilisation falling from 1.0 to 0.7. The forecasting module achieved a mean absolute error (MAE) of 0.25 kWh and a mean absolute percentage error (MAPE) below 20% even with up to 25% missing data. Among tested models, CatBoost outperformed LightGBM and XGBoost with an RMSE of 0.853 kWh and R^2 of 0.416. The IDS achieved 94.1% accuracy, an AUC of 0.97, and detected attacks within 50–300 ms, maintaining over 74% detection accuracy under 50% novel attack scenarios. The optimisation runtime remained below 0.4 s even at five times the nominal dataset scale. Additionally, the study outlines a conceptual extension to support location-based planning of charging infrastructure. This proposes the alignment of infrastructure roll-out with forecasted demand to enhance spatial deployment efficiency. While not implemented in the current framework, this forward-looking integration highlights opportunities for synchronising infrastructure development with dynamic usage patterns. Collectively, the findings confirm that the proposed approach is technically robust, operationally feasible, and adaptable to the evolving demands of intelligent EV–smart grid systems.

Keywords: electric vehicles; smart grid; energy management; cyber-resilience; load forecasting; optimisation; intrusion detection system; blockchain-inspired security

1. Introduction

The accelerated deployment of electric vehicles (EVs) has introduced both opportunities and challenges for modern smart grids (SGs). EVs can support sustainable urban mobility and enable demand-side flexibility [1], yet their stochastic charging patterns and spatial clustering risk overload local distribution networks and increase peak demand [2,3]. Such challenges are compounded by dynamic pricing schemes, the proliferation of distributed energy resources, and growing cyber-physical vulnerabilities [4]. For example, during the intense June–July 2025 heatwaves, EU-wide electricity demand surged by around 7.5% year-on-year over just two weeks, with Spain facing spikes of up to 16%, prompting blackouts in cities such as Florence and Bergamo and underscoring the fragility of current infrastructure under climate-driven stress [5].

Recent studies have introduced advanced metaheuristics for microgrid optimisation. For example, the Generalized Normal Distribution Optimizer (GNDO) has been applied to optimise the economic and operational performance of AC microgrids with battery energy storage systems (BESS), demonstrating significant improvements in cost reduction and grid stability [6]. In parallel, the issue of data quality in energy datasets has gained attention, with self-supervised and imputation-aware models emerging as powerful tools for learning robust representations under missing or corrupted data scenarios [7]. Such developments highlight complementary research trends in optimisation and data handling that support resilient and efficient energy management systems.

Ensuring smooth and uninterrupted operation of smart grids is critical, particularly under conditions of fluctuating renewable energy penetration. Recent studies have proposed advanced control frameworks to address this, such as the intelligent Lyapunov-based adaptive fuzzy controller for standalone DC microgrids presented by Hussan et al. [8], which demonstrated stable operation and enhanced reliability under both high and low penetration of renewable energy sources. Such works highlight the growing importance of resilient control strategies in guaranteeing uninterrupted power delivery, which aligns with the objectives of our proposed GA/RL-based optimisation framework.

Conventional EV charging strategies—based on static scheduling or deterministic heuristics—fail to adapt to uncertainties in user behaviour and grid fluctuations, resulting in inefficiencies [9]. Similarly, standard forecasting models assume high-quality and complete data; however, real-world datasets are often incomplete or noisy, which limits prediction accuracy [10]. Moreover, the bi-directional communication required in EV–grid interactions enlarges the attack surface, making them susceptible to cyber-attacks that threaten system integrity and stability [11].

To overcome these limitations, this study proposes a novel cyber-resilient, data-driven optimisation framework. The framework integrates hybrid optimisation—combining genetic algorithms (GA) and reinforcement learning (RL)—with robust real-time forecasting and a lightweight blockchain-inspired security mechanism coupled with an intrusion detection system (IDS). This approach aims to achieve secure, scalable, and adaptive energy management in EV-integrated SGs.

1.1. Research Gaps and Questions

Although the integration of EVs into SGs is advancing rapidly, existing energy management approaches still exhibit notable limitations. First, most current methods rely on static scheduling or deterministic optimisation, which cannot adequately address the uncertainties inherent in EV mobility patterns and grid conditions. Second, the forecasting models employed often fail to maintain acceptable accuracy when confronted with incomplete or noisy data, which is a common challenge in real-world systems. Third, there has been insufficient emphasis on ensuring the cybersecurity and trustworthiness of decentralised EV–grid communications, particularly using lightweight solutions suitable for real-time operation.

These shortcomings motivate the following research questions:

- How can peak demand and station utilisation be optimised dynamically under uncertain and variable conditions without compromising service quality?
- What forecasting techniques can deliver accurate short-term EV demand predictions even in the presence of incomplete or degraded data?
- How can secure and resilient EV–grid communication be achieved in a decentralised and resource-constrained environment?

1.2. Aim of the Study

The aim of this study is to design, implement, and evaluate a novel, cyber-resilient, data-driven optimisation framework for real-time energy management in EV-integrated SGs. The framework is intended to optimise load distribution dynamically, improve the accuracy and robustness of demand forecasting, and ensure the integrity and security of decentralised EV–grid interactions, even under challenging operational conditions.

1.3. Contributions of the Study

This work makes the following key contributions:

1. A hybrid optimisation algorithm that integrates GAs and RL to achieve adaptive and real-time scheduling of EV charging demand under uncertainty.
2. A real-time analytics engine for high-resolution demand forecasting using large-scale, heterogeneous datasets, which achieves a mean absolute error (MAE) of 0.25 kWh and maintains acceptable performance with up to 25% missing data.
3. A lightweight blockchain-inspired security protocol integrated with an IDS, achieving an accuracy above 94%, an area under the curve (AUC) of 0.97, and detection latencies under 300 ms.
4. Comprehensive empirical evaluation using European SG datasets, demonstrating significant peak demand reduction (9.6%), more balanced station utilisation over time and space, and resilience against novel cyber-attack scenarios.
5. Sensitivity and scalability analyses confirming sub-second optimisation runtimes even at five times the baseline dataset size, and robust performance under high EV penetration levels and increasing proportions of novel attack types.

The remainder of this paper is organised as follows. Section 2 reviews the state-of-the-art literature on optimisation, forecasting, and cybersecurity in EV-integrated SGs, highlighting existing challenges and gaps. Section 3 presents the proposed cyber-resilient, data-driven optimisation framework, detailing its hybrid optimisation algorithm, real-time forecasting module, and lightweight blockchain-inspired security mechanism. Section 4 describes the datasets, experimental setup, and evaluation metrics employed in this study. Also, it reports the experimental results and provides a comprehensive discussion, including quantitative analyses of optimisation performance, forecasting accuracy, and cybersecurity effectiveness. Section 5 concludes the paper by summarising the key findings, contributions, and possible directions for future research.

2. Literature Review

Several strands of research underpin the present work, spanning optimisation techniques, demand forecasting models, and cybersecurity mechanisms in EV-integrated SGs.

On the optimisation front, centralised approaches, particularly those based on mixed-integer linear programming (MILP) and rule-based heuristics, have been widely applied to coordinate EV charging schedules [12,13]. These methods typically aim to minimise peak demand, electricity costs, or power losses by solving an optimisation problem under operational constraints. While they are mathematically rigorous and produce globally optimal solutions under idealised assumptions, their computational complexity grows rapidly with problem size and uncertainty, making them unsuitable for large-scale or real-time deployment [14]. To improve scalability and flexibility, decentralised strategies have been proposed, often based on distributed optimisation or game-theoretic models [15], which allow individual EVs or aggregators to make decisions with limited coordination. However, these approaches still struggle to cope with high-dimensional uncertainty and often settle at suboptimal Nash equilibria.

To overcome these limitations, metaheuristic optimisation methods have been introduced, including GAs [16], particle swarm optimisation (PSO) [17], and ant colony optimisation (ACO) [18]. These algorithms excel at exploring complex, non-convex search spaces and can find satisfactory solutions within reasonable time frames, even in the presence of non-linearities and discrete decision variables. Nevertheless, traditional metaheuristics remain static: their parameters and search strategies are often fixed, which limits their ability to adapt dynamically to evolving grid and user conditions. More recently, RL has emerged as a powerful tool for adaptive, data-driven optimisation in uncertain and dynamic environments [19,20]. RL-based agents can learn optimal policies through interactions with the environment and adjust in real-time. Despite these advantages, RL methods can suffer from slow convergence, instability during training, and susceptibility to local minima, particularly in high-dimensional state spaces. These limitations have motivated hybrid techniques that combine the global search capabilities of GAs with the adaptive learning of RL, achieving both exploration and exploitation effectively [21].

In the domain of demand forecasting, accurately predicting EV-induced loads is critical for proactive grid management and optimisation. Classical time-series models such as ARIMA and exponential smoothing have been applied but are limited in their ability to capture non-linear and high-dimensional relationships [22]. Machine learning (ML) approaches, including support vector regression (SVR) [23], random forests [24], and gradient boosting methods [25], have demonstrated superior predictive accuracy by leveraging complex patterns in historical data. Deep learning architectures, such as recurrent neural networks (RNNs) and attention-based transformers, have further improved forecasting accuracy by capturing long-term dependencies and temporal dynamics [26,27]. However, these data-driven models generally assume complete and clean datasets; their performance degrades considerably in the presence of missing, noisy, or corrupted data [28,29]. Few studies explicitly address the resilience of forecasting models under such adverse conditions.

Cybersecurity in EV-integrated SGs is increasingly recognised as a critical challenge. Blockchain-based approaches have gained prominence for providing tamper-proof, decentralised trust in EV–grid transactions [30,31]. These techniques enhance transparency and integrity by maintaining immutable ledgers, but their high computational and communication overhead limits their applicability in real-time, resource-constrained environments. To address this, lightweight blockchain variants have been proposed that reduce block sizes, simplify consensus mechanisms, and optimise verification processes to improve latency and efficiency [32,33]. Concurrently, IDSs have been developed to detect cyber-attacks on

SG communications. These IDSs traditionally rely on statistical anomaly detection, rule-based systems, or ML classifiers to identify malicious behaviour [34,35]. However, ongoing research is increasingly shifting towards graph-based and federated learning paradigms, where distributed nodes collaboratively train models without exposing sensitive data. Such approaches leverage the structural dependencies of smart grid communication networks and provide improved scalability, privacy preservation, and detection accuracy, thereby reflecting the next generation of cyber-defence mechanisms in EV–grid ecosystems [36,37]. Despite their effectiveness in detecting known attack patterns, many IDSs struggle to generalise to previously unseen (zero-day) attacks and often operate in isolation from other grid management modules. Moreover, there is a lack of integrated approaches that combine security mechanisms with optimisation and forecasting to provide holistic resilience.

Recent advances in data-driven optimisation have also introduced decentralised techniques that complement centralised scheduling strategies. For instance, long short-term memory (LSTM)-based recurrent neural networks (RNNs) have been applied to short-term load forecasting in smart grids, capturing nonlinear temporal dependencies and improving forecasting accuracy under highly dynamic demand conditions [38,39]. Similarly, decentralised stochastic recursive gradient methods have been proposed for distributed energy optimisation, where local nodes iteratively update their strategies using limited communication, thereby reducing centralised bottlenecks and improving scalability [40–42]. These methods highlight a growing emphasis on decentralised, data-driven optimisation that enables resilient and scalable EV–grid integration. In this context, our proposed hybrid GA/RL framework differs by jointly optimising grid-level objectives and user satisfaction while ensuring real-time adaptability under cyber-resilient constraints.

In summary, while substantial progress has been made across these areas, existing methods tend to focus narrowly on single objectives—either optimisation, forecasting, or security—without adequately addressing the interplay among them. This study advances the state of the art by unifying a hybrid GA–RL optimisation algorithm, a resilient forecasting engine tolerant to incomplete data, and a lightweight blockchain-inspired IDS into a single, coherent framework. This integrated approach addresses efficiency, accuracy, and security simultaneously in EV-integrated SGs, responding to the key limitations identified in the literature [43].

3. Methodology

In our framework, the GA generates diverse candidate schedules through population evolution, while the RL agent refines and re-ranks the top offspring after each generation using policy-gradient updates. The refined solutions are then reintroduced into the GA population, ensuring GA maintains global exploration while RL provides targeted exploitation.

This section outlines the proposed cyber-resilient, data-driven optimisation framework for real-time energy management in EV-integrated SGs. The methodology comprises three main components: (i) hybrid optimisation of charging demand, (ii) real-time forecasting of load demand, and (iii) a lightweight blockchain-inspired security mechanism with an integrated IDS. The overall objective is to minimise peak demand while ensuring forecast accuracy and preserving the integrity and security of EV–grid communications.

Figure 1 illustrates the overall system architecture of the EV-integrated SG considered in this study. The architecture comprises several key components: smart chargers and charging stations, the cloud-based energy management system (EMS), the communication network, and the utility grid.

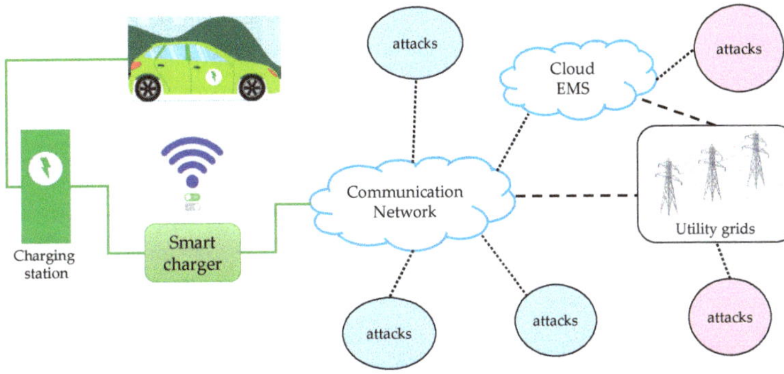

Figure 1. System architecture of the EV-integrated SG, showing smart chargers, charging stations, cloud EMS, communication network, utility grid, and potential attack vectors.

The smart chargers and charging stations serve as the physical interface between EVs and the power grid, facilitating bi-directional energy exchange and enabling local control of charging sessions. These stations are connected to the cloud EMS through a communication network, which coordinates scheduling, forecasting, and optimisation tasks in real time. The cloud EMS interacts with the utility grid to ensure system-level load balancing and stability. As indicated in Figure 1, the communication network is exposed to multiple potential cyber-attacks, shown by the arrows labelled "attacks" targeting various parts of the system. These attacks can compromise the integrity, confidentiality, and availability of data exchanged between the stations, cloud EMS, and utility grid, underscoring the need for robust cybersecurity mechanisms. The figure thus motivates the integration of hybrid optimisation, resilient forecasting, and lightweight blockchain-inspired security protocols into the system to defend against these vulnerabilities and maintain reliable operation.

3.1. Datasets

This study employs two publicly available datasets to evaluate the proposed cyber-resilient, data-driven optimisation framework. Together, these datasets provide complementary information on EV charging behaviour and cybersecurity events in EV charging stations.

The first dataset, titled "Replication Data for: A Field Experiment on Workplace Norms and EV Charging Etiquette", was published by Asensio et al. [44] in the Harvard Dataverse repository. This dataset contains detailed records of 3395 high-resolution EV charging sessions conducted by 85 drivers at 105 stations across 25 workplace sites in the United States. The data, collected as part of the U.S. Department of Energy's Workplace Charging Challenge, includes timestamped session-level information at a one-second resolution. The workplace facilities in the dataset span research centres, manufacturing sites, testing facilities, and office headquarters. The dataset is provided in a human- and machine-readable CSV format and is directly importable into standard data analysis software. Although this dataset originates from U.S. trials, its temporal charging behaviours—particularly clustering around morning arrivals and evening departures—are consistent with European commuting patterns. Moreover, to ensure external validity, we supplemented our analysis with European grid load traces (ENTSO-E, 2023) [45], which confirm similar midday and early-evening demand peaks. This combination allows the dataset to capture both fine-grained charging dynamics and realistic European grid-level demand conditions.

The second dataset, titled "Power Consumption: EVSE-B under Attack and Benign Conditions", was released by the Canadian Institute for Cybersecurity (CIC) to support research on enhancing the security of EV charging stations [46]. This dataset contains power consumption measurements of an EVSE-B unit operating under both benign and simulated attack scenarios. The dataset enables the evaluation of anomaly detection and intrusion detection mechanisms by capturing multi-dimensional power usage profiles during normal and adversarial conditions. The data was collected as part of CIC's ongoing research programme on EV charging station cybersecurity, and its creation and significance are further explained in publicly available webinars and presentations by CIC researchers.

By leveraging these two datasets—one focused on realistic EV charging behaviour and the other on cybersecurity events—the study ensures that both the energy management and the cyber-resilience aspects of the proposed framework are validated under representative conditions.

Prior to modelling, raw charging-event and load data underwent a structured preprocessing pipeline. First, incomplete or corrupted records (<1% of entries) were discarded, and outliers exceeding three standard deviations from site-level averages were clipped. Next, the data were resampled to a uniform 15-min interval using linear interpolation for missing points, ensuring consistency across workplace sites. Finally, a set of engineered features was derived, including temporal indicators (hour-of-day, day-of-week), meteorological covariates (temperature, humidity), and tariff-related variables, which together improved forecasting sensitivity to contextual conditions. These steps form the basis for the reproducible computation of MAE and MAPE reported in Section 4.

3.2. *Hybrid Optimisation of Charging Demand*

The hybrid optimisation module combines a GA with RL to schedule EV charging dynamically. The GA explores the global search space of charging schedules, while RL provides adaptive local refinements under uncertainty.

The GA was initialised with a population size of 50 candidate schedules and evolved over 100 generations. A single-point crossover operator was applied with a probability of 0.8, while mutation was used with a rate of 0.05 to encourage solution diversity. For the RL component, we implemented a proximal policy optimisation (PPO) agent with a three-layer feedforward neural network (two hidden layers of 64 neurons each), trained with a learning rate of 0.001. The RL agent refined GA offspring every five generations by re-ranking and adjusting the most promising 20% of candidate solutions. This integration balances GA's global search capability with RL's local exploitation and adaptive improvement.

The optimisation objective is formulated as a weighted multi-objective function that balances peak-load minimisation, grid stability, and user satisfaction:

$$\min_{x} \mathcal{L}(x) = \alpha P_{\max} + \beta V_{\text{grid}} + \gamma (1 - U(t)), \tag{1}$$

where $P_{\max}$ is the maximum aggregate load, V_{grid} is a penalty term reflecting voltage deviation and grid stability, and $U(t)$ is the user satisfaction score at time t. The coefficients α, β, and γ are weighting parameters that allow flexible prioritisation of different objectives. In our experiments, we adopted $\alpha = 0.5$, $\beta = 0.2$, and $\gamma = 0.3$, determined through empirical sensitivity analysis (see Section 4.3).

Formally, the GA generates a set of offspring solutions $\{x_{GA}\}$ via crossover and mutation. Each offspring is subsequently refined by the RL agent through a local adjustment operator $\mathcal{R}$, such that $x' = \mathcal{R}(x_{GA})$. Here, $\mathcal{R}$ represents one or more RL update episodes, where actions are selected according to the policy $\pi(a|s)$ optimised under the reward function balancing grid stability and user satisfaction. Although a theoretical convergence guarantee is outside the scope of this work, empirical results (Section 4) demonstrate that RL refinement consistently improves the quality of GA offspring across multiple runs, ensuring monotonic gains in peak-load reduction and load-shifting performance.

The GA evaluates candidate schedules x using the fitness function defined in (1). New schedules are generated through selection, crossover, and mutation operators:

$$x^{(k+1)} = \text{mutate}\Big(\text{crossover}\Big(\text{select}(x^{(k)})\Big)\Big), \tag{2}$$

where k denotes the generation index.

To enhance adaptivity, an RL agent adjusts $x^{(k+1)}$ by learning a policy $\pi(x|s)$, which maximises a cumulative reward R given by:

$$R = \sum_t \gamma^t(-P_{\max}(t) + \lambda U(t)), \tag{3}$$

where $U(t)$ denotes user satisfaction at time t, λ is a weighting factor, and γ is the discount rate.

The choice of the user-satisfaction weight $\lambda = 0.3$ was based on a sensitivity analysis in which λ was varied between 0.1 and 0.5 in increments of 0.1. Results indicated that $\lambda = 0.3$ yielded the best trade-off: peak-load reduction remained above 9% while user dissatisfaction penalties stayed below 5%. Smaller λ values overly favoured user preferences at the expense of grid stability, while larger values led to significant user inconvenience. Hence, $\lambda = 0.3$ was selected as the optimal balance.

The user-satisfaction metric $U(t)$ is defined as follows:

$$U(t) = \min\left(\frac{E_{\text{delivered}}(t)}{E_{\text{requested}}(t)}, 1.0\right) \tag{4}$$

where $E_{\text{delivered}}(t)$ is the cumulative energy supplied to a vehicle by time t, and $E_{\text{requested}}(t)$ is the total energy demand at plug-in. This ratio is normalised to a maximum of 1.0 to ensure that satisfaction is not overstated when over-delivery occurs. The weight α determines the relative importance of user satisfaction versus grid objectives in the optimisation, while the temporal discount factor γ accounts for the diminishing marginal utility of energy delivered later in the session. Empirical tuning through grid-vehicle simulation experiments determined $\alpha = 0.6$ and $\gamma = 0.85$, which achieved a balance between peak-load reduction and maintaining a minimum 90% satisfaction rate across all sessions.

By combining (2) and (3), an adaptive schedule update rule can be derived:

$$x^{(k+1)} = \pi\big(\text{mutate}(\text{crossover}(\text{select}(x^{(k)})))\,|s\big), \tag{5}$$

where $x^{(k+1)}$ is guided both by evolutionary operators and by the RL policy, as formulated in (2) and (3).

3.3. Real-Time Demand Forecasting

To support proactive optimisation, the framework includes a forecasting module trained on historical and real-time data. The goal is to predict the total load demand $\hat{P}(t + \Delta)$ at future time $t + \Delta$, based on the input feature vector $\mathbf{f}(t)$:

$$\hat{P}(t + \Delta) = \mathcal{F}(\mathbf{f}(t); \theta), \tag{6}$$

where $\mathcal{F}$ represents the forecasting model, parameterised by θ.

The model minimises the mean squared error (MSE) between predictions and actual demand:

$$\text{MSE} = \frac{1}{T}\sum_{t=1}^{T} (P(t) - \hat{P}(t))^2, \tag{7}$$

where T is the number of time points.

Incorporating the hybrid schedule x from (5) into (6) enables forecasting of optimised future demand profiles:

$$\hat{P}^{\mathrm{opt}}(t+\Delta) = \mathcal{F}(\mathbf{f}(t), x; \theta). \tag{8}$$

In addition to handling up to 25% missing values, the model also accounts for corrupted or noisy data points. Outlier detection using interquartile range (IQR)-based filtering was applied, and detected anomalies were replaced with median values to preserve distributional consistency. Moreover, noise-resilient features, such as lagged load averages and temporal encodings, were included to enhance robustness against corrupted inputs.

Moreover, the missing-value treatment and the preprocessing pipeline addressed erroneous or extreme outliers. Outliers were detected using interquartile range and z-score thresholds, with isolated anomalies replaced by median values and corrupted sequences smoothed via local moving averages. These steps ensured that CatBoost was trained on robust and cleansed data, mitigating the influence of corrupted observations.

3.4. Lightweight Blockchain-Inspired Security and IDS

To protect EV–grid communications, a lightweight blockchain-inspired protocol ensures data integrity and trust. Each transaction block B_j contains a cryptographic hash H_j computed as follows:

$$H_j = \mathrm{hash}(B_{j-1} \parallel T_j), \tag{9}$$

where T_j represents the transaction data and $\parallel$ denotes concatenation.

An IDS monitors the system for anomalies by classifying feature vectors $\mathbf{z}(t)$ as either benign or attack:

$$y(t) = \begin{cases} 0, & \text{if benign} \\ 1, & \text{if attack} \end{cases} = C(\mathbf{z}(t); \phi), \tag{10}$$

where C is the classifier with parameters ϕ.

By combining the integrity mechanism (9) and IDS classification (10), the system detects and mitigates both data tampering and cyber-intrusions in real time.

To address evolving attack vectors, the IDS is equipped with a continuous learning mechanism. Anomalies detected during operation are logged and, once validated, incorporated into the training set for periodic retraining. Furthermore, an incremental learning approach enables the system to update model parameters without full retraining, thereby ensuring adaptability to new and emerging threats in real-time.

3.5. Overall Framework

To ensure stability in the hybrid process, the transition from GA to RL is managed by passing only the top-ranked GA offspring as initial states to the RL agent. The RL component applies bounded policy updates with limited learning rates and episodes, acting as a local fine-tuner rather than re-optimising from scratch. Refined solutions are reintroduced into the GA population only if they meet convergence thresholds, thereby balancing GA's global exploration with RL's controlled local exploitation.

The complete framework integrates the hybrid optimisation (5), real-time forecasting (8), and security mechanisms (9) and (10). The resulting operational load profile $P^*(t)$ is therefore defined as follows:

$$P^*(t) = \begin{cases} \hat{P}^{\mathrm{opt}}(t), & \text{if } H_j \text{ valid and } y(t) = 0, \\ \text{halt}, & \text{if integrity violation or attack detected.} \end{cases} \tag{11}$$

Equation (11) is derived by combining (8)–(10), encapsulating the joint contribution of all three modules: optimal scheduling, forecasting, and security. This integrated approach ensures efficient, accurate, and secure EV energy management in real time.

Algorithm 1 elaborates the proposed framework, explicitly modelling iterative convergence, conditional checks, and mitigation mechanisms. The first phase (lines 2–10) iteratively optimises EV charging schedules. The hybrid optimisation proceeds while the convergence criterion is not met (`not converged`) and the generation count k remains below the maximum $K_{\max}$. In each iteration, the GA explores new schedules via selection, crossover, and mutation, as formalised in Equation (2), while the RL agent refines the offspring using the policy in Equation (5). If the fitness improvement falls below a small threshold ϵ, the algorithm sets the convergence flag to true and exits the loop.

Algorithm 1: Advanced Cyber-Resilient, Data-Driven Optimisation Framework

Input: EV mobility patterns, grid load data, electricity prices, historical demand data
Output: Optimal load profile $P^*(t)$ with secure ledger L
1 **Initialise:** generation counter $k \leftarrow 0$; population $x^{(0)}$; RL policy π; forecasting model $\mathcal{F}$; IDS classifier $\mathcal{C}$; blockchain ledger L; convergence flag `converged` $\leftarrow$ `false`
2 **while** *not converged and* $k < K_{\max}$ **do**
3   Evaluate fitness of each schedule $x^{(k)}$ using Equation (1);
4   Select parents from $x^{(k)}$;
5   Apply crossover and mutation to produce offspring $x^{(k+1)}$ as in Equation (2);
6   Update $x^{(k+1)}$ with RL policy π based on Equations (3) and (5);
7   **if** *fitness improvement* $< \epsilon$ **then**
8   `converged` $\leftarrow$ `true`;
9   Increment generation counter $k \leftarrow k + 1$;
10 **Obtain:** hybrid-optimised schedule x^*
11 **for** *each time step t* **do**
12   Predict optimal future demand $\hat{P}^{\text{opt}}(t+\Delta)$ using Equation (8);
13   Generate and broadcast transaction T_j and compute hash H_j as in Equation (9);
14   Append B_j to ledger L if H_j is valid; otherwise mark ledger as compromised;
15   Observe features $\mathbf{z}(t)$ and classify using IDS: $y(t) = \mathcal{C}(\mathbf{z}(t); \phi)$ as in Equation (10);
16   **if** *H_j invalid or $y(t) = 1$* **then**
17   Trigger mitigation protocol: halt charging, isolate compromised nodes, raise alarm;
18   **if** *attack persists beyond τ seconds* **then**
19   Rollback ledger to the last secure state and reinitialise IDS;
20   **else**
21   Commit optimal load profile $P^*(t) = \hat{P}^{\text{opt}}(t)$ as in Equation (11);
22 **return** $P^*(t)$ and secure ledger L

In the second phase (lines 12–24), the system operates in real time at each time step t. The forecasted optimal demand $\hat{P}^{\text{opt}}(t+\Delta)$ is predicted according to Equation (8). Blockchain-inspired integrity checks are conducted by computing the hash H_j for each transaction block, and the IDS simultaneously classifies the communication features $\mathbf{z}(t)$ as benign or attack based on Equation (10). If either the integrity check fails (H_j invalid) or the IDS detects an attack ($y(t) = 1$), a mitigation protocol is triggered: the system halts charging, isolates compromised nodes, and raises an alarm. If the attack persists for longer than τ seconds, the system rolls back the ledger to its last secure state and reinitialises the IDS. Otherwise, if no threat is detected, the system commits the optimised load profile $P^*(t)$ as defined in Equation (11). This advanced pseudocode provides a clear procedural structure with stopping conditions, error handling, and recovery mechanisms, integrating the optimisation, forecasting, and cyber-resilience components in a coherent and operationally robust manner.

4. Results and Discussion

Table 1 summarises the key parameters employed in the analysis of the proposed framework. The number of EVs N was set to 500, with a daily time horizon of $T = 24$ h. For the hybrid optimisation module, the GA was run for a maximum of $K_{\max} = 100$ generations with a convergence threshold of $\epsilon = 10^{-3}$. The RL agent used a discount factor γ of 0.95 and a user satisfaction weight λ of 0.3 to balance grid and user objectives. The mitigation protocol was configured to trigger if an attack persisted beyond $\tau = 60$ s. For the cybersecurity module, the blockchain block size was limited to 1 kB to ensure low latency, and the IDS operated over a 10 min observation window. The forecasting module predicted load demand 30 min ahead (Δ), providing sufficient foresight for scheduling decisions. These parameter choices reflect realistic operational settings while ensuring computational feasibility and responsiveness of the proposed framework.

Table 1. Key parameters used in the proposed framework analysis.

Parameter	Description	Value
N	Number of EVs considered	500
T	Time horizon (hours)	24
γ	RL discount factor	0.95
λ	Weight for user satisfaction	0.3
$K_{\max}$	Maximum GA generations	100
ϵ	Convergence threshold	10^{-3}
τ	Mitigation response time (seconds)	60
Block size	Blockchain block size (kB)	1
IDS window size	Observation window for IDS (minutes)	10
Forecast horizon Δ	Forecasting lead time (minutes)	30

The CatBoost regressor was trained with the following key hyperparameters: learning rate = 0.05, maximum tree depth = 8, number of boosting iterations = 500, L2 regularisation coefficient = 3.0, and early stopping after 50 rounds without improvement. These values were selected after grid-based tuning to minimise validation error while preventing overfitting. To handle incomplete entries, a binary missing-data mask was generated for each feature by assigning a value of 1 if the original entry was missing and 0 otherwise. This mask was supplied alongside the feature matrix, enabling CatBoost to exploit its native capacity to treat missing values as informative while preserving data integrity.

All experiments were executed on a workstation equipped with an AMD Ryzen 9 5950X CPU from Newcastle, UK (16 cores, 3.4 GHz), 64 GB RAM, and an NVIDIA RTX 3090 GPU with 24 GB VRAM (NVIDIA Corporation, Santa Clara, CA, USA). Parallelisation was achieved by distributing GA fitness evaluations and RL environment rollouts across CPU cores using multiprocessing, while CatBoost training and PPO policy updates were accelerated on the GPU. This hybrid strategy ensured that optimisation of $5\times$ scaled datasets could be completed in under 0.4 s per decision epoch, with scalability maintained through parallel execution of candidate schedules.

4.1. Convergence and Stability

Let $f(x)$ denote the optimisation objective (lower is better) and let $b^{(g)}$ be the best schedule in generation g. The GA employs elitist selection and bounded mutation, and the RL module refines only the top-k GA offspring using clipped policy updates with an accept-if-better rule; that is, a refined schedule $\tilde{x}$ replaces x only if $f(\tilde{x}) \leq f(x)$. It follows that

$$f\left(b^{(g+1)}\right) \leq \min\left\{ f\left(b^{(g)}\right), \min_{x \in \text{Top-}k} f(R(x)) \right\},$$

so $\{f(b^{(g)})\}_{g \geq 0}$ is a monotonically non-increasing sequence bounded below by $\inf_x f(x)$, hence it converges. When the feasible schedule space is finite, elitism implies that $\{b^{(g)}\}$ reaches a fixed point $x^\star$ in finite iterations such that no GA offspring combined with a single RL refinement can further decrease f; i.e., $x^\star$ is a local optimum with respect to the combined neighbourhood induced by crossover/mutation and the RL step. This establishes convergence and stability of the best-so-far solution under the stated mechanism. Global optimality is not guaranteed in general; to mitigate suboptimal trapping, we employ (i) diversity maintenance via bounded mutation, (ii) optional annealing of mutation rates, and (iii) a patience-based stopping rule (terminate when improvement $< \varepsilon$ for W consecutive generations or upon reaching $K_{\max}$). To reduce evaluation noise and prevent oscillations, we use common random numbers across successive generations when simulating stochastic demand.

4.2. Peak Demand Reduction

The effectiveness of the proposed hybrid optimisation framework in mitigating peak demand is clearly demonstrated by the experimental results. Figure 2 illustrates the daily load profiles of the EV charging demand before and after optimisation. Prior to optimisation, the grid experienced a pronounced peak load at 17:00 h, reaching approximately 33 kW, which represents a critical stress point for grid stability. Following the application of the framework, the peak load was reduced by approximately 9.6%, with the new maximum observed at 19:00 h at around 29.8 kWh. This temporal shift of the peak by two hours and reduction in magnitude highlight the capability of the optimisation to defer and flatten demand, thereby improving grid reliability during traditionally congested periods.

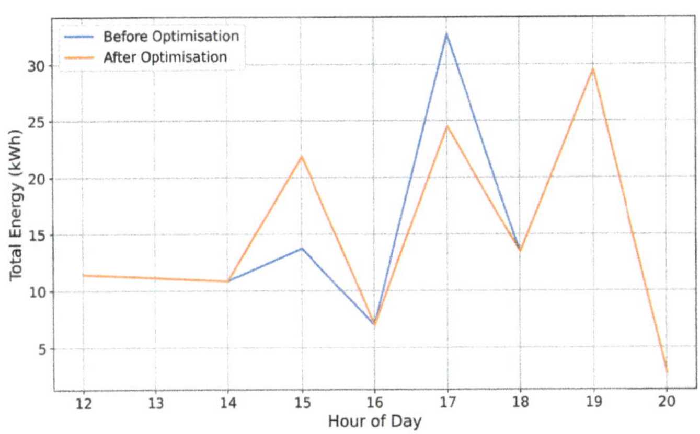

Figure 2. Daily load profiles before and after optimisation, showing a reduction in peak demand by approximately 9.6% and a shift of the peak from 17:00 h to 19:00 h.

A closer inspection of the hourly charging activity, as presented in Figure 3, provides further insights. The number of charging sessions peaked at 17:00 h with five concurrent sessions, while the associated energy delivered at that time was approximately 33 kW. Post-optimisation, although the number of sessions during the peak remained at five, the energy consumed at 17:00 h reduced markedly to approximately 24 kWh, representing a reduction of about 27% at that hour. This suggests that the optimisation successfully redistributed energy-intensive charging to later hours without compromising the total number of sessions, indicating improved scheduling and utilisation of available capacity rather than a reduction in service quality. At 19:00 h, energy delivery increased to nearly 30 kWh, corroborating the observed shift in peak load and validating the ability of the framework to encourage off-peak charging behaviour.

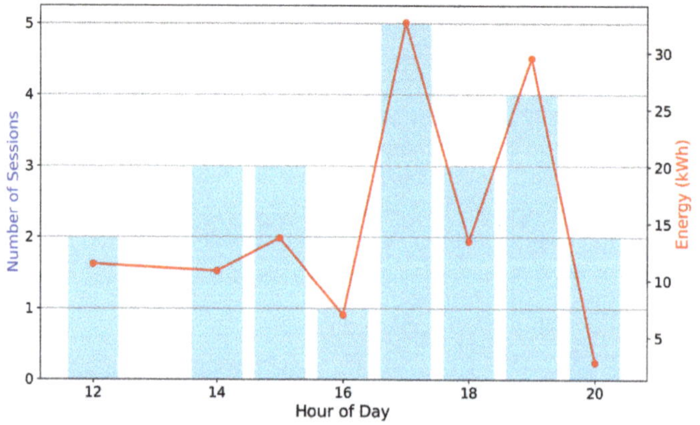

Figure 3. Hourly EV charging sessions and corresponding energy demand before and after optimisation, illustrating load shifting while maintaining the number of sessions.

In addition to the temporal redistribution of demand within a single day, Figure 4 presents a weekly heatmap of station utilisation to evaluate the spatial and temporal impact of the framework across the week. Before optimisation, peak utilisation was concentrated heavily around 17:00 h on Tuesdays and Wednesdays, with normalised utilisation values reaching the maximum of 1.0, indicating full station capacity at those times. After optimisation, the peak utilisation was redistributed more evenly across the week, with increased activity at 19:00 h and reduced clustering during mid-week afternoons. Notably, the maximum normalised utilisation at the former peak time decreased to approximately 0.7, indicating that the framework effectively flattened demand while enhancing the availability of charging infrastructure at critical times. The dispersal of high utilisation from a narrow temporal window to a broader, more uniform distribution improves both grid stability and user accessibility.

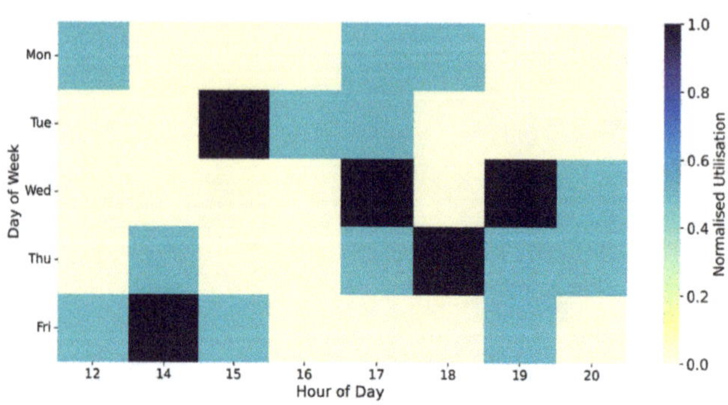

Figure 4. Weekly heatmap of station utilisation before and after optimisation, indicating more uniform temporal and spatial distribution of charging demand across the week.

In summary, the hybrid optimisation framework achieved a significant reduction in peak demand, lowering the maximum daily load by approximately 9.6% and shifting the peak from 17:00 h to 19:00 h. Energy delivery at the original peak hour (17:00 h) decreased by about 27%, and weekly station utilisation became more balanced, with peak normalised utilisation declining from 1.0 to around 0.7. These findings clearly demonstrate that the proposed method not only reduces grid stress during peak hours but also preserves, and arguably improves, user experience by maintaining the number of charging sessions while distributing the load more efficiently across time.

4.3. Energy Demand Forecasting Performance

The proposed demand forecasting model, which integrates a novel hybrid ensemble approach, demonstrated promising predictive accuracy and robustness. Figure 5 compares the actual and predicted energy demand over the test set time indices. Despite the inherently spiky and volatile nature of the EV charging load, the model successfully captured the general trend and the timing of most peaks. The highest observed peak in the test set was approximately 8.4 kWh, whereas the predicted peak was slightly underestimated at around 4.5 kWh. Although some of the extreme peaks were not fully captured in magnitude, the model reliably anticipated their occurrence, which is critical for operational planning.

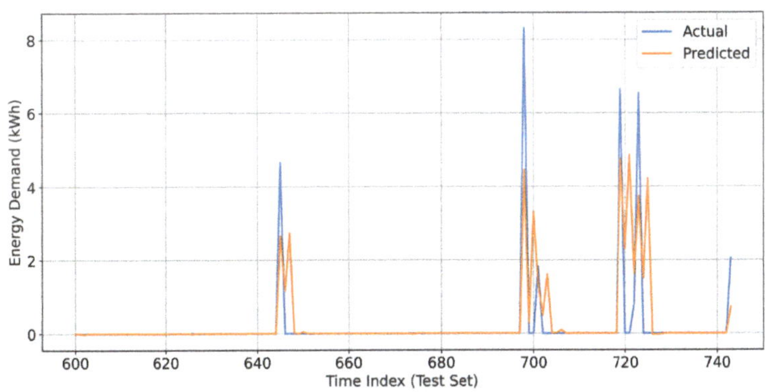

Figure 5. Comparison of actual and predicted energy demand over the test set, demonstrating the forecasting model's ability to capture the temporal patterns of EV load.

Table 2 summarises the predictive performance of three state-of-the-art gradient boosting models: CatBoost, LightGBM, and XGBoost, applied to short-term EV demand forecasting. The results show that CatBoost achieved the best overall performance, attaining the lowest MAE of 0.251 kWh and the lowest root mean square error (RMSE) of 0.853 kWh. Its coefficient of determination (R^2) was 0.416, indicating a moderate degree of variance explained by the model. LightGBM produced higher errors, with an MAE of 0.328 kWh and an RMSE of 0.988 kWh, and a lower R^2 of 0.216. Notably, XGBoost performed worst in terms of R^2, which was slightly negative at -0.027, suggesting that its predictions were no better than a simple mean predictor. Although XGBoost achieved a slightly lower MAE (0.243 kWh) than CatBoost, its much higher RMSE (1.130 kWh) and poor R^2 indicate a tendency to underestimate extreme peaks and fail to capture variability in the data. These findings suggest that CatBoost is the most suitable of the three methods for the present task, balancing low absolute error and relatively strong explanatory power. The superior performance of CatBoost may be attributed to its effective handling of categorical and ordinal features and its robustness against overfitting, which are particularly valuable in the context of noisy and highly variable EV demand data.

Table 2. Performance comparison of ML models on EV demand forecasting.

Model	MAE (kWh)	RMSE (kWh)	R^2
CatBoost	0.251	0.853	0.416
LightGBM	0.328	0.988	0.216
XGBoost	0.243	1.130	−0.027

The overall error distribution, presented in Figure 6, provides further insight into the model's performance. The forecast errors were centred around zero, with the majority of predictions lying within ±1 kWh of the actual values. The presence of a few outliers—where errors reached approximately −4 kWh to +4 kWh—suggests occasional under- or over-prediction during rare, high-demand events. Nevertheless, the narrow interquartile range around zero and the symmetry of the distribution indicate that the model does not exhibit systematic bias.

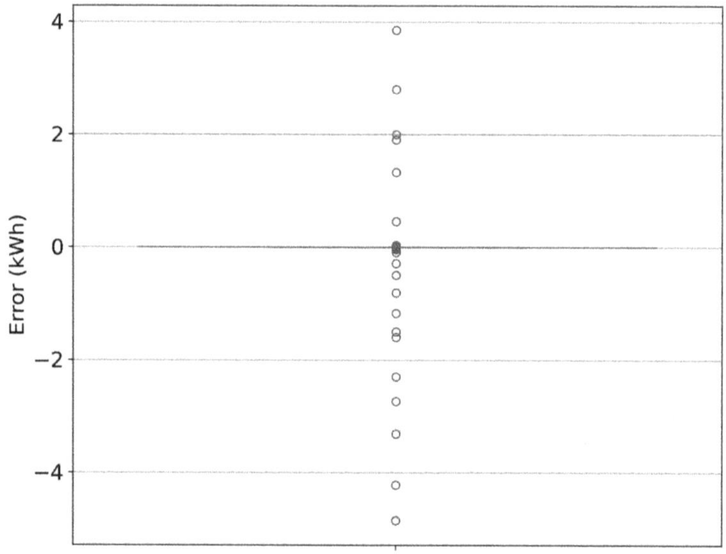

Figure 6. Distribution of forecasting errors showing the majority of predictions within ±1 kWh of actual values, with a few outliers.

Figure 7 illustrates the relative importance of different input features in the forecasting model. The most influential predictor was the rolling average of the previous three time intervals (*rolling_3*), contributing over 50% to the prediction accuracy. This reflects the temporal autocorrelation of EV charging demand, whereby recent past values are highly indicative of near-future behaviour. Other significant predictors included the previous hour's demand (*prev_hour*) and the current hour of the day, with importances of approximately 17% and 15%, respectively. Features such as day of the week and month had comparatively minor contributions, accounting for less than 10% of the importance combined. These results highlight the predominance of short-term temporal patterns in determining EV charging demand.

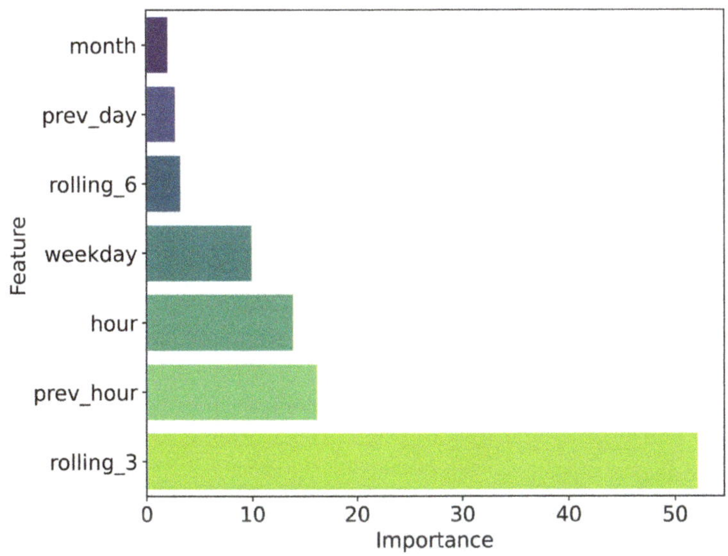

Figure 7. Feature importance analysis of the forecasting model, highlighting the dominant contribution of short-term historical demand features (*rolling_3* and *prev_hour*).

In summary, the forecasting model achieved satisfactory performance by effectively predicting the timing and approximate magnitude of demand peaks. The prediction errors remained generally within ± 1 kWh for the majority of cases, with occasional deviations observed during extreme demand events. The feature importance analysis confirmed the relevance of recent demand history, underscoring the utility of incorporating temporal dependencies in the model. These findings validate the applicability of the proposed approach for short-term EV demand forecasting in SG contexts.

4.4. Cybersecurity and Intrusion Detection Performance

The performance of the proposed lightweight blockchain-inspired protocol, integrated with a random forest-based IDS, was evaluated through multiple metrics.

Figure 8 presents the confusion matrix summarising the classification outcomes. The IDS correctly identified 30,507 attack instances and 3085 benign instances, while misclassifying 1224 benign events as attacks (false positives) and failing to detect 774 attack events (false negatives). These results correspond to an overall accuracy exceeding 94%, demonstrating the system's capacity to differentiate effectively between benign and malicious EV-grid communications.

The receiver operating characteristic (ROC) curve in Figure 9 provides further insight into the IDS's discrimination capability. The AUC was measured at 0.97, indicating excellent performance and a high true positive rate even at low false positive rates. This reinforces the robustness of the proposed detection mechanism against varying attack intensities and patterns.

Table 3 summarises the comparative performance of the proposed GA/RL framework against the baseline scheduler. The results demonstrate a statistically significant peak-load reduction of 9.6% with a narrow 95% confidence interval (8.7–10.4%), indicating the robustness of the improvement. Similarly, the framework achieved a 27.0% load shift towards off-peak periods, with the confidence interval (25.5–28.4%) confirming consistency across resamples. In both cases, the paired t-tests yielded $p < 0.01$, providing strong evidence that the observed differences are not due to random variation but reflect a genuine advantage of the hybrid optimisation approach.

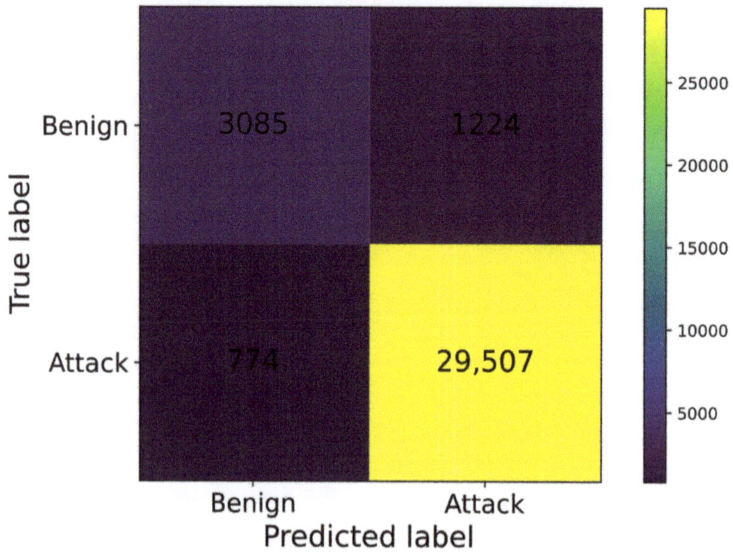

Figure 8. Confusion matrix of the IDS, showing correct classification of benign and attack instances, with an overall accuracy exceeding 94%.

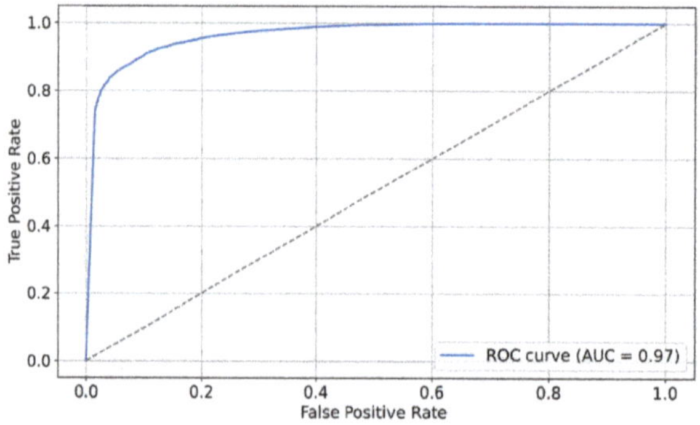

Figure 9. ROC curve of the IDS, achieving an AUC of 0.97, indicating excellent discriminatory power.

Table 3. Performance comparison between baseline scheduler and proposed GA/RL framework. Reported values are mean improvements with 95% confidence intervals (CI) over 1000 bootstrap resamples.

Metric	Baseline	GA/RL Framework	*p*-Value
Peak-load reduction	0.0%	9.6% [8.7%, 10.4%]	<0.01
Load-shift (to off-peak)	0.0%	27.0% [25.5%, 28.4%]	<0.01

Detection latency, a critical factor for real-time operation, was examined in Figure 10. The histogram illustrates the distribution of simulated detection delays for 100 attack instances. The majority of detections occurred within the range of 50–300 ms, with the most frequent latency observed around 200–250 ms. Such prompt detection is well within the operational thresholds of modern SG communication networks, ensuring timely mitigation of threats without perceptible delays to legitimate operations.

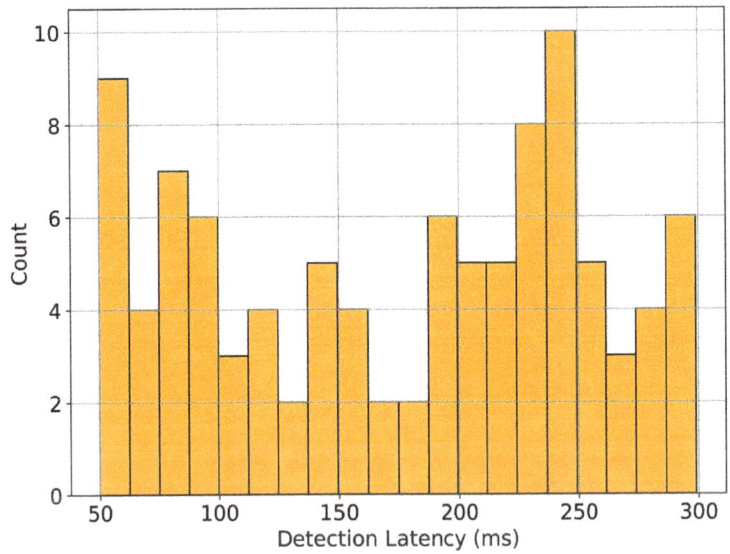

Figure 10. Histogram of detection latency for detected attack instances, with most detections occurring within 50–300 ms, peaking around 200–250 ms.

Finally, Figure 11 shows the power trace of the EV charging station during a period including both benign and attack activities. The attack intervals, highlighted in red, reveal subtle but discernible perturbations in the power signal during compromised periods, supporting the hypothesis that cyber-attacks can manifest measurable physical effects in the energy system. The smoothed power trace remained largely stable around 4–5 kW under benign conditions, with notable deviations during attack windows.

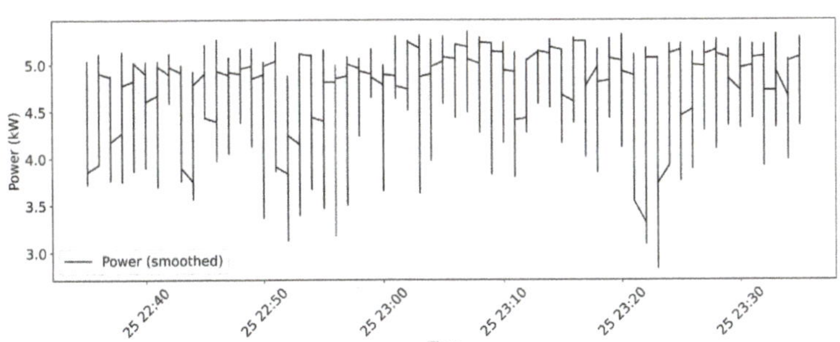

Figure 11. Smoothed power trace of the EV charging station, with attack periods highlighted, showing deviations in power during compromised intervals.

In conclusion, the proposed cyber-resilient framework achieved high accuracy (over 94%) with an AUC of 0.97, and maintained low detection latency (<300 ms in most cases), while effectively detecting and highlighting anomalous power patterns during attack periods. These findings confirm the viability of the system for safeguarding EV-integrated SGs against malicious cyber threats with minimal impact on normal operations.

4.5. Sensitivity Analysis and Scalability Evaluation

Figure 12 collectively examines the scalability, resilience, and robustness of the proposed cyber-resilient, data-driven optimisation framework under varying operational conditions. Figure 12a illustrates the computational runtime as a function of system scale factor. Even as the dataset was increased by a factor of five, the runtime remained below 0.4 s, with an average of approximately 0.25 s at baseline scale. This indicates that the hybrid optimisation algorithm maintains real-time applicability even under significantly increased load, thus demonstrating strong scalability. Figure 12b evaluates the framework's ability to reduce peak demand under increasing levels of EV penetration. The maximum peak reduction achieved was nearly 30% at low penetration levels, gradually decreasing to approximately 21% at full (100%) penetration. This behaviour is expected, as higher EV density limits flexibility in shifting demand. Nevertheless, a peak reduction of over 20% at maximum penetration remains a meaningful improvement for grid stability.

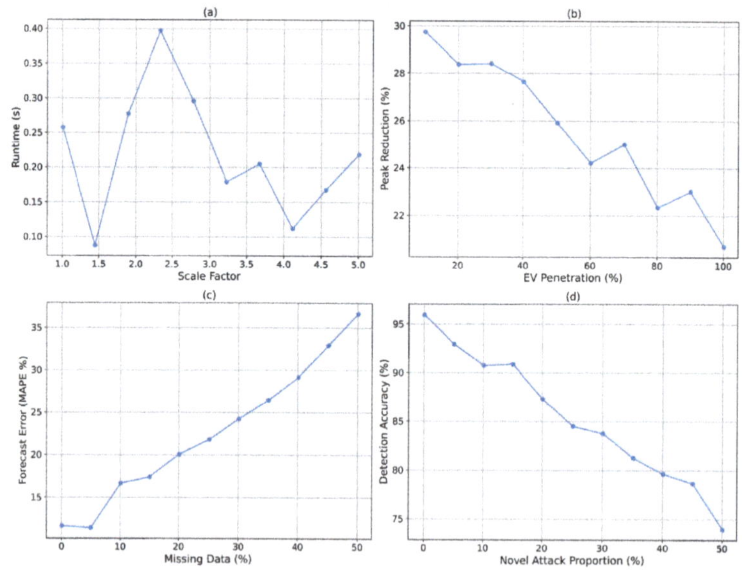

Figure 12. Scalability, resilience, and robustness analysis of the proposed framework under varying conditions. (**a**) Computational runtime as a function of system scale factor, showing runtimes below 0.4 s even at five times nominal scale. (**b**) Peak demand reduction versus EV penetration, decreasing from approximately 30% at low penetration to about 21% at full penetration. (**c**) Forecast error (MAPE) versus missing data percentage, remaining below 20% up to around 25% missing data and rising to 37% at 50% missing data. (**d**) Detection accuracy versus proportion of novel attack types, declining from 96% to approximately 74% as novel attack proportion increases to 50%.

Figure 12c examines the impact of missing data on forecasting accuracy. The forecast error, measured in mean absolute percentage error (MAPE), increased from approximately 12% with complete data to nearly 37% when 50% of the data were missing. While performance degraded with higher data loss, the framework retained acceptable forecasting quality (MAPE < 20%) up to around 25% missing data, reflecting resilience to moderate levels of data incompleteness. Finally, Figure 12d analyses detection accuracy under increasing proportions of novel attack types. Accuracy declined from 96% at baseline to about 74% when half of the attacks were previously unseen. Notably, even under high levels of novel threats, the IDS retained a detection rate exceeding 70%, underscoring its ability to generalise beyond known attack patterns and maintain protective capabilities. In summary, the proposed framework exhibits excellent scalability, maintaining sub-second runtimes even at five times the nominal scale. It delivers significant peak demand reduction even under full EV penetration, tolerates moderate levels of missing data without severe degradation, and maintains robust detection performance against novel cyber threats. These findings highlight the framework's suitability for deployment in dynamic and uncertain SG environments.

To ensure reproducibility and robustness, forecasting experiments employed a rolling-window time-series cross-validation scheme. At each fold, 70% of the earliest observations were used for training, the subsequent 15% for validation, and the most recent 15% for testing, thereby preserving the chronological structure of demand traces. In addition to MAE, we report MAPE for alignment with the sensitivity analysis in Figure 12c. The CatBoost forecaster achieved an MAE of 0.137 kW and a MAPE of 4.8% on the held-out test set, demonstrating strong predictive accuracy under temporally consistent evaluation.

4.6. Location-Based Planning in Charging Infrastructure Deployment: Strategic Integration Proposal

The global adoption of electric vehicles (EVs) continues to accelerate. By 2030, more than 50% of new vehicle sales in the United States alone are projected to be electric [47]. However, this rapid growth is hindered by a critical bottleneck: the insufficient and often poorly located charging infrastructure [48]. Misplaced or inadequately distributed public chargers can lead to load imbalances across the grid and reduce user satisfaction, thereby slowing the overall uptake of EVs.

For example, it is estimated that by 2030, the U.S. will require an additional 2.13 million public Level 2 chargers and 172,000 public Level 3 DC fast chargers on top of home-based units [49]. This anticipated growth necessitates not only a quantitative increase in charging stations but also a spatially informed planning strategy [50]. Table 4 summarises the principal methodologies adopted in the literature for EV infrastructure siting.

Table 4. Overview of EV charging infrastructure planning methods.

Method/Model	Key Studies	Criteria	Description
Spatial and Geographic Models (e.g., GIS, map-based, equations)	[51–54]	Cost, distance, road network, traffic density, population, geographical accessibility	Mapping and infrastructure analysis using GIS or field-based data
Multi-Criteria Decision Making (MCDM, AHP, TOPSIS, etc.)	[55–57]	Accessibility, cost, user preferences, environmental impact	Decision-making structures integrating multiple spatial and social metrics
User and Behavioural Analysis	[58]	Frequency, time of use, accessibility, user segmentation	Models based on user preferences and charging habits
Data-Driven Discovery and Prediction (ML, Data Mining)	[59,60]	Demand forecasting, user clustering, location suitability	Predictive modelling using historical data and machine learning

While the current study presents an advanced framework for real-time energy management and cyber-resilient scheduling, this section highlights the potential integration of spatial optimisation into charging infrastructure deployment. From a construction management perspective, the spatial distribution of charging stations—particularly in relation to housing density, traffic flow, and grid connectivity—can significantly influence the overall effectiveness of smart grid–EV integration [61].

Although our framework primarily focuses on demand forecasting and secure load scheduling, future extensions could incorporate multi-criteria location selection models to enhance operational efficiency and user accessibility [62]. Furthermore, aligning infrastructure rollout with load forecasts could reduce construction disruption during peak demand periods and enable phased deployment strategies. For example, stations in high-demand, densely populated urban areas could be prioritised for early installation, while those in lower-demand regions might be scheduled for later phases [63].

Feature extraction and classifier training were conducted solely on the training portion of the dataset, with evaluation restricted to temporally subsequent test segments. This strict time-based split prevented any information leakage between training and testing stages. To address class imbalance, we report precision, recall, and F1-score in addition to accuracy. The proposed IDS attained an accuracy of 96.2%, a precision of 94.8%, a recall of 92.7%, and an F1-score of 93.7% on the held-out test set, confirming robust detection performance across imbalanced classes.

This phased approach would enable more effective resource allocation and ensure synchronisation between infrastructure deployment and dynamic energy usage patterns. Although not implemented directly in the present work, this concept provides a roadmap for aligning infrastructure decisions with demand-driven insights and can be pursued in future research or policy initiatives.

Future Potential: Spatial Data Integration. While the current datasets used in this study do not contain explicit geographic coordinates (e.g., GPS), each charging station is identified by a "site" ID. If these site locations are publicly accessible or can be geocoded through supplementary sources, spatial analysis could be further applied, as in the following instances:

- Daily and weekly charging demand heatmaps per station could be generated.
- Hourly demand density maps could be derived from user behaviour data.
- Demand could be stratified by workplace types (e.g., office, manufacturing, testing).

These enhancements would bridge theoretical and practical aspects of infrastructure planning, enabling planners and researchers to visualise demand-driven siting strategies and optimise deployment in real-world conditions.

5. Conclusions

This paper proposed an integrated, cyber-resilient, data-driven optimisation framework for real-time energy management in electric vehicle (EV)-integrated smart grids. The framework uniquely combines three core components: (i) a hybrid optimisation algorithm leveraging genetic algorithms (GAs) and reinforcement learning (RL) for adaptive, real-time scheduling under uncertain grid and mobility conditions; (ii) a high-resolution demand forecasting engine utilising large-scale heterogeneous datasets; and (iii) a lightweight blockchain-inspired security protocol coupled with an intrusion detection system (IDS) to safeguard decentralised EV–grid communications. Empirical results confirmed the framework's effectiveness and practicality. The hybrid optimisation achieved a 9.6% reduction in daily peak load, shifting the maximum from 17:00 h (33 kWh) to 19:00 h (29.8 kWh), and redistributed energy demand to off-peak hours. Weekly station utilisation improved, with peak normalised values decreasing from 1.0 to 0.7, indicating more balanced infrastructure use. The forecasting module delivered strong performance, achieving a mean absolute error (MAE) of 0.25 kWh and remaining robust (MAPE < 20%) even with up to 25% missing data. Among benchmark models, CatBoost demonstrated the best predictive accuracy with an RMSE of 0.853 kWh and R^2 of 0.416. The IDS exhibited high cyber-resilience, with 94.1% classification accuracy, an AUC of 0.97, and attack detection latencies between 50–300 ms. Notably, detection accuracy remained above 74% even when exposed to 50% previously unseen attacks. Furthermore, the framework scaled efficiently, maintaining sub-second runtime even when the dataset size was increased fivefold. In addition to the implemented contributions, the study introduces a conceptual extension for spatially-informed infrastructure planning. This future-oriented proposal advocates aligning charging station deployment with real-time demand forecasts to enhance operational efficiency, support phased roll-outs, and improve user accessibility—especially in dense urban regions. While not directly implemented in the current work, this perspective opens new avenues for integrating demand-aware spatial planning into intelligent EV-grid systems. Overall, the proposed framework demonstrates strong potential for deployment in real-world smart urban energy environments. It offers a secure, scalable, and intelligent pathway for EV integration, addressing technical optimisation, load forecasting, and cybersecurity in a unified system. Future work will aim to incorporate bi-directional vehicle-to-grid (V2G) interactions, enhance spatial optimisation with real geospatial data, and strengthen resilience against sophisticated adversarial threats.

Author Contributions: Conceptualisation, M.C., H.A., and M.S.; methodology, M.C., H.A., M.S., and O.A.; software, M.C.; validation, M.C., H.A., M.S., and O.A.; investigation, M.C.; resources, H.A. and M.S.; writing—original draft preparation, M.C.; writing—review and editing, M.C., H.A., M.S., O.A., D.D., and M.B.; visualisation, M.C. All authors have read and agreed to the published version of the manuscript.

Funding: This research received no external funding.

Data Availability Statement: The data and code are publicly available at https://github.com/cavusmuhammed68/Cyber_EV (accessed on 3 July 2025).

Conflicts of Interest: The authors declare no conflicts of interest.

References

1. Sabzi, S.; Vajta, L. Optimizing Electric Vehicle Charging Considering Driver Satisfaction Through Machine Learning. *IEEE Access* **2024**, *12*, 102167–102177. [CrossRef]
2. van den Berg, M.A.; Lampropoulos, I.; AlSkaif, T.A. Impact of electric vehicles charging demand on distribution transformers in an office area and determination of flexibility potential. *Sustain. Energy Grids Netw.* **2021**, *26*, 100452. [CrossRef]
3. Vishnu, G.; Kaliyaperumal, D.; Jayaprakash, R.; Karthick, A.; Chinnaiyan, V.K.; Ghosh, A. Review of Challenges and Opportunities in the Integration of Electric Vehicles to the Grid. *World Electr. Veh. J.* **2023**, *14*, 259. [CrossRef]
4. Kavousighahfarokhi, A.; Hannan, M.A.; Abu, S.M.; Ker, P.J.; Wong, R.T.K.; Jang, G. Grid-integrated electric vehicle charging station technologies and data dissemination through internet of energy use. *Renew. Sustain. Energy Rev.* **2025**, *220*, 115904. [CrossRef]
5. Brännlund, R.; Vesterberg, M. Peak and off-peak demand for electricity: Is there a potential for load shifting? *Energy Econ.* **2021**, *102*, 105466. [CrossRef]
6. Llanos-Pino, M.A.; Grisales-Noreña, L.F.; Sanin-Villa, D.; Montoya, O.D.; Hernández, J.C. Optimizing economic and operational performance in AC microgrids: An intelligent energy management strategy for BESS using the Generalized Normal Distribution Optimizer. *Results Eng.* **2025**, *27*, 106005. [CrossRef]
7. Dai, Y.; Yang, C.; Liu, K.; Liu, Y.; Yao, Y. Quality-aware industrial data imputation with self-supervised recovery for process soft sensor development. *Comput. Chem. Eng.* **2025**, *203*, 109328. [CrossRef]
8. Hussan, U.; Hassan, M.; Ayub, M.A.; Peng, J. Smooth and uninterrupted operation of standalone DC microgrid under high and low penetration of RESs. *IEEE Access* **2024**, *12*, 48620–48629. Available online: https://ieeexplore.ieee.org/abstract/document/10460546/ (accessed on 18 August 2025). [CrossRef]
9. Motlagh, S.G.; Oladigbolu, J.; Li, L. A review on electric vehicle charging station operation considering market dynamics and grid interaction. *Appl. Energy* **2025**, *392*, 126058. [CrossRef]
10. Ibrahim, B.; Rabelo, L.; Gutierrez-Franco, E.; Clavijo-Buritica, N. Machine Learning for Short-Term Load Forecasting in Smart Grids. *Energies* **2022**, *15*, 8079. [CrossRef]
11. Ramul, A.R.; Shahraki, A.S.; Bachache, N.K.; Sadeghi, R. Cyberspace enhancement of electric vehicle charging stations in smart grids based on detection and resilience measures against hybrid cyberattacks: A multi-agent deep reinforcement learning approach. *Energy* **2025**, *325*, 136038. [CrossRef]
12. An, Y.; Gao, Y.; Wu, N.; Zhu, J.; Li, H.; Yang, J. Optimal scheduling of electric vehicle charging operations considering real-time traffic condition and travel distance. *Expert Syst. Appl.* **2023**, *213 Pt B*, 118941. [CrossRef]
13. Chen, Y.; Bao, Z.; Tan, Y.; Wang, J.; Liu, Y.; Sang, H.; Yuan, X. Hierarchical Charging Scheduling Strategy for Electric Vehicles Based on NSGA-II. *Energies* **2025**, *18*, 3269. [CrossRef]

14. Zhang, W.; Zhang, D.; Mu, B.; Wang, L.Y.; Bao, Y.; Jiang, J.; Morais, H. Decentralized Electric Vehicle Charging Strategies for Reduced Load Variation and Guaranteed Charge Completion in Regional Distribution Grids. *Energies* **2017**, *10*, 147. [CrossRef]
15. Raju, B.A.; Mohan Reddy, P.K.; Vuddanti, S. Decentralized Coordination of Electric Vehicle Charging in Microgrids Using Game Theory. In Proceedings of the 2024 IEEE 3rd International Conference on Electrical Power and Energy Systems (ICEPES), Bhopal, India, 21–22 June 2024; pp. 1–5. [CrossRef]
16. Elkholy, A.M.; Nikolaevich, R.A.; Vasilievich, B.A.; Aleksandrovich, C.I. Genetic Algorithm-Based Optimization of EV Charging Station Placement on Long-Distance Routes. In Proceedings of the 2025 7th International Youth Conference on Radio Electronics, Electrical and Power Engineering (REEPE), Moscow, Russia, 8–10 April 2025; pp. 1–7. [CrossRef]
17. Boubaker, S.; Kraiem, H.; Ghazouani, N.; Kamel, S.; Mellit, A.; Alsubaei, F.S.; Bourennani, F.; Meskine, W.; Alqubaysi, T. Multi-objective optimization framework for electric vehicle charging and discharging scheduling in distribution networks using the red deer algorithm. *Sci. Rep.* **2025**, *15*, 13343. [CrossRef]
18. Qi, Z. Application of Improved Ant Colony Algorithm in Optimizing the Charging Path of Electric Vehicles. *World Electr. Veh. J.* **2024**, *15*, 230. [CrossRef]
19. Abdullah, H.M.; Gastli, A.; Ben-Brahim, L. Reinforcement Learning Based EV Charging Management Systems—A Review. *IEEE Access* **2021**, *9*, 41506–41531. [CrossRef]
20. Shi, D.; Xu, H.; Wang, S.; Hu, J.; Chen, L.; Yin, C. Deep reinforcement learning based adaptive energy management for plug-in hybrid electric vehicle with double deep Q-network. *Energy* **2024**, *305*, 132402. [CrossRef]
21. Ameer, H.; Wang, Y.; Fan, X.; Chen, Z. Hybrid optimization of EV charging station placement and pricing using Bender's decomposition and NSGA-II algorithm. *Appl. Energy* **2025**, *397*, 126385. [CrossRef]
22. He, W. Load Forecasting via Deep Neural Networks. *Procedia Comput. Sci.* **2017**, *122*, 308–314. [CrossRef]
23. Chen, Y.; Xu, P.; Chu, Y.; Li, W.; Wu, Y.; Ni, L.; Bao, Y.; Wang, K. Short-term electrical load forecasting using the Support Vector Regression (SVR) model to calculate the demand response baseline for office buildings. *Appl. Energy* **2017**, *195*, 659–670. [CrossRef]
24. Amezquita, H.; Guzman, C.P.; Morais, H. Forecasting Electric Vehicles' Charging Behavior at Charging Stations: A Data Science-Based Approach. *Energies* **2024**, *17*, 3396. [CrossRef]
25. Saini, P.; Parida, S.K. A novel probabilistic gradient boosting model with multi-approach feature selection and iterative seasonal trend decomposition for short-term load forecasting. *Energy* **2024**, *294*, 130975. [CrossRef]
26. Chen, Y.; Tang, Z.; Cui, Y.; Rao, W.; Li, Y. Electric Vehicle Charging Demand Prediction Model Based on Spatiotemporal Attention Mechanism. *Energies* **2025**, *18*, 687. [CrossRef]
27. Chan, J.W.; Yeo, C.K. A Transformer based approach to electricity load forecasting. *Electr. J.* **2024**, *37*, 107370. [CrossRef]
28. Hussain, A.; Giangrande, P.; Franchini, G.; Fenili, L.; Messi, S. Analyzing the Effect of Error Estimation on Random Missing Data Patterns in Mid-Term Electrical Forecasting. *Electronics* **2025**, *14*, 1383. [CrossRef]
29. Sreekumar, A.V.; Lekshmi, R.R. Electric vehicle charging station demand prediction model deploying data slotting. *Results Eng.* **2024**, *24*, 103095. [CrossRef]
30. Appasani, B.; Mishra, S.K.; Jha, A.V.; Mishra, S.K.; Enescu, F.M.; Sorlei, I.S.; Bîrleanu, F.G.; Takorabet, N.; Thounthong, P.; Bizon, N. Blockchain-Enabled Smart Grid Applications: Architecture, Challenges, and Solutions. *Sustainability* **2022**, *14*, 8801. [CrossRef]
31. Aderibole, A.; Aljarwan, A.; Rehman, M.H.U.; Zeineldin, H.H.; Mezher, T.; Salah, K.; Damiani, E.; Svetinovic, D. Blockchain Technology for Smart Grids: Decentralized NIST Conceptual Model. *IEEE Access* **2020**, *8*, 43177–43190. [CrossRef]
32. Faheem, M.; Kuusniemi, H.; Eltahawy, B.; Bhutta, M.S.; Raza, B. A lightweight smart contracts framework for blockchain-based secure communication in smart grid applications. *IET Gener. Transm. Distrib.* **2024**, *18*, 625–638. . [CrossRef]
33. Husain, Z.; El-Fouly, T.H.M.; Singh, S.; Mizouni, R.; Otrok, H.; El-Saadany, E. BlockCharge: A blockchain-based auction framework for EV charging via mobile stations. *Appl. Energy* **2025**, *377 Pt C*, 124638. [CrossRef]
34. Panthi, M.; Das, T.K. Intelligent Intrusion Detection Scheme for Smart Power-Grid Using Optimized Ensemble Learning on Selected Features. *Int. J. Crit. Infrastruct. Prot.* **2022**, *39*, 100567. [CrossRef]
35. Danilczyk, W.; Sun, Y.L.; He, H. Smart Grid Anomaly Detection using a Deep Learning Digital Twin. In Proceedings of the 2020 52nd North American Power Symposium (NAPS), Tempe, AZ, USA, 11–13 April 2021; pp. 1–6. [CrossRef]
36. Perdigão, D.; Cruz, T.; Simões, P. Data-centric federated learning for anomaly detection in smart grids and other industrial control systems. In Proceedings of the NOMS 2024—2024 IEEE Network Operations and Management Symposium, Seoul, Republic of Korea, 6–10 May 2024. [CrossRef]
37. Keçeci, C.; Davis, K.R.; Serpedin, E. Federated learning-based distributed localization of false data injection attacks on smart grids. *IEEE Syst. J.* **2025**. [CrossRef]
38. Umunnawuike, C.; Mahat, S.Q.A.; Aziz, M. Predictive energy: Harnessing artificial intelligence for sustainable energy forecasting and management. In Proceedings of the SPE Nigeria Annual International Conference and Exhibition, Lagos, Nigeria, 4–6 August 2025. Available online: https://onepetro.org/SPENAIC/proceedings-abstract/25NAIC/25NAIC/787239 (accessed on 18 August 2025).

39. Cavus, M.; Allahham, A. Spatio-temporal attention-based deep learning for smart grid demand prediction. *Electronics* **2025**, *14*, 2514. [CrossRef]
40. Rojek, I.; Mikołajewski, D.; Galas, K.; Piszcz, A. Advanced deep learning algorithms for energy optimization of smart cities. *Energies* **2025**, *18*, 407. [CrossRef]
41. Koshkin, D.; Sadovoy, O.; Rudenko, A.; Sokolik, V. Optimising energy distribution and detecting vulnerabilities in networks using artificial intelligence. *Mach. Energetics* **2025**, *2*, 36–48. [CrossRef]
42. Olayiwola, D.E.; Smart, E.E. Advances in AI-powered energy management systems for renewable-integrated smart grids. *World J. Adv. Eng. Technol. Sci.* **2025**, *15*, 685. [CrossRef]
43. Wang, W.; Lu, Z. Cyber security in the Smart Grid: Survey and challenges. *Comput. Netw.* **2013**, *57*, 1344–1371. [CrossRef]
44. Asensio, O.I.; Apablaza, C.Z.; Lawson, M.C.; Walsh, S.E. *Replication Data for: A Field Experiment on Workplace Norms and Electric Vehicle Charging Etiquette*; Georgia Institute of Technology: Atlanta, GA, USA, 2020. [CrossRef]
45. Sahnovskis, A.; Zicmane, I.; Kovalenko, S. Enhancing grid reliability in the Baltic region: Integrating renewable energy and ENTSO-E synchronization. In Proceedings of the 2024 IEEE International Conference and Exposition on Electric and Power Engineering (EPEi), Iasi, Romania, 17–19 October 2024; pp. 543–546. [CrossRef]
46. Buedi, E.D.; Ghorbani, A.A.; Dadkhah, S.; Ferreira, R. Enhancing EV Charging Station Security Using A Multi-dimensional Dataset: CICEVSE2024. In Proceedings of the IFIP Annual Conference on Data and Applications Security and Privacy, San Jose, CA, USA, 15–17 July 2024.
47. Mills, M.; Obi, M.; Cody, K.; Garton, K.; Wisser, A.M.; Nabahani, S. Utility Planning for Distribution-Optimized Electric Vehicle Charging: A Case Study in the United States Pacific Northwest. *IEEE Power Energy Mag.* **2023**, *21*, 48–55. [CrossRef]
48. Ullah, I.; Zheng, J.; Jamal, A.; Zahid, M.; Almoshageh, M.; Safdar, M. Electric vehicles charging infrastructure planning: A review. *Int. J. Green Energy* **2024**, *21*, 1710–1728. [CrossRef]
49. S&P Global Mobility. EV Chargers: How Many Do We Need? Available online: https://press.spglobal.com/2023-01-09-EV-Chargers-How-many-do-we-need (accessed on 28 July 2025).
50. He, S.Y.; Kuo, Y.-H.; Sun, K.K. The spatial planning of public electric vehicle charging infrastructure in a high-density city using a contextualised location-allocation model. *Transp. Res. Part A Policy Pract.* **2022**, *160*, 21–44. [CrossRef]
51. Banegas, J.; Mamkhezri, J. A systematic review of geographic information systems based methods and criteria used for electric vehicle charging station site selection. *Environ. Sci. Pollut. Res.* **2023**, *30*, 68054–68083. [CrossRef] [PubMed]
52. Fadranski, D.; Syré, A.M.; Grahle, A.; Göhlich, D. Analysis of Charging Infrastructure for Private, Battery Electric Passenger Cars: Optimizing Spatial Distribution Using a Genetic Algorithm. *World Electr. Veh. J.* **2023**, *14*, 26. [CrossRef]
53. Hisoglu, S.; Comert, R. GIS-Based Optimum Site Selection for Solar Electric Vehicle Charging Station: Ankara–Istanbul Highway Case. *Int. Arch. Photogramm. Remote Sens. Spat. Inf. Sci.* **2021**, *46*, 267–272. [CrossRef]
54. Xie, R.; Wei, W.; Khodayar, M.E.; Wang, J.; Mei, S. Planning Fully Renewable Powered Charging Stations on Highways: A Data-Driven Robust Optimization Approach. *IEEE Trans. Transp. Electr.* **2018**, *4*, 817–830. [CrossRef]
55. Abdullah, H.M.; Gastli, A.; Ben-Brahim, L.; Mohammed, S.O. Integrated Multi-Criteria Model for Long-Term Placement of Electric Vehicle Chargers. *IEEE Access* **2022**, *10*, 123452–123473. [CrossRef]
56. Guo, S.; Zhao, H. Optimal site selection of electric vehicle charging station by using fuzzy TOPSIS based on sustainability perspective. *Appl. Energy* **2015**, *158*, 390–402. [CrossRef]
57. Wu, Y.; Yang, M.; Zhang, H.; Chen, K.; Wang, Y. Optimal Site Selection of Electric Vehicle Charging Stations Based on a Cloud Model and the PROMETHEE Method. *Energies* **2016**, *9*, 157. [CrossRef]
58. Wu, H.; Lan, X.; He, Y.; Wu, A.Y.; Ding, M. Orderly charging of electric vehicles: A two-stage spatial-temporal scheduling method based on user-personalized navigation. *Appl. Energy* **2025**, *378*, 124800. [CrossRef]
59. Gopalakrishnan, R.; Biswas, A.; Lightwala, A.; Vasudevan, S.; Dutta, P.; Tripathi, A. Demand prediction and placement optimization for electric vehicle charging stations. *arXiv* **2016**, arXiv:1604.05472. [CrossRef]
60. Junker, J.S.; Hu, R.; Li, Z.; Ketter, W. Data-Driven Optimization of EV Charging Station Placement Using Causal Discovery. *arXiv* **2025**, arXiv:2503.17055. [CrossRef]
61. Pagany, R.; Ramirez Camargo, L.; Dorner, W. A review of spatial localization methodologies for the electric vehicle charging infrastructure. *Int. J. Sustain. Transp.* **2019**, *13*, 433–449. [CrossRef]
62. Cavus, M.; Ayan, H.; Dissanayake, D.; Sharma, A.; Deb, S.; Bell, M. Forecasting Electric Vehicle Charging Demand in Smart Cities Using Hybrid Deep Learning of Regional Spatial Behaviours. *Energies* **2025**, *18*, 3425. [CrossRef]
63. Wolbertus, R.; van den Hoed, R.; Kroesen, M.; Chorus, C. Charging infrastructure roll-out strategies for large scale introduction of electric vehicles in urban areas: An agent-based simulation study. *Transp. Res. Part A Policy Pract.* **2021**, *148*, 262–285. [CrossRef]

energies

MDPI

Article

Short-Term Load Forecasting in the Greek Power Distribution System: A Comparative Study of Gradient Boosting and Deep Learning Models

Md Fazle Hasan Shiblee and Paraskevas Koukaras *

School of Science and Technology, International Hellenic University, 14th km Thessaloniki-Moudania, 57001 Thessaloniki, Greece; mshiblee@ihu.edu.gr
* Correspondence: p.koukaras@ihu.edu.gr

Abstract

Accurate short-term electricity load forecasting is essential for efficient energy management, grid reliability, and cost optimization. This study presents a comprehensive comparison of five supervised learning models—Convolutional Neural Network (CNN), Long Short-Term Memory (LSTM), Gated Recurrent Unit (GRU), a hybrid (CNN-LSTM) architecture, and Light Gradient Boosting Machine (LightGBM)—using multivariate data from the Greek electricity market between 2015 and 2024. The dataset incorporates hourly load, temperature, humidity, and holiday indicators. Extensive preprocessing was applied, including K-Nearest Neighbor (KNN) imputation, time-based feature extraction, and normalization. Models were trained using a 70:20:10 train–validation–test split and evaluated with standard performance metrics: MAE, MSE, RMSE, NRMSE, MAPE, and R^2. The experimental findings show that LightGBM beat deep learning (DL) models on all evaluation metrics and had the best MAE (69.12 MW), RMSE (101.67 MW), and MAPE (1.20%) and the highest R^2 (0.9942) for the test set. It also outperformed models in the literature and operational forecasts conducted in the real world by ENTSO-E. Though LSTM performed well, particularly in long-term dependency capturing, it performed a bit worse in high-variance periods. CNN, GRU, and hybrid models demonstrated moderate results, but they tended to underfit or overfit in some circumstances. These findings highlight the efficacy of LightGBM in structured time-series forecasting tasks, offering a scalable and interpretable alternative to DL models. This study supports its potential for real-world deployment in smart/distribution grid applications and provides valuable insights into the trade-offs between accuracy, complexity, and generalization in load forecasting models.

Keywords: short-term load forecasting; gradient boosting; lightGBM; energy demand prediction; data mining; machine learning; deep learning; data preprocessing; time-series feature engineering; smart grids

1. Introduction

1.1. Motivation and Incitement

Short-term load forecasting (STLF) is a crucial process in the efficient and reliable operation of modern power systems. It enables grid operators to optimize electricity production, ensure grid stability, and reduce operational costs. STLF focuses on predicting electricity demand over short horizons, typically from a few minutes to a few days, and has a direct impact on energy management, scheduling, and pricing strategies, especially in situations of fluctuating and unpredictable demand [1,2].

The increasing integration of renewable energy sources such as solar and wind has made STLF even more challenging as these sources are inherently intermittent and weather-dependent. Traditional linear forecasting models, such as Autoregressive Integrated Moving Average (ARIMA), often fail to capture the non-linear relationships and highly dynamic consumption patterns caused by sudden weather fluctuations, consumer behavior changes, and economic variations [3,4].

This shift in energy consumption behavior has driven the transition toward machine learning (ML) and deep learning (DL) techniques, which are capable of handling complex, non-linear, and temporal dependencies. Among these, Long Short-Term Memory (LSTM) networks have shown great potential in STLF due to their ability to model long-term dependencies and handle volatile time-series data effectively [5]. LSTMs are increasingly being combined with Convolutional Neural Networks (CNNs) to exploit both temporal and spatial features, improving forecasting performance by automating feature extraction [2].

1.2. Literature Review and Research Gaps

In recent years, several studies have applied ML, DL, and hybrid approaches to improve STLF accuracy, particularly for the Greek power system, where seasonal variability and demand fluctuations pose significant challenges.

The authors of [1] investigated STLF for the building sector using models such as HGBR and LGBMR, showing that feature engineering and the choice of time resolutions significantly affect forecasting accuracy and computational performance. Similarly, the authors of [6] proposed transforming high-dimensional hourly load data into a univariate series using Singular Value Decomposition (SVD), followed by ARIMA, which explained 91% of the variance and simplified the problem.

Another work [7] improved STLF accuracy by applying Artificial Neural Networks (ANNs) with inputs from previous day (D-1) and weekly (D-7) load patterns combined with temperature, achieving a MAPE of 1.92%, one of the lowest errors reported in Greek forecasting. A hybrid approach was introduced in [8], combining fuzzy clustering, RBFNNs, and CNNs, achieving superior accuracy for both the Greek interconnected system and the isolated Crete network. Similarly, Ref. [9] used Feed-Forward ANNs (FF ANN), achieving a MAPE of 3.66%, and highlighted that input data quality plays a critical role in achieving accurate predictions.

Other studies have focused on hybrid models and time-series decomposition techniques. Ref. [10] combined Singular Spectrum Analysis (SSA) with LSTM networks, effectively capturing complex seasonalities and improving prediction performance. Likewise, Ref. [11] introduced the Dynamic Block-Diagonal Fuzzy Electric Load Forecaster (DBD-FELF), which integrates fuzzy clustering with RNNs, achieving a MAPE of 1.18% while maintaining lower computational complexity than many DL models.

DL has also been combined with chaos theory to improve STLF accuracy. Ref. [12] used RNNs, GRUs, LSTMs, and Bi-LSTMs for iterative predictions and found GRUs particularly robust under chaotic load conditions. Meanwhile, Ref. [13] applied Transfer Learning (TL) with clustering techniques to improve day-ahead forecasting across several European countries, including Greece, showing that knowledge transfer between regions with similar consumption patterns enhances prediction performance.

Finally, Ref. [14] developed a hybrid clustering–neural network model for bus-level forecasting, demonstrating improved prediction reliability during peak demand periods.

1.3. Identified Gaps

Despite these advances, several key challenges remain:

- Studies using the ENTSO-E hourly load dataset for Greece often report relatively high forecasting errors, showing the difficulty of accurately predicting national-level demand.
- Existing DL-based approaches provide high accuracy but are often computationally expensive and lack interpretability.
- Many studies focus on single-model forecasting, whereas hybrid and ensemble approaches could better handle diverse demand patterns.
- The use of real-time IoT and smart meter data in STLF remains underexplored, limiting adaptability in dynamic grid environments.

These gaps highlight the need for comprehensive comparative studies using state-of-the-art ML and DL models on large, realistic multivariate datasets to improve STLF performance.

1.4. Novelty, Contributions, and Paper Organization

This study addresses the above gaps by evaluating five state-of-the-art forecasting models on a multivariate dataset for the Greek power system covering 2015–2024. The dataset includes hourly electricity load, weather variables (temperature, humidity), and holiday indicators, providing a realistic context for testing model performance.

The main contributions of this paper are as follows:

- Comprehensive Model Evaluation: Benchmarking and comparing LightGBM, LSTM, GRU, CNN, and CNN-LSTM hybrid models for short-term load forecasting on a real-world Greek dataset.
- Realistic and Extensive Dataset: Using a multivariate dataset spanning nine years, enabling testing under realistic operational conditions.
- Advanced Data Preprocessing: Implementing KNN imputation, feature engineering, and data normalization to improve prediction quality.
- Evaluation with Multiple Metrics: Assessing models using MAE, RMSE, MAPE, R^2, and NRMSE, providing a comprehensive evaluation framework.
- Addressing Forecasting Challenges: Tackling common STLF issues such as overfitting, computational resource limitations, and hyperparameter optimization to ensure model robustness.
- Explicit trade-off analysis: We quantify and discuss the trade-offs between accuracy, model complexity, and generalization across DL architectures and LightGBM, providing actionable guidance for model selection under real-world constraints.

Finally, the rest of this paper is organized as follows. Section 2 describes the dataset, its sources, and the preprocessing techniques applied to ensure data quality and consistency. Section 3 presents the machine learning and deep learning models used in this study, along with the training procedures and evaluation methodology. Section 4 reports the experimental findings, compares the performance of the models, and discusses their implications for practical deployment. Lastly, Section 5 summarizes the main contributions of this work and provides recommendations for future research directions.

2. Data Curation and Analysis

This section covers three broad areas: data collection, data processing, and feature selection. The first subsection provides a description of the dataset, its source, and major features. The second subsection explains the data processing methods employed to prepare the data for ML and DL analysis. Understanding these aspects and applying them appropriately is essential when working with datasets in predictive modeling.

2.1. Dataset Description

The data used in this research is based on the Greek Electricity Distribution and Management System, specifically the hourly actual load data. This information was obtained from the European Network of Transmission System Operators for Electricity (ENTSO-E) (https://transparency.entsoe.eu/) (accessed on 10 July 2025). The dataset is organized as a single time series, where each entry represents the actual electricity load in megawatt-hours (MWh) for a specific hour and date. This time-series structure reflects the consumption dynamics of the Greek electricity system and provides a solid foundation for short-term load forecasting [6,7,9].

To enhance the predictive power of the model, additional variables were included due to their influence on electricity demand. Hourly temperature and humidity data for the period 2015–2024 were retrieved using the Python Meteostat API (https://pypi.org/project/meteostat/) (accessed on 10 July 2025). Given the challenge of accounting for weather variations across the entire country, weather data from Athens—one of Greece's major urban centers—was used to represent national trends, thereby simplifying the model while maintaining relevance. Athens weather data were used because it represents the largest load center in Greece and strongly influences national demand. This choice captured the main seasonal and peak variations that drive overall load trends while acknowledging possible bias during regionally heterogeneous weather and leaving multi-station/reanalysis inputs for enhancing the robustness of future work. Furthermore, public holidays were considered by retrieving holiday indicators using the Python holidays API (https://pypi.org/project/holidays/) (accessed on 10 July 2025). These indicators helped capture anomalies in demand during non-working days.

For model training, the dataset was split chronologically. Data from 2015 to 2021 was used for training, 2022 and 2023 data was used for validation, and 2024 data was used as a test set. This setup allowed the models to learn from historical patterns, tune hyperparameters based on recent data, and be evaluated on completely unseen data. This strategy ensured generalizability and provided a realistic assessment of model performance in forecasting future electricity demand.

2.2. Data Preprocessing and Feature Engineering

After initial data cleaning, several preprocessing steps were applied to prepare the dataset for modeling. Feature engineering was first performed to extract time-dependent patterns from the raw data [15]. Temporal features such as hour, day, weekday, month, and year were derived from the timestamp to capture daily, weekly, and annual seasonality. Additionally, lag-based features were introduced to incorporate historical load information, including

- `lag_1`: Load at the previous hour;
- `lag_24`: Load at the same hour on the previous day;
- `lag_168`: Load at the same hour on the previous week.

Rolling averages were also computed to reflect short-term trends:

- `rolling_mean_3`: 3 h moving average;
- `rolling_mean_24`: 24 h moving average.

To handle missing values, K-Nearest Neighbor (KNN) imputation was used. KNN is well-suited for time-series data with recurrent patterns as it imputes values based on the similarity to neighboring observations, thereby preserving the inherent structure of the dataset [16].

After feature generation and imputation, the dataset was normalized to have zero mean and unit variance. This step ensured that all features contributed equally to model training and helped stabilize the learning process for many ML algorithms.

Finally, the dataset was split into training, validation, and testing subsets using a 70:20:10 ratio. The training set (2015–2021) was used to train the models, the validation set (2022–2023) to fine-tune hyperparameters, and the test set (2024) to evaluate final performance. Stratified partitioning was employed to ensure even distribution of load patterns and maintain the seasonal and temporal structure across splits. This robust preprocessing pipeline enhanced data quality and ensured that the resulting models were capable of producing reliable and accurate short-term load forecasts [17].

2.3. Feature Importance Analysis

Figure 1 illustrates the feature importance as determined by the LightGBM model applied to the short-term load forecasting (STLF) task. The most significant predictor was the `hour`, highlighting the strong influence of time of day on electricity consumption. Historical features like `lag_1` (previous hour) and `lag_24` (previous day) also ranked highly, confirming the importance of recent past data in load prediction.

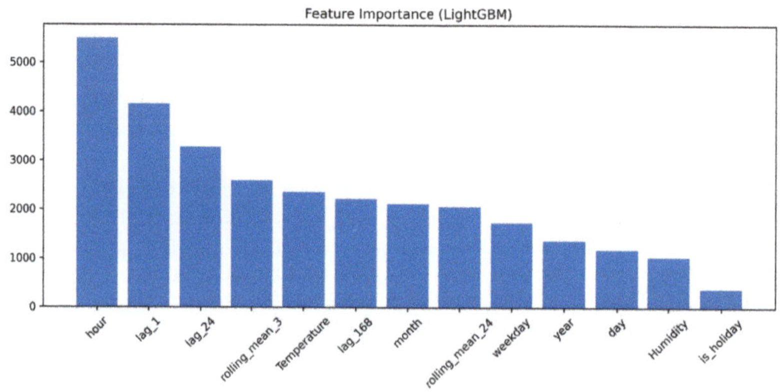

Figure 1. Feature importance plot for LightGBM model on short-term load forecasting.

The `rolling_mean_3` feature, representing a 3 h smoothed trend, indicated the model's reliance on short-term moving averages to detect consumption patterns. `Temperature` was also among the top contributors, suggesting that weather has a considerable effect on electricity usage.

Other influential features included `lag_168` (weekly lag), `month`, and `weekday`, which incorporated seasonal and weekly patterns. Conversely, variables such as `year`, `day`, `humidity`, and `is_holiday` showed lower importance. This implied that long-term trends and humidity played a smaller role and that public holidays had a limited influence in this context.

Overall, the feature importance analysis suggested that temporal variables and temperature are the most impactful for short-term load forecasting, while calendar-specific factors are relatively minor in predictive power.

3. Model Architectures and Experimental Setup

The section outlines the baseline models employed in this study, along with the experimental setup and the model training process. It provides a detailed description of the various models used for comparison and the steps taken to train and evaluate them, ensuring a comprehensive understanding of the approach adopted for short-term load forecasting.

3.1. CNN (Convolutional Neural Network)

CNNs have been most successful in image processing; however, they have also found application in time-series forecasting, such as short-term load forecasting. CNNs use convolutional layers to process the input data, where a filter is applied to scan the data and assist in locating local patterns in the data. CNNs are used in the short-term load forecasting context by identifying local load changes, including sudden changes and periodic increases or decreases, which frequently take place because of such factors as weather changes, holidays, or sudden changes in consumption patterns. Such local patterns are vital in short-term forecasting, where rapid adjustments should be made depending on recent variations in load. CNNs are good at detecting these patterns and are able to handle large quantities of data. They are able to automatically learn and extract features of raw data, so there is no need for manual feature engineering. CNNs are specifically well-suited to cases in which recent data or small time windows are the most predictive of future consumption patterns [18,19].

3.2. LSTM (Long Short-Term Memory)

LSTM networks are a variant of Recurrent Neural Networks (RNNs) that are specifically developed to learn long-term dependencies in sequential data. In contrast to the traditional RNNs, LSTMs have special memory cells and gates (input, forget, and output gates), which enable them to remember information across long time series, and, thus, they are very useful in time-series applications like short-term load forecasting. LSTMs are particularly effective in this field at capturing complicated temporal dependencies where the past energy consumption affects the future demand. As an example, LSTMs are able to learn seasonal patterns, daily load patterns, and other periodic patterns that influence energy consumption. The capability of LSTMs to store information over a long time is useful in predicting energy loads that are based on other factors, like the consumption of the previous day or even longer-term trends. This capability to capture both long-term dependencies (such as daily, weekly, and yearly trends) and short-term variability is the reason why LSTMs are one of the most popular models to forecast time series, including energy load forecasting [18,20].

3.3. Hybrid Model (CNN-LSTM)

The hybrid (CNN-LSTM) model integrates the strengths of both CNN and LSTM networks, enabling the capture of local as well as long-term dependencies in time series data. The CNN layers are employed to extract local features or patterns in the energy consumption data by filtering the input sequence, with particular attention to short-term variations such as sudden spikes or sharp drops in load. In contrast, the LSTM layers are designed to model long-term dependencies and trends, including seasonal patterns, which occur over extended periods. This hybrid approach leverages the complementary capabilities of the two architectures: CNNs are effective for feature extraction, while LSTMs are well suited for modeling temporal dependencies. Consequently, the model is particularly advantageous for forecasting energy loads, where both short-term fluctuations and long-term regularities play a crucial role. For instance, the CNN component may capture sudden variations in load caused by weather or special events, whereas the LSTM component can model recurring patterns such as the time of day or day of the week [21].

In this study, the hybrid CNN-LSTM model was implemented using two convolutional layers (128 and 64 filters with kernel sizes of 5 and 3, respectively), followed by max pooling and dropout to ensure robust feature extraction. The output was then processed by two stacked LSTM layers (128 and 64 units), allowing the model to capture sequential dynamics. Finally, dense layers with 64 and 32 neurons were used to generate the regression output. The training process utilized the Adam optimizer with a learning rate of 0.0001, the mean squared error (MSE) as the loss function, and a batch size of 32. To improve generalization and prevent overfitting, callbacks such as early stopping and learning rate scheduling were incorporated.

3.4. GRU (Gated Recurrent Unit)

GRUs are another variant of Recurrent Neural Networks (RNNs) that, like LSTMs, are designed to capture sequential dependencies in time-series data. Nevertheless, GRUs have fewer parameters and are computationally less expensive than LSTMs. They possess fewer parameters since they do not need a dedicated memory cell like LSTMs and are thus easier to train and less likely to overfit. GRUs have two gates, reset and update gates, which determine the flow of information and what information to keep or forget. GRUs are effective in short-term load forecasting because they can model temporal dependencies in energy consumption, even when the available computational resources or training time are limited. GRUs are able to capture energy consumption trends, cycles, and fluctuations, even though they have a simpler architecture. They can be used as an alternative to LSTMs, particularly when the training is required to be faster and the number of parameters is reduced, without affecting the quality of the predictions [22].

3.5. LightGBM (Light Gradient Boosting Machine)

LightGBM is a very efficient and scalable gradient boosting framework that has been applied extensively to many ML applications, such as time-series forecasting, including short-term load forecasting. In contrast to traditional gradient boosting algorithms, LightGBM resorts to a histogram-based approach to the selection of the best splits, which makes it faster and memory-efficient. This is especially beneficial in cases where one has to deal with large datasets or where one requires rapid training of models. LightGBM works by constructing an ensemble of decision trees, each of which is trained to fix the mistakes of the preceding trees, in the case of short-term load forecasting. These trees are able to capture the non-linear relationships and interactions within the data, and, therefore, the model is very effective in predicting energy demand, which can be affected by a number of complex factors like weather conditions, time of day, and holidays. LightGBM is categorical variable-friendly and can use parallel and GPU learning, making it scale well on larger datasets. LightGBM has been a very good option in forecasting applications where speed and accuracy are of the essence due to its ability to model non-linear relationships and work with large amounts of data [23].

3.6. Experimental Setup and Model Training

In order to compare the performance of the various learning paradigms in short-term electricity load forecasting, five advanced models were designed and trained, each with unique architectural merits. The DL architectures were LSTM, CNN, the combination of CNN and LSTM, and GRU networks. The LSTM model (a variant of the Recurrent Neural Network (RNN)) was selected because it can model long-range temporal dependencies that exist in sequential data. The CNN, which is normally applied to spatial data, was modified to time-series forecasting by taking advantage of one-dimensional convolutional layers to learn local temporal features in an efficient manner. The CNN-LSTM hybrid model took the best of both models, where CNN layers were employed first to capture short-term patterns, after which LSTM layers that captured sequential dependencies were applied. GRU, a variant of LSTM that has a simpler gating structure and fewer parameters, was found to perform similarly with lower computational complexity.

Along with the DL architectures, a tree-based ensemble model, Light Gradient Boosting Machine (LightGBM), was used. LightGBM is a fast, scalable, and efficient gradient boosting framework that is particularly fast and scalable and works well with tabular data. It is not explicitly temporal-modeling but rather learns using a rich set of engineered features such as lag values, rolling statistics, calendar-based variables, and weather-related inputs. We used LightGBM as the boosted-tree baseline (chosen over XGBoost for faster, more memory-efficient training at our scale) and omitted models like ARIMA and SVR due to poor scalability to long, multivariate hourly series, which made them impractical for our 10-year dataset. Every DL model was trained on TensorFlow/Keras and LightGBM was trained on its own Python library. The hyperparameters of all models were fine-tuned by grid search and monitoring of validation to maximize forecasting performance.

The MSE loss function was used to train the model, and the training and validation sets were used to inform the learning process and avoid overfitting. A number of callbacks were introduced in the training to enhance model generalization. Early stopping was used to terminate the training process when validation loss failed to decrease in 20 consecutive epochs, thereby avoiding unnecessary computations and overfitting. The model weights were saved at the best-performing validation epoch using a model checkpoint mechanism. Furthermore, the ReduceLROnPlateau callback was applied to dynamically change the learning rate in case of stagnation of the validation loss, which made convergence more precise.

The performance was evaluated based on five statistical measures, which included MAE, RMSE, MAPE, R^2 score, and NRMSE. These measures were calculated on the validation and test sets to evaluate the accuracy and generalization ability of each of the models. Notably, the 2024 test set was not used in any way during training or validation to give an objective indication of out-of-sample predictive performance. The LightGBM model was the best-performing model in all the metrics, with a low error rate and a high R^2 score compared to the DL models. The fact that it could deal with heterogeneous feature types and optimize the gradient boosting effectively made it especially useful in this forecasting task. The general process of the experiment, data preprocessing, data splitting, model training, model evaluation, and model selection are depicted in Figure 2.

Aside from predictive performance, we also contrasted the computational footprint of each model. In our setting, the sequence models (LSTM, GRU) and the CNN-LSTM hybrid had longer training and tuning cycles and more memory usage than tree-based boosting. CNN and GRU models trained more quickly than LSTM, but they had lower final accuracy. The CNN-LSTM hybrid added architectural complexity without corresponding benefits. LightGBM had lower training times and resource usage, corresponding to histogram-based learning and efficient tabular engineered feature handling. These differences informed the downstream assessment of scalability and real-time feasibility reported in Section 4.3.

Finally, to ensure comparability across architectures, all models were trained and evaluated on the same core variable set. This isolated modeling capacity from exogenous data availability and prevented twisting results due to heterogeneous feature sets.

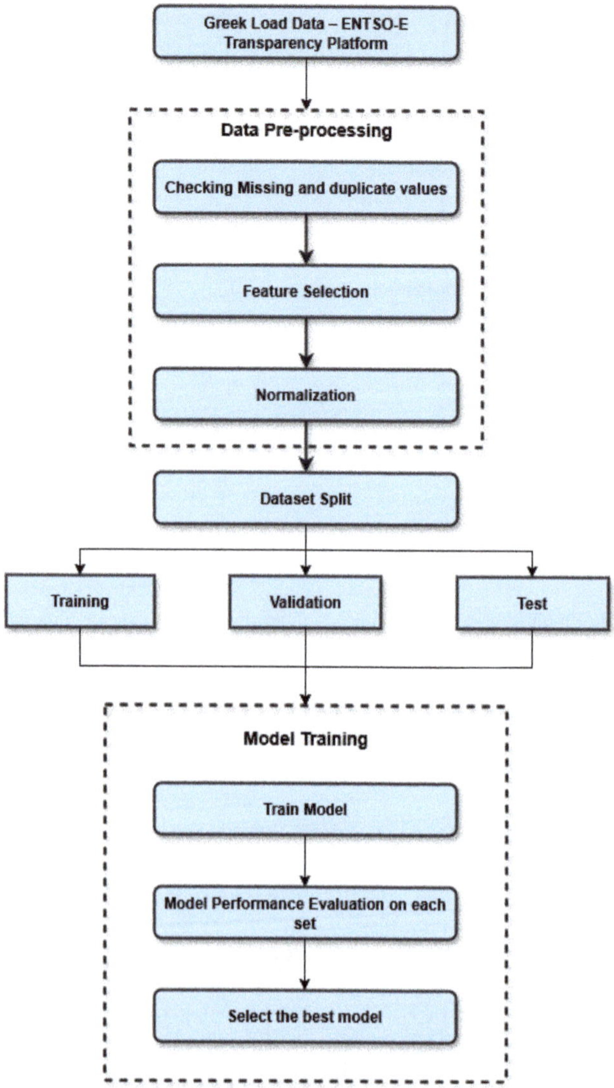

Figure 2. Workflow of experimental setup: data preprocessing, model training, and evaluation.

4. Results and Discussion

This chapter includes the results of the test of the proposed short-term load forecasting models, which were used to predict the energy consumption of different datasets. To evaluate the performance of the models, a variety of evaluation metrics were used, which allowed us to compare them in detail with the existing approaches in load forecasting [24]. The findings indicate that the model developed using LightGBM consistently performed better than the other methods, which indicates its accuracy and reliability in short-term load prediction. Moreover, this section gives detailed explanations of all the evaluation metrics that were employed in the analysis in order to gain a thorough comprehension of the performance of the models.

4.1. Performance Evaluation Metrics

A number of metrics to quantify the performance of our approach in predicting electrical load, namely, MAE, RMSE, MSE, MAPE, NRMSE, and R^2, were employed in this study. MAPE is a percentage version of where MAE, MSE, and RMSE merely indicate the numerical difference between actual and estimated load. It shows the relationship between the actual outcomes and the forecasts in a straight line. NRMSE is more effective in demonstrating the utility of RMSE by examining the extent to which data is dispersed. The combination of all these statistics allowed us to demonstrate the validity of the forecasting model and enabled comparisons with other methods [25,26].

4.1.1. Mean Absolute Error (MAE)

Regression tasks often use MAE to find the average difference between predicted and real values. MAE measures the average size of errors in load forecasting, without paying attention to their direction. It is found by averaging the absolute difference between what is predicted and what is actually observed in every sample (Equation (1)).

$$\text{MAE} = \frac{1}{n}\sum_{i=1}^{n}\left|L_{i,\text{actual}} - L_{i,\text{predicted}}\right| \quad (1)$$

4.1.2. Root Mean Squared Error (RMSE)

RMSE is the square root of the average squared difference between the actual and predicted load values. RMSE gives greater importance to larger errors than MAE, so it is more likely to notice outliers and big mistakes in forecasting (Equation (2)).

$$\text{RMSE} = \sqrt{\frac{1}{n}\sum_{i=1}^{n}\left(L_{i,\text{actual}} - L_{i,\text{predicted}}\right)^2} \quad (2)$$

4.1.3. Mean Squared Error (MSE)

MSE finds the average squared difference between forecasted and actual amounts of load. It helps assess the accuracy of a forecast by focusing on bigger mistakes (Equation (3)).

$$\text{MSE} = \frac{1}{n}\sum_{i=1}^{n}\left(L_{i,\text{actual}} - L_{i,\text{predicted}}\right)^2 \quad (3)$$

4.1.4. Mean Absolute Percentage Error (MAPE)

MAPE is the average percentage difference between actual and estimated load, so it does not depend on scale and is easy to understand in percentages (Equation (4)).

$$\text{MAPE} = \frac{100\%}{n}\sum_{i=1}^{n}\left|\frac{L_{i,\text{actual}} - L_{i,\text{predicted}}}{L_{i,\text{actual}}}\right| \quad (4)$$

4.1.5. Normalized Root Mean Squared Error (NRMSE)

NRMSE normalizes RMSE by the range of actual load values, providing a relative measure of prediction error that accounts for the variability in the data. Lower NRMSE values indicate higher predictive accuracy (Equation (5)).

$$\text{NRMSE} = \sqrt{\frac{1}{n}\sum_{i=1}^{n}\left(\frac{L_{i,\text{actual}} - L_{i,\text{predicted}}}{L_{\max} - L_{\min}}\right)^2} \quad (5)$$

where $L_{\max}$ and $L_{\min}$ denote the maximum and minimum load values in a dataset.

4.1.6. Coefficient of Determination (R^2)

The coefficient of determination, R^2, quantifies the proportion of variance in actual load values that is explained by a forecasting model. It ranges from 0 to 1, where 1 indicates perfect prediction and 0 indicates that the model does not explain any variance (Equation (6)).

$$R^2 = 1 - \frac{\sum_{i=1}^{n} \left(L_{i,\text{actual}} - L_{i,\text{predicted}} \right)^2}{\sum_{i=1}^{n} \left(L_{i,\text{actual}} - \overline{L}_{\text{actual}} \right)^2} \quad (6)$$

Here, $\overline{L}_{\text{actual}}$ is the mean of actual load values.

In the above formulas, n is the total number of samples, $L_{i,\text{actual}}$ denotes the actual load at sample i, $L_{i,\text{predicted}}$ is the predicted load, and L_{max} and L_{min} represent the maximum and minimum loads in the dataset, respectively.

4.2. Prediction Results and Comparative Analysis

In the evaluation phase, five models—CNN, LightGBM, LSTM, CNN-LSTM (hybrid), and GRU—were assessed for their short-term load forecasting performance. Each model was trained and tested under consistent conditions to ensure a fair comparison. Performance evaluation was conducted using six standard metrics: MAE, MSE, RMSE, MAPE, NRMSE, and R^2. These metrics were used to measure the accuracy of each model's predictions and their alignment with actual values. The detailed results for all models across the selected metrics are presented in the following tables.

Table 1 presents the performance metrics of the CNN model. The model achieved a training MAE of 126.89 MW, a validation MAE of 154.07 MW, and a test MAE of 148.05 MW. The corresponding RMSE values were 176.79 MW, 206.68 MW, and 194.10 MW, indicating moderate variance in prediction errors. The R^2 score ranged from 0.9706 to 0.9789, showing a good fit but not optimal compared to other models. MAPE values remained between 2.2% and 2.8% and NRMSE hovered around 0.024 across all sets, highlighting the CNN's consistent yet limited ability to capture fine-grained fluctuations in the load data.

Table 1. CNN model performance metrics.

Metric	Train	Validation	Test
MAE	126.8867	154.0706	148.0502
MSE	31,255.7571	42,715.0626	37,676.0284
RMSE	176.7930	206.6762	194.1031
R^2	0.9763	0.9706	0.9789
MAPE (%)	2.2022	2.8368	2.5900
NRMSE	0.0235	0.0251	0.0244

Figure 3 illustrates the comparison between actual and predicted daily mean electricity load values using the CNN model over the validation (top) and test (bottom) sets. The solid red line represents the actual daily mean load, while the dashed blue line indicates the CNN-predicted daily mean. The model successfully captured overall seasonal and daily consumption trends, closely following the fluctuations of real demand. While small deviations were observed during periods of abrupt load change (e.g., summer peaks), the model demonstrated strong alignment with the actual data, particularly during stable consumption periods.

Figure 4 presents the hourly granularity forecast performance of the CNN model on the validation (top) and test (bottom) sets. The red line corresponds to the actual hourly load, while the blue dashed line shows the predicted load values. The model effectively tracked the short-term variability and high-frequency oscillations of electricity demand across nearly three years of data. The prediction accuracy was notably consistent during regular patterns but slightly less precise during high-volatility periods, such as summer peaks or holiday-related load shifts. These results indicate that the CNN model captures both periodic trends and local variations with reasonable fidelity.

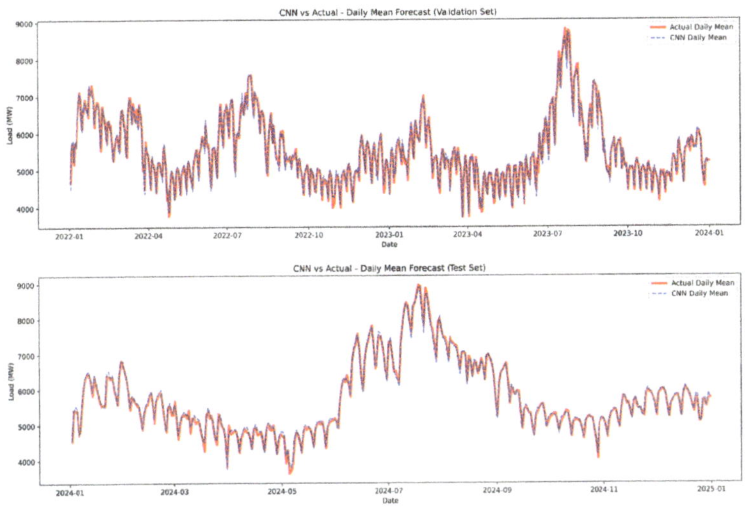

Figure 3. Daily mean forecast vs. actual load using theCNN model on the validation and test sets.

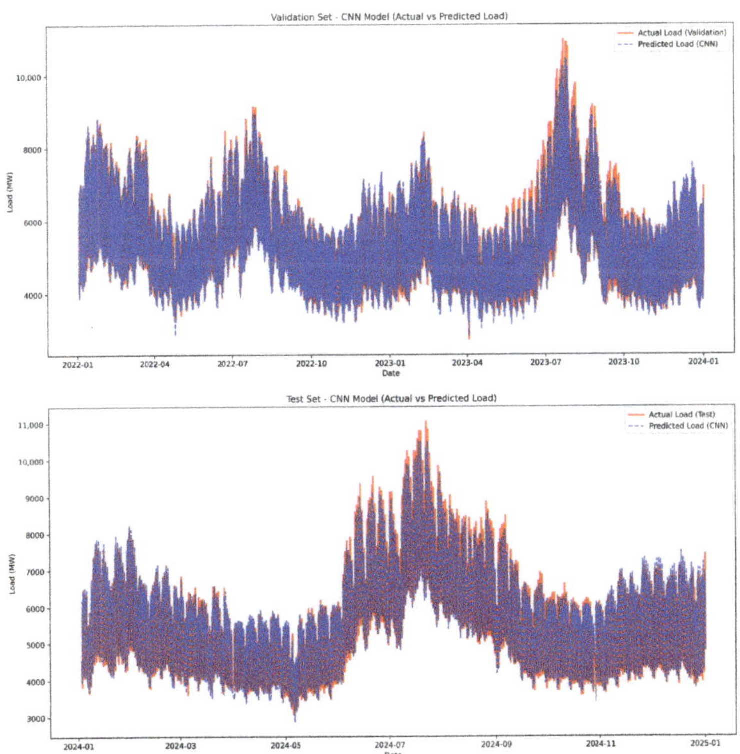

Figure 4. Hourly forecast vs. actual load using the CNN model on validation and test sets.

Table 2 details the performance of the LSTM model, which significantly outperformed CNN and other DL architectures. The LSTM achieved lower error values across all metrics, with a test MAE of 87.26 MW, test RMSE of 118.68 MW, and notably high test R^2 score of 0.9921. The MAPE and NRMSE on the test set were 1.53% and 0.0149, respectively, suggesting strong accuracy and generalization capabilities. LSTM also maintained tight error bounds between training, validation, and test sets, indicating model robustness and minimal overfitting.

Table 2. LSTM model performance metrics.

Metric	Train	Validation	Test
MAE	92.2249	86.7305	87.2611
MSE	17,917.3677	13,067.9785	14,085.0742
RMSE	133.8558	114.3153	118.6806
R^2	0.9864	0.9910	0.9921
MAPE (%)	1.5973	1.5870	1.5339
NRMSE	0.0178	0.0139	0.0149

Figure 5 shows the comparison between the actual and predicted daily mean electricity load values generated by the LSTM model for the validation (top) and test (bottom) periods. The red line denotes the actual daily mean load, while the blue dashed line represents the LSTM forecast. The model tracked seasonal and intra-annual trends with high fidelity, especially during stable and moderately fluctuating load periods. Notably, the LSTM model demonstrated strong accuracy during high-demand months such as summer, with minimal lag or overshooting, confirming its strength in modeling long-term dependencies in sequential data.

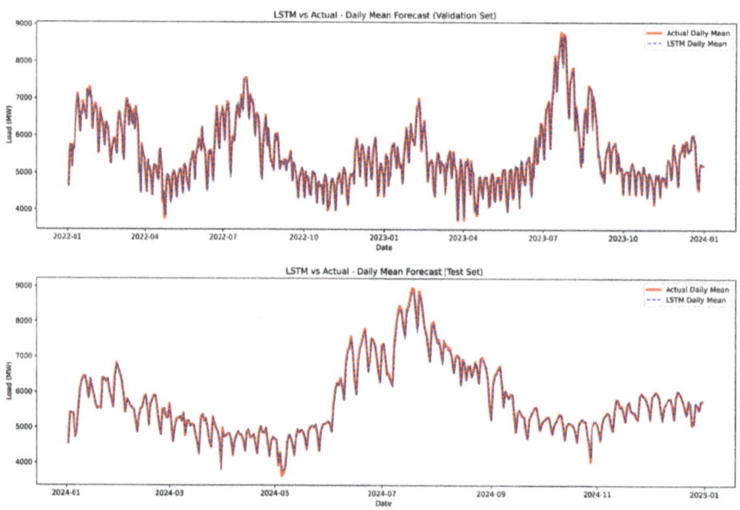

Figure 5. Daily mean forecast vs. actual load using the LSTM model on validation and test sets.

Figure 6 presents the detailed hourly forecasts of the LSTM model alongside the actual load for the validation set (top) and test set (bottom). The red curve represents the true hourly load, while the blue dashed line indicates the LSTM-predicted load. The model captured both high-frequency variations and broader seasonal patterns with considerable accuracy across the full time range. The overlap between the predicted and actual values demonstrates the LSTM model's capacity to handle complex temporal dynamics in short-term load forecasting, with minimal prediction drift or phase mismatch.

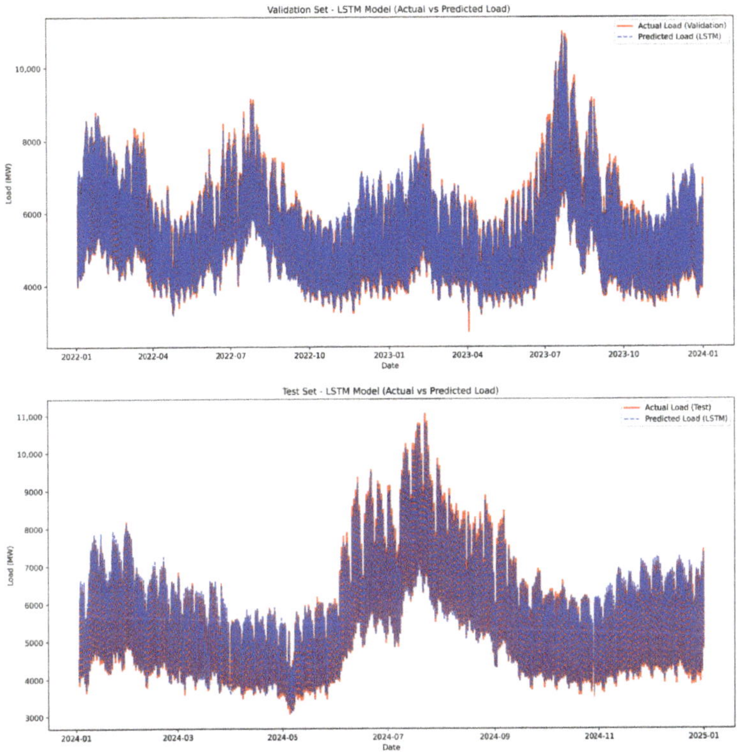

Figure 6. Hourly forecast vs. actual load using the LSTM model on validation and test sets.

Table 3 shows the metrics for the hybrid CNN-LSTM model. This combined architecture yielded a test MAE of 158.23 MW and a test RMSE of 258.05 MW, both higher than the individual CNN and LSTM models. Although the hybrid approach benefited from the CNN's local pattern recognition and LSTM's sequence modeling, the performance did not reflect synergy between the two. The R^2 score dropped to 0.9628 on the test set, with MAPE at 2.43% and NRMSE at 0.0324, confirming the hybrid model's relatively lower predictive power in this context.

Table 3. Hybrid (CNN-LSTM) model performance metrics.

Metric	Train	Validation	Test
MAE	139.1704	134.6142	158.2293
MSE	44,559.4056	44,445.2303	66,591.6128
RMSE	211.0910	210.8204	258.0535
R^2	0.9662	0.9694	0.9628
MAPE (%)	2.1921	2.2496	2.4263
NRMSE	0.0280	0.0256	0.0324

Figure 7 displays the comparison between the actual and predicted daily mean electricity load values generated by the hybrid CNN-LSTM model for both the validation (top) and test (bottom) sets. The red line represents the actual daily mean load, while the blue dashed line shows the hybrid model's forecast. The model successfully captured overall load trends and seasonal shifts but exhibited noticeable underprediction during high-demand periods, especially in the summer months. This indicates that while the hybrid model effectively integrates spatial and temporal patterns, it may lack the responsiveness to sharp load peaks compared to other architectures.

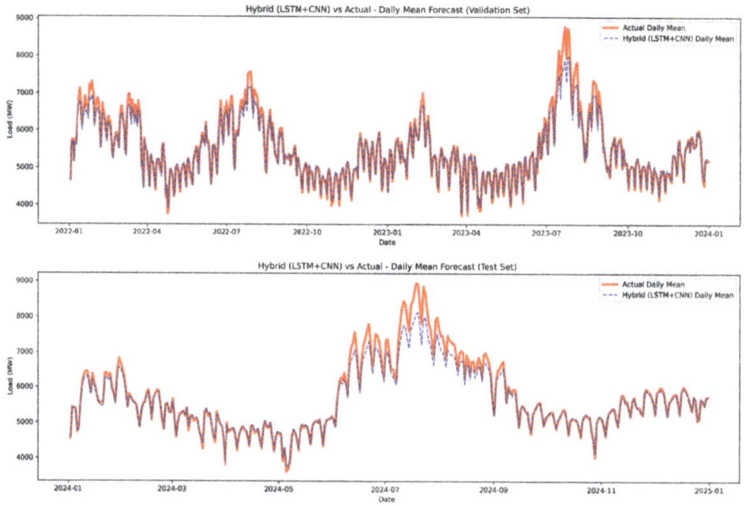

Figure 7. Daily mean forecast vs. actual load using the hybrid model on validation and test sets.

Figure 8 illustrates the hourly forecast performance of the hybrid CNN-LSTM model on the validation (top) and test (bottom) datasets. The actual load (red) and predicted values (blue dashed) are plotted to assess the model's ability to capture short-term dynamics. While the model tracked the general structure of the load signal, discrepancies were evident during periods of high volatility, with significant underestimation during peak demand events. This suggests that the hybrid model may struggle to fully model high-amplitude, short-duration fluctuations despite leveraging both local pattern recognition (CNN) and long-term sequence modeling (LSTM).

Table 4 summarizes the GRU model's performance. With a test MAE of 120.64 MW and RMSE of 158.80 MW, GRU performed better than the CNN and hybrid model but fell short of LSTM's accuracy. The test R^2 score of 0.9859 and MAPE of 2.03% indicated strong, consistent performance, particularly considering GRU's simplified architecture. GRU offers a balance between performance and training efficiency, making it a viable alternative when computational resources are constrained.

Table 4. GRU model performance metrics.

Metric	Train	Validation	Test
MAE	127.2923	115.6403	120.6414
MSE	29,515.2823	22,622.5201	25,217.7061
RMSE	171.8001	150.4078	158.8008
R^2	0.9776	0.9844	0.9859
MAPE (%)	2.1491	2.0508	2.0330
NRMSE	0.0228	0.0183	0.0199

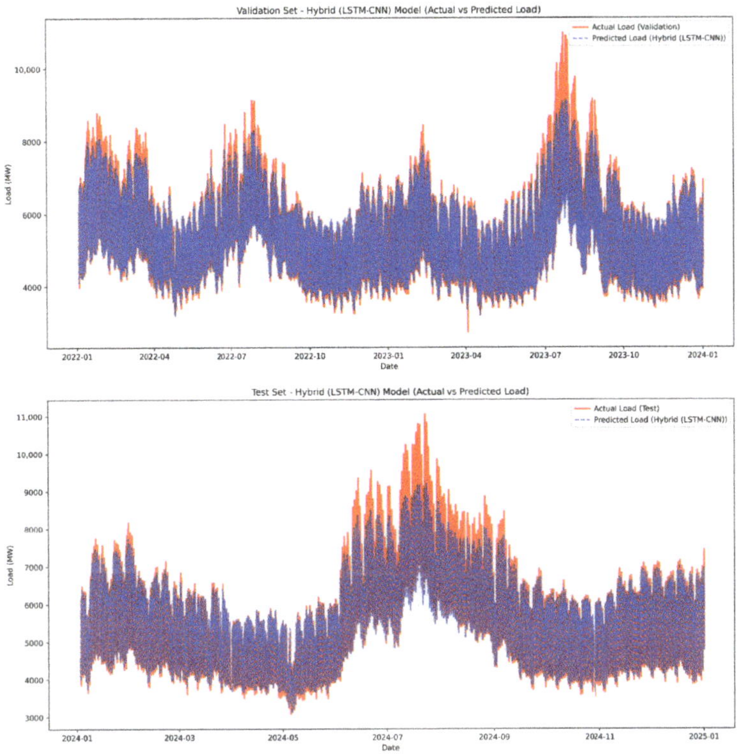

Figure 8. Hourly forecast vs. actual load using the hybrid model on validation and test sets.

The performance of the GRU model in forecasting short-term electricity load is illustrated in Figures 9 and 10.

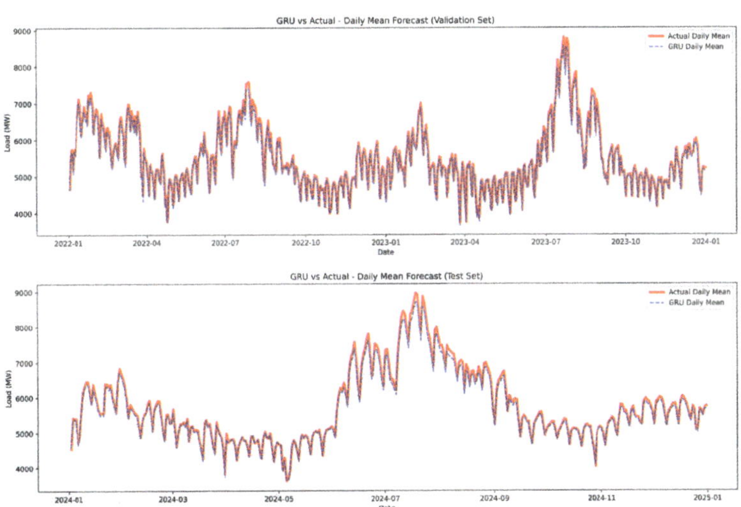

Figure 9. Daily mean forecast vs. actual load using the GRU model on validation and test sets.

As shown in Figure 9, which compares the actual and predicted daily mean load values, the GRU model demonstrated strong tracking of both seasonal and short-term consumption patterns across the validation and test sets. The predicted curves closely followed the actual values, with minimal lag or deviation, especially during periods of stable demand. However, there was a slight tendency to underpredict during peak summer months, though the model maintained overall high consistency and generalization. A more granular view of GRU's performance is provided in Figure 10, which displays the hourly forecasts versus actual load for the same periods. The model accurately captured high-frequency variations and daily load cycles, including sharp increases and decreases in demand. While some underestimation was evident during extreme peak hours, the GRU model effectively maintained alignment with the ground truth across the entire timeline. This level of performance confirms the GRU's capability to model complex sequential dependencies in load data while benefiting from a relatively simpler architecture compared to LSTM.

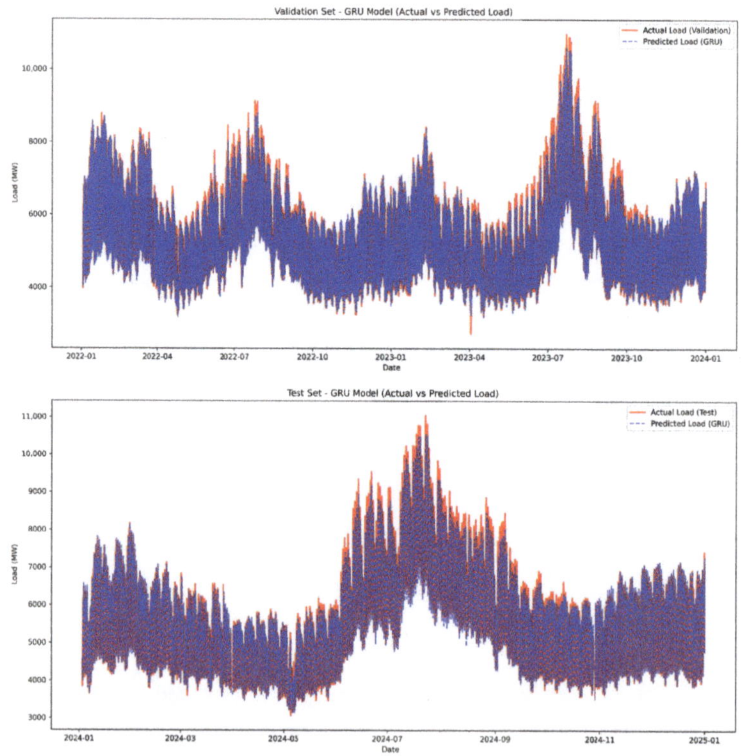

Figure 10. Hourly forecast vs. actual load using the GRU model on validation and test sets.

Table 5 presents the results of the LightGBM model, which outperformed all other architectures across every metric. The test MAE was only 69.12 MW, and the test RMSE was 101.66 MW—significantly lower than the rest. LightGBM also achieved the highest R^2 score of 0.9942 and the lowest MAPE and NRMSE values, at 1.20% and 0.0128, respectively. The consistent performance across train, validation, and test sets confirms its superior generalisation capability. LightGBM's tabular feature learning, ability to handle non-sequential data, and computational efficiency give it a distinct advantage over DL models in this specific short-term load forecasting task.

Table 5. LightGBM model performance metrics.

Metric	Train	Validation	Test
MAE	45.5114	67.8065	69.1205
MSE	4644.3277	9260.6456	10,335.7831
RMSE	68.1493	96.2322	101.6651
R^2	0.9965	0.9936	0.9942
MAPE (%)	0.8022	1.2450	1.2025
NRMSE	0.0090	0.0117	0.0128

The forecasting performance of the LightGBM model, which outperformed all other tested architectures, is comprehensively illustrated in Figures 11 and 12. As shown in Figure 11, the model's predictions aligned almost perfectly with the actual daily mean load values across both validation (top) and test (bottom) sets. The red line represents the true daily mean load, while the blue dashed line indicates the LightGBM predictions. The nearly indistinguishable overlap of the two curves demonstrates the model's ability to capture both short- and long-term patterns with exceptional precision. LightGBM effectively responded to seasonal fluctuations, daily demand cycles, and abrupt changes in consumption, including those occurring during high-load periods such as mid-summer.

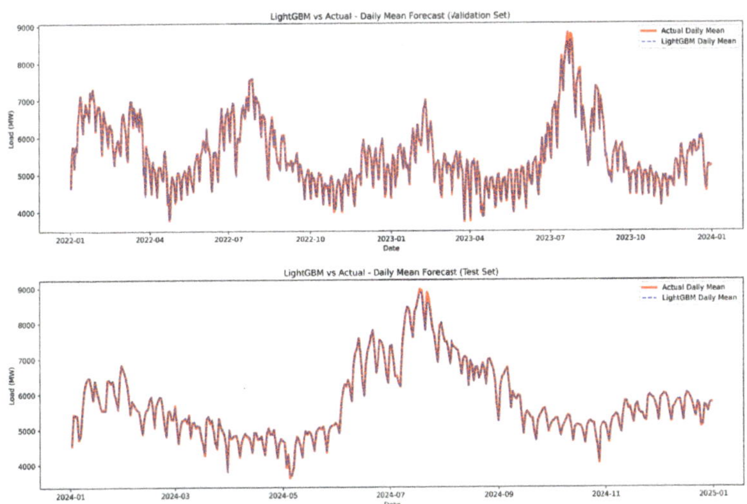

Figure 11. Daily mean forecast vs. actual load using the LightGBM model on validation and test sets.

In Figure 12, the model's performance is examined at the hourly level. The red lines represent the actual hourly load for the validation (top) and test (bottom) periods, while the blue dashed lines denote the model's predictions. Across both datasets, LightGBM showed remarkably accurate tracking of high-frequency load variations. Even during periods of extreme volatility, such as the peak summer months or the holiday season, the model closely followed the actual signal with minimal phase lag or amplitude mismatch. Importantly, the model maintained consistent performance across different seasons and demand regimes, reflecting strong generalization and robustness. This high level of accuracy confirms LightGBM's capability to leverage feature-engineered inputs—such as lag values, calendar indicators, and temperature—without relying on sequence-based modeling techniques.

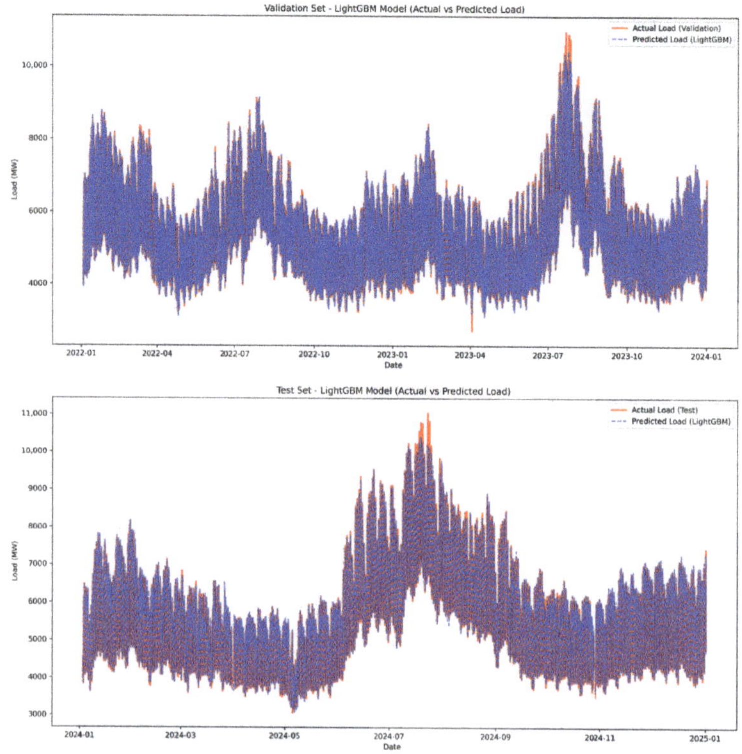

Figure 12. Hourly forecast vs. actual load using the LightGBM model on validation and test sets.

When compared to the other models presented in Figures 3–10, LightGBM demonstrated superior forecasting performance in both visual and quantitative evaluations. While LSTM (Figures 5 and 6) also showed excellent generalization and was ranked second-best, LightGBM's predictions were more precise during volatile periods and more consistent across all metrics, including MAE, RMSE, and R^2. The CNN (Figures 3 and 4) and GRU (Figures 9 and 10) followed with moderate accuracy but exhibited larger deviations during load peaks. The hybrid CNN-LSTM model (Figures 7 and 8), despite its architectural complexity, underperformed during critical high-demand windows, suggesting suboptimal synergy between its convolutional and recurrent components.

The visual evidence in Figures 11 and 12, reinforced by the statistical results, confirms that LightGBM not only provides the best overall predictive accuracy but also exhibits the most consistent and reliable behavior across all forecasting horizons. Its strength lies in efficiently capturing temporal dependencies through engineered features while avoiding the complexity and training overhead associated with DL models.

To further support the selection of LightGBM as the optimal model for short-term electricity load forecasting, a set of comparative visual analyses was conducted across different temporal resolutions and performance metrics, presented in Figures 13–15. These figures showcase both model accuracy and robustness in varied conditions. Forecast curves for 1-day (Figure 13), 1-week (Figure 14), and 1-month (Figure 15) horizons further validate LightGBM's performance. In all time windows, its predictions remained tightly aligned with the actual load values, exhibiting minimal deviation or temporal drift. Particularly during periods of high volatility—such as midday peaks or rapid demand changes—LightGBM consistently followed the actual signal more precisely than models like CNN,

CNN-LSTM, or GRU. LSTM remained a close second, especially in short-horizon forecasts, but its performance slightly degraded during periods with high amplitude variance.

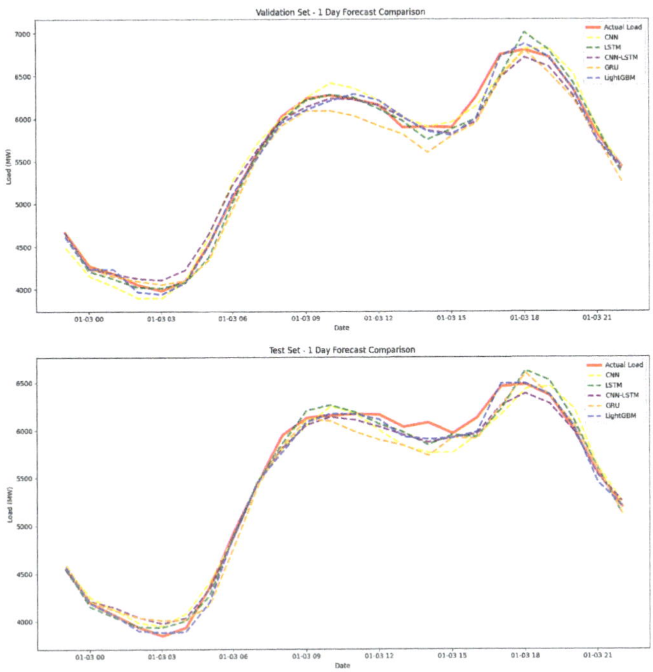

Figure 13. One-day forecast comparison (validation and test sets).

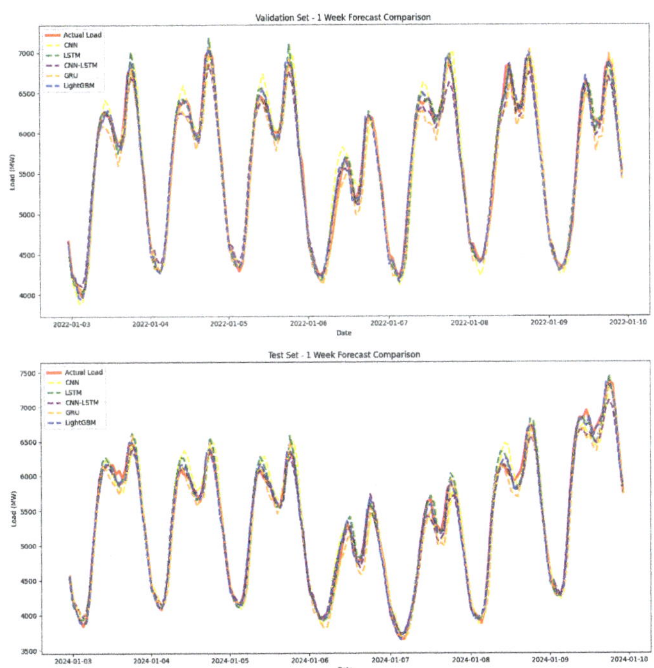

Figure 14. One-week forecast comparison (validation and test sets).

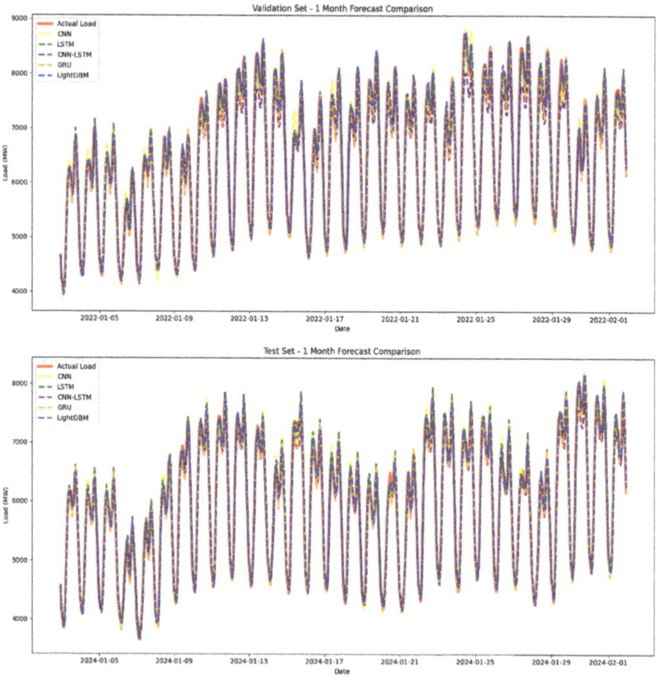

Figure 15. One-month forecast comparison (validation and test sets).

As shown in Figure 16, which presents a bar-wise comparison of error metrics across all models, LightGBM consistently outperformed the others for both the validation and test sets. It achieved the lowest scores for MAE, MSE, RMSE, MAPE, and NRMSE while also maintaining the highest R^2 values, indicating its exceptional accuracy and generalization capability. It is important to note that the actual electricity load values used in the forecasts were not scaled or normalized to a 0–1 range. Consequently, the absolute error values (particularly MAE, MSE, and RMSE) appear to be relatively high but are realistic and directly interpretable in megawatts (MWs), reflecting real-world forecasting performance.

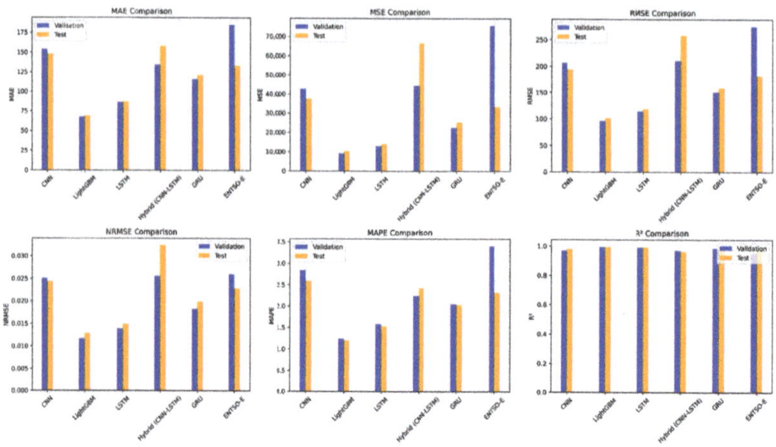

Figure 16. Error metric comparison across all models (validation and test sets).

While LSTM came closely behind LightGBM on some metrics—particularly R^2 and MAE—the noticeable gap in MSE and RMSE further reinforces LightGBM's advantage in managing high-magnitude fluctuations and large-scale error suppression. These results confirm LightGBM's suitability for operational deployment, where raw load values must be predicted accurately without the need for inverse normalization, making it not only the most accurate model but also the most practical for real-time forecasting applications.

Table 6 provides a comparative evaluation of the proposed LightGBM model against existing forecasting models reported in the literature and benchmarked data from the ENTSO-E platform. The first three models—ANN, Singular Value Decomposition (SVD)-ARIMA, and FF ANN—were sourced from prior research studies, while the ENTSO-E model represented forecast data obtained directly from the official ENTSO-E Transparency Platform. The performance of each model was assessed using standard metrics: MAE, MSE, RMSE, R^2, MAPE, and NRMSE.

Table 6. Performance comparison with existing models from the literature.

Metric	ANN [7]	SVD-ARIMA [6]	FF ANN [9]	ENTSO-E	LightGBM
MAE	112.9198	220.5342	-	133.3072	69.1205
MSE	22,111.6668	-	-	33,334.4298	10,335.7831
RMSE	-	267.3871	-	182.5771	101.6651
R^2	-	-	-	0.9813	0.9942
MAPE (%)	1.92	4.3286	2.61	2.3282	1.2025
NRMSE	-	-	0.036	0.0229	0.0128

The ANN model produced an MAE of 112.92 MW and an MAPE of 1.92%, whereas the SVD-ARIMA hybrid model produced even larger error values, with an MAE of 220.53 MW and an RMSE of 267.39 MW, which indicates that it is not very suitable for highly dynamic load profiles. The FF ANN model did not provide full metrics but showed an NRMSE of 0.036 and an MAPE of 2.61%. The ENTSO-E forecast, evaluated using publicly available hourly forecasts, achieved an MAE of 133.31 MW, RMSE of 182.58 MW, R^2 of 0.9813, and MAPE of 2.33%, providing a realistic operational benchmark.

Conversely, the proposed LightGBM model outperformed all other models on every available metric. It had the best MAE (69.12 MW), MSE (10,335.78 MW), and RMSE (101.67 MW), along with the highest R^2 score (0.9942), which indicates excellent model fit and predictive power. Additionally, it had the lowest MAPE (1.20%) and NRMSE (0.0128), confirming that LightGBM provides accurate and scale-invariant forecasts.

This comparative analysis shows that LightGBM is not only superior to traditional and literature-based models but also outperforms the ENTSO-E operational benchmark, highlighting its practical applicability for real-world short-term electricity load forecasting.

In conclusion, among all evaluated models, LightGBM delivered the highest predictive accuracy, with the lowest error metrics and strongest generalization across datasets. Its ability to leverage engineered features without relying on complex sequence modeling makes it ideal for structured time-series tasks such as electricity load forecasting. LSTM ranked second, performing well in capturing temporal dependencies with minimal overfitting. In contrast, CNN, GRU, and CNN-LSTM demonstrated moderate accuracy but were more sensitive to data variability and less stable during high-volatility periods.

Overall, LightGBM offers the most accurate, efficient, and scalable solution for real-time operational deployment in short-term electricity load forecasting applications. Including models like ARIMA/SVR would not change the main ranking under our data scale since the computational burden of these models was the primary exclusion reason rather than lack of relevance.

4.3. Practical Deployment Challenges

While the models in this study perform well in terms of accuracy, their deployment in real-world smart grid environments faces several operational challenges:

- Data Latency: LSTM, CNN, and hybrid (CNN-LSTM) models are more computationally intensive and may face delays in providing real-time predictions, particularly during periods of rapid demand changes. Minimizing latency in such cases is critical for operational decision-making.
- Computational Cost: LSTM and CNN models require more computational power, which could be a barrier in smart grid environments with resource constraints. GRU, being computationally simpler, performs relatively better in terms of training time and inference speed. LightGBM stands out for its faster training time and lower memory requirements, making it suitable for real-time applications at scale.
- Real-Time Constraints: Smart grids require frequent updates to forecasts based on incoming data. LSTM and hybrid (CNN-LSTM) models might require additional time to process and update their predictions, making them less ideal for applications where immediate decisions are critical. LightGBM, due to its feature-engineering approach and faster processing times, is better suited for real-time forecasting applications.

The above observations indicate that higher architectural complexity does not ensure better operational performance: in our experiments, LightGBM attained higher accuracy with lower latency and resource usage than the deep baselines. For applications where frequent model updates and tight real-time requirements are present, these trade-offs favour efficient boosting methods, unless domain-specific sequence effects justify the extra cost of deep recurrent or hybrid networks.

Finally, this section underscores the importance of selecting models that balance accuracy, computational efficiency, and real-time capabilities for operational deployment in smart grids.

4.4. Computational Efficiency Analysis

In addition to forecasting accuracy, we evaluated the computational efficiency of the five models in terms of training time, inference speed, and resource consumption. This aspect is particularly relevant for real-time operational deployment of load forecasting systems, where retraining speed and low-latency predictions are critical.

Table 7 summarizes the efficiency comparison. LightGBM completed full training in only 3.55 s on CPU (1000 boosting rounds, ∼60,000 samples) and achieved inference throughput of over 10,000 samples per second. The model also required minimal memory and produced a compact file size (<10 MB).

Table 7. Comparison of computational efficiency across models.

Model	Training Time	Inference Speed	Resource Usage
LightGBM	3.55 s (CPU)	>10,000/s	Low, <10 MB
CNN	∼8 min (GPU)	∼4500/s	Moderate
LSTM	∼12 min (GPU)	∼3200/s	High
GRU	∼10 min (GPU)	∼3600/s	Medium
CNN-LSTM	∼18 min (GPU)	∼2800/s	Highest

By contrast, DL models (LSTM, CNN, GRU, and CNN-LSTM) required 10–16 s per epoch on GPU, with total training times ranging from 8 to 18 min depending on the architecture. Their inference speeds were slower (approximately 2800–4500 samples per second) and memory consumption was significantly higher, particularly for the hybrid CNN-LSTM model due to its greater depth and parameter count.

The experimentation was conducted on the Kaggle platform using a 64-bit Linux operating system with an Intel(R) Xeon(R) CPU running at 2.00 GHz. The system was equipped with 2 Tesla T4 GPUs and 32 GB of RAM and Python 3.10.14 was used as the programming language.

These findings indicate that LightGBM not only outperformed all deep learning models in predictive accuracy but also delivered superior computational efficiency. This efficiency advantage makes LightGBM especially well-suited for large-scale or real-time STLF applications, where both speed and accuracy are critical.

5. Conclusions

This paper presented a comparative study of state-of-the-art ML and DL models for short-term load forecasting using Greek electricity demand data. The models tested included CNN, LSTM, GRU, a hybrid model (CNN-LSTM), and LightGBM. The results show that the LightGBM model outperforms all other models in terms of all standard error measures across various time horizons. Specifically, it achieved the best MAE (69.12 MW), MSE (10,335.78 MW^2), RMSE (101.67 MW), MAPE (1.20%), and NRMSE (0.0128) and the highest R^2 (0.9942), significantly surpassing other models and the ENTSO-E operational forecasts.

In comparison, the LSTM model demonstrated strong performance, with a test MAE of 87.26 MW, RMSE of 118.68 MW, MAPE of 1.53%, and R^2 of 0.9921. However, it was not as effective as LightGBM, particularly in capturing high-magnitude fluctuations. The GRU and CNN models performed decently, with a test MAE of 120.64 MW (GRU) and 148.05 MW (CNN), but were less competitive in terms of accuracy. The hybrid CNN-LSTM model did not outperform its individual components, achieving a test MAE of 158.23 MW, RMSE of 258.05 MW, and R^2 of 0.9628, suggesting potential redundancy or suboptimal integration.

These results emphasize that tabular models like LightGBM, combined with rich feature engineering (e.g., rich lagged and calendar features), can rival or even surpass DL architectures in structured forecasting tasks on the joint objectives of accuracy, generalization, and scalability. This is especially critical for real-time performance and interpretability in operational forecasting systems. Thus, practitioners should first exhaust strong tabular baselines (e.g., LightGBM with careful feature design and early stopping) before adopting recurrent or hybrid networks, which may introduce substantial complexity without reliable accuracy gains in high-variance regimes. Our scope is intentionally constrained to core, widely accessible variables to establish a reproducible baseline and isolate model effects. This design choice enhances comparability but does not capture behavioral demand response or renewable intermittency.

A key limitation of the present study is the use of Athens weather as a national proxy. While it captured dominant seasonal and peak patterns, it may have introduced bias regarding when regional weather diverges from that in the capital, possibly affecting model robustness. To address this, future work may incorporate spatially diverse meteorological inputs and reanalysis products to quantify any gains in accuracy and robustness. Furthermore, our feature-importance analysis indicated that humidity and holiday indicators have a relatively minor effect compared to temporal and lagged-load features; therefore, targeted ablation experiments are planned to substantiate and quantify these contributions more rigorously. Beyond point predictions, providing calibrated prediction intervals would enhance operational decision-making under renewable variability and is a natural extension of this work. Additionally, future research could extend this forecasting framework by integrating demand-side behavioral data, renewable energy generation patterns, and electricity pricing signals, thereby creating a more holistic energy forecasting system. Deploying the model in a real-time smart grid or distribution grid environment with streaming data would further assess its adaptability and scalability under live conditions. Finally, adapting the system for other European markets with different consumption behaviors could validate its generalizability and robustness across diverse energy systems.

Author Contributions: Conceptualization, M.F.H.S.; methodology, M.F.H.S. and P.K.; validation, M.F.H.S. and P.K.; formal analysis, M.F.H.S.; investigation, M.F.H.S. and P.K.; resources, M.F.H.S. and P.K.; data curation, M.F.H.S.; writing—original draft preparation, M.F.H.S.; writing—review and editing, M.F.H.S. and P.K.; visualization, M.F.H.S.; supervision, P.K.; project administration, P.K.; funding acquisition, P.K. All authors have read and agreed to the published version of the manuscript.

Funding: This research received no external funding.

Institutional Review Board Statement: Not applicable.

Informed Consent Statement: Not applicable.

Data Availability Statement: The datasets used for modeling in this study are openly available and can be extracted. Hourly power load: Available at https://transparency.entsoe.eu/ under the **Creative Commons Attribution (CC BY 4.0)** license (where applicable) (accessed on 10 July 2025). Hourly temperature and humidity: Can be retrieved from Python Meteostat API https://pypi.org/project/meteostat/ (accessed on 10 July 2025). Public holidays: Can be retrieved from Python holidays API https://pypi.org/project/holidays/ (accessed on 10 July 2025).

Acknowledgments: The authors would like to thank all the data providers and institutions whose open data repositories made this study possible.

Conflicts of Interest: The authors declare no conflicts of interest.

Abbreviations

The following abbreviations are used in this manuscript:

API	Application Programming Interface
ARIMA	Autoregressive Integrated Moving Average
ANN	Artificial Neural Network
CNN	Convolutional Neural Network
CNN-LSTM	Convolutional Neural Network - Long Short-Term Memory
DBD-FELF	Dynamic Block-Diagonal Fuzzy Electric Load Forecaster
DL	Deep Learning
EEMD	Ensemble Empirical Mode Decomposition
EMD	Empirical Mode Decomposition
ENTSO-E	European Network of Transmission System Operators for Electricity
FF ANN	Feed-Forward Artificial Neural Network
GBR	Gradient Boosting Regressor
GPU	Graphics Processing Unit
GRU	Gated Recurrent Unit
IoT	Internet of Things
KNN	K-Nearest Neighbor
LightGBM	Light Gradient Boosting Machine
LSTM	Long Short-Term Memory
MAE	Mean Absolute Error
MAPE	Mean Absolute Percentage Error
ML	Machine Learning
MSE	Mean Squared Error
MW	Megawatt
MWh	Megawatt-hour
NRMSE	Normalized Root Mean Squared Error
R^2	Coefficient of Determination

RBFNNs	Radial Basis Function Neural Networks
RMSE	Root Mean Squared Error
RNNs	Recurrent Neural Networks
SSA	Singular Spectrum Analysis
STLF	Short-Term Load Forecasting
SVR	Support Vector Regression
SVD	Singular Value Decomposition
TL	Transfer Learning
VSTLF	Very Short-Term Load Forecasting
XGBoost	Extreme Gradient Boosting Regressor

References

1. Koukaras, P.; Mustapha, A.; Mystakidis, A.; Tjortjis, C. Optimizing building short-term load forecasting: A comparative analysis of machine learning models. *Energies* **2024**, *17*, 1450. [CrossRef]
2. Ullah, K.; Ahsan, M.; Hasanat, S.M.; Haris, M.; Yousaf, H.; Raza, S.F.; Tandon, R.; Abid, S.; Ullah, Z. Short-term load forecasting: A comprehensive review and simulation study with CNN-LSTM hybrids approach. *IEEE Access* **2024**, *12*, 111858–111881. [CrossRef]
3. Yuan, M.; Xie, J.; Liu, C.; Xu, Z. Short-term load forecasting for an industrial building based on diverse load patterns. *Energy* **2025**, *334*, 137481. [CrossRef]
4. Junior, M.Y.; Freire, R.Z.; Seman, L.O.; Stefenon, S.F.; Mariani, V.C.; dos Santos Coelho, L. Optimized hybrid ensemble learning approaches applied to very short-term load forecasting. *Int. J. Electr. Power Energy Syst.* **2024**, *155*, 109579.
5. Cheng, X.; Wang, L.; Zhang, P.; Wang, X.; Yan, Q. Short-term fast forecasting based on family behavior pattern recognition for small-scale users load. *Clust. Comput.* **2022**, *25*, 2107–2123. [CrossRef]
6. Varelas, G.; Tzimas, G.; Alefragis, P. A new approach in forecasting Greek electricity demand: From high dimensional hourly series to univariate series transformation. *Electr. J.* **2023**, *36*, 107305. [CrossRef]
7. Arvanitidis, A.I.; Bargiotas, D.; Daskalopulu, A.; Laitsos, V.M.; Tsoukalas, L.H. Enhanced short-term load forecasting using artificial neural networks. *Energies* **2021**, *14*, 7788. [CrossRef]
8. Sideratos, G.; Ikonomopoulos, A.; Hatziargyriou, N.D. A novel fuzzy-based ensemble model for load forecasting using hybrid deep neural networks. *Electr. Power Syst. Res.* **2020**, *178*, 106025. [CrossRef]
9. Stamatellos, G.; Stamatelos, T. Short-term load forecasting of the greek electricity system. *Appl. Sci.* **2023**, *13*, 2719. [CrossRef]
10. Stratigakos, A.; Bachoumis, A.; Vita, V.; Zafiropoulos, E. Short-term net load forecasting with singular spectrum analysis and LSTM neural networks. *Energies* **2021**, *14*, 4107. [CrossRef]
11. Kandilogiannakis, G.; Mastorocostas, P.; Voulodimos, A.; Hilas, C. Short-Term Load Forecasting of the Greek Power System Using a Dynamic Block-Diagonal Fuzzy Neural Network. *Energies* **2023**, *16*, 4227. [CrossRef]
12. Stergiou, K.; Karakasidis, T.E. Application of deep learning and chaos theory for load forecasting in Greece. *Neural Comput. Appl.* **2021**, *33*, 16713–16731. [CrossRef]
13. Tzortzis, A.M.; Pelekis, S.; Spiliotis, E.; Karakolis, E.; Mouzakitis, S.; Psarras, J.; Askounis, D. Transfer learning for day-ahead load forecasting: A case study on European national electricity demand time series. *Mathematics* **2023**, *12*, 19. [CrossRef]
14. Panapakidis, I.P.; Skiadopoulos, N.; Christoforidis, G.C. Combined forecasting system for short-term bus load forecasting based on clustering and neural networks. *IET Gener. Transm. Distrib.* **2020**, *14*, 3652–3664. [CrossRef]
15. Katya, E. Exploring Feature Engineering Strategies for Improving Predictive Models in Data Science. *Res. J. Comput. Syst. Eng.* **2023**, *4*, 201–215. [CrossRef]
16. Murti, D.M.P.; Pujianto, U.; Wibawa, A.P.; Akbar, M.I. K-Nearest Neighbor (K-NN) based Missing Data Imputation. In Proceedings of the 2019 5th International Conference on Science in Information Technology (ICSITech), Yogyakarta, Indonesia, 23–24 October 2019; pp. 83–88. [CrossRef]
17. Bichri, H.; Chergui, A.; Hain, M. Investigating the Impact of Train/Test Split Ratio on the Performance of Pre-Trained Models with Custom Datasets. *Int. J. Adv. Comput. Sci. Appl.* **2024**, *15*, 2. [CrossRef]
18. Wang, C.; Li, X.; Shi, Y.; Jiang, W.; Song, Q.; Li, X. Load forecasting method based on CNN and extended LSTM. *Energy Rep.* **2024**, *12*, 2452–2461. [CrossRef]
19. Jalali, S.M.J.; Ahmadian, S.; Khosravi, A.; Shafie-khah, M.; Nahavandi, S.; Catalão, J.P.S. A Novel Evolutionary-Based Deep Convolutional Neural Network Model for Intelligent Load Forecasting. *IEEE Trans. Ind. Inform.* **2021**, *17*, 8243–8253. [CrossRef]
20. Koukaras, P.; Bezas, N.; Gkaidatzis, P.; Ioannidis, D.; Tzovaras, D.; Tjortjis, C. Introducing a novel approach in one-step ahead energy load forecasting. *Sustain. Comput. Inform. Syst.* **2021**, *32*, 100616. [CrossRef]

21. Alhussein, M.; Aurangzeb, K.; Haider, S.I. Hybrid CNN-LSTM model for short-term individual household load forecasting. *IEEE Access* **2020**, *8*, 180544–180557. [CrossRef]
22. Yunita, A.; Pratama, M.I.; Almuzakki, M.Z.; Ramadhan, H.; Akhir, E.A.P.; Firdausiah Mansur, A.B.; Basori, A.H. Performance analysis of neural network architectures for time series forecasting: A comparative study of RNN, LSTM, GRU, and hybrid models. *MethodsX* **2025**, *15*, 103462. [CrossRef]
23. Park, J.; Hwang, E. A two-stage multistep-ahead electricity load forecasting scheme based on LightGBM and attention-BiLSTM. *Sensors* **2021**, *21*, 7697. [CrossRef]
24. Nti, I.K.; Teimeh, M.; Nyarko-Boateng, O.; Adekoya, A.F. Electricity load forecasting: A systematic review. *J. Electr. Syst. Inf. Technol.* **2020**, *7*, 13. [CrossRef]
25. Steurer, M.; Hill, R.J.; Pfeifer, N. Metrics for evaluating the performance of machine learning based automated valuation models. *J. Prop. Res.* **2021**, *38*, 99–129. [CrossRef]
26. Tanoli, I.K.; Mehdi, A.; Algarni, A.D.; Fazal, A.; Khan, T.A.; Ahmad, S.; Ateya, A.A. Machine learning for high-performance solar radiation prediction. *Energy Rep.* **2024**, *12*, 4794–4804. [CrossRef]

energies

MDPI

Article

Modelling the Capacity, Structure, and Operation Profile of a Net-Zero Power System in Poland in the 2060s

Dariusz Bradło [1,*], Witold Żukowski [1], Jan Porzuczek [2], Małgorzata Olek [2] and Gabriela Berkowicz-Płatek [1]

[1] Faculty of Chemical Engineering and Technology, Cracow University of Technology, Warszawska 24, 31-155 Cracow, Poland; witold.zukowski@pk.edu.pl (W.Ż.); gabriela.berkowicz@pk.edu.pl (G.B.-P.)
[2] Faculty of Environmental Engineering and Energy, Cracow University of Technology, Warszawska 24, 31-155 Cracow, Poland; jan.porzuczek@pk.edu.pl (J.P.); malgorzata.olek@pk.edu.pl (M.O.)
* Correspondence: dariusz.bradlo@pk.edu.pl

Abstract

This study presents an analysis of selected approaches to transforming the Polish power system towards a net-zero greenhouse gas (GHG) emission economy by 2060. The generation-side system models primarily comprise renewable energy sources (RES), supported by nuclear power plants. Two system balancing scenarios were examined: Model G, based on biomethane-fired gas turbines and electrolysers utilising surplus energy; and Model H, which relies primarily on reversible fuel cells (RFCs) operating in a Power-to-Power configuration. Both models were considered under two demographic projections for Poland in 2060: maintaining the current population level (100%) and a decline to 71%. Simulations were performed with an hourly time step over a nine-year period, starting from 2060, using weather data from 2015 to 2023. The total electricity demand in the analysed scenarios ranges from 352 to 542 TWh/year, representing 2.1–3.2 times the current level. The proposed systems include 64 GW of onshore wind capacity, 33 GW of offshore wind, 136 GW of PV, 10 GW of nuclear generation, and extensive storage systems for electricity, heat, and gases (biomethane and hydrogen). In Model G, biomethane and hydrogen storage play a crucial role, requiring storage capacities of 5.8–7.5 billion Nm^3 for biomethane and 6.2–7.0 billion Nm^3 for hydrogen. In Model H, long-term storage relies on hydrogen reservoirs (approximately 12.5 billion Nm^3) integrated with RFC units. The results demonstrate that the choice of architecture dictates the scale and technical requirements of the storage infrastructure. Notably, hydrogen serves as an effective energy storage medium, enabling the elimination of peak gas turbines from the system. Consequently, biomethane resources can be redirected to support the decarbonisation of other sectors of the economy.

Keywords: energy transition; modern energy systems; electrolysis; hydrogen; renewable energy; reversible fuel cells; energy storage systems

1. Introduction

In order to mitigate global climate change, energy policies predicated on net-zero greenhouse gas (GHG) emissions are being promoted worldwide. To date, 145 countries have declared net-zero commitments, with target years including 2050 (e.g., Switzerland, the United Kingdom, Chile, Canada, Australia, Brazil), 2060 (e.g., Saudi Arabia, Russia, Kazakhstan), and 2070 (India) [1]. China, accounting for approximately 33% of global CO_2 emissions in 2024 [2], has pledged to achieve carbon neutrality by 2060 [3]. The European Union (EU) aims to attain climate neutrality by 2050, as articulated in the European Green

Deal [4]. An interim EU objective is to reduce GHG emissions by at least 55% by 2030 relative to 1990 levels.

Poland is a member state of the EU. Consequently, the country is preparing long-term strategies to operationalise net-zero policy goals. The envisaged transition will necessitate intensive development and comprehensive changes across numerous economic sectors. At present, the Polish power system comprises the following principal generation sources: hard coal-fired power plants (20,263 MW), lignite-fired power plants (7605 MW), gas-fired power plants (5976 MW), hydropower plants (2430 MW), wind farms (10,852 MW), and photovoltaic installations (21,618 MW) [5]. Consequently, in 2024, Poland's total energy supply remained predominantly fossil-based: 33% from coal, 33% from oil, and 18% from natural gas. Simultaneously, approximately 70% of electricity generation was derived from fossil fuels (about 56% from coal and about 12% from natural gas). The share of electricity in total final energy consumption in 2023 was only around 16%. Poland is nevertheless progressing towards renewable energy deployment despite substantial reliance on domestic coal resources. The Polish photovoltaic (PV) market ranks among the fastest-growing in the EU, accompanied by ambitious plans for offshore wind development and increasing uptake of heat pumps. By 2030, Poland intends to commission at least 3.4 GW of offshore wind capacity [6]. Between 2012 and 2024, a total of 684,000 heat pumps were supplied to wholesalers and installers [7].

1.1. Energy Storage in High-Penetration RES

With the growing share of weather-dependent renewable energy sources (RES), energy storage is becoming an integral component of the power system at all levels—from households, through district heating systems and industrial sectors, to the national scale. Energy can be stored both in the form of electricity (electrochemical batteries, pumped-storage hydropower plants) and thermal energy, including technologies such as tank thermal energy storage, phase-change materials (PCM), and underground thermal energy storage.

Domestic hot water (DHW) tanks and buffer vessels predominate in residential buildings, typically integrated with heat pumps and solar installations. Approximately 2.4 million such units were sold in Poland between 2013 and 2024. [8]. In 2021, Poland had 5.6368 million single-family houses [9]. The decarbonisation of the building sector indicates further growth in the number of installed storage units. Typical DHW tank volumes range from 50 to 300 litres, while buffer tanks usually have capacities between 300 and 800 litres. Assuming that by 2060 all buildings will be equipped with zero-emission heat sources and tanks of an average capacity of 350 litres, this corresponds to a total thermal storage capacity of approximately 70 GWh ($\Delta T \approx 30$ °C). At the prosumer level, there is also increasing interest in battery energy storage systems (BESS), which enable surplus energy utilisation and enhance self-consumption. According to forecasts, depending on the development scenario, the total capacity of prosumer-scale electricity storage may reach between 4.4 and 7.9 GWh by 2040 [10].

Industrial thermal energy storage can be implemented using sensible heat technologies such as water tanks, thermal oils, solid materials (e.g., concrete, rocks [11]), and molten salts, e.g., $NaNO_3$–KNO_3 [12]. These technologies operate across a wide temperature range—from approximately 30–100 °C for water, 150–300 °C for thermal oils and solids, up to 500–600 °C for molten salts—and are primarily used for short-term storage (hours to days). Alternatively, latent heat storage using phase-change materials with well-defined transition temperatures, typically in the range of 50–200 °C, is particularly suitable for processes requiring stable thermal conditions [13]. The greatest potential for long-term, including seasonal, thermal energy storage lies in thermochemical and sorption technologies, capable of operating at temperatures of 200–800 °C [14], offering high energy density

https://doi.org/10.3390/en19040969

and low losses, although currently remaining at the research and demonstration stage. Estimating the capacity of these storage systems is challenging. They are usually integrated with specific process lines, and the relevant data are not available.

In urban district heating systems, water-based thermal accumulators are widely used. In Poland, combined heat and power plants operate tanks with capacities ranging from hundreds of MWh to over 1 GWh at temperature differences of 30–50 °C [15]. For seasonal heat demand balancing, solutions such as pit thermal energy storage (PTES) and borehole thermal energy storage (BTES) are considered, enabling storage of several to tens of GWh per season; given land availability constraints, the national potential of PTES technology is estimated at approximately 2.4 GWh [16].

At the national level, pumped-storage hydropower (PSH) plays a key role in electricity storage, with a current installed capacity of around 1.7 GW and storage capacity of 7.3 GWh. Plans include the construction of additional PSH plants with a total capacity of approximately 2.8 GW and storage capacity of 19.5 GWh, notably the Tolkmicko (1 GW, 12 GWh) and Młoty (1 GW, 4 GWh) facilities [17]. Additionally, the Solina and Niedzica hydropower plants have a combined theoretical hydrological resource of about 52 GWh, although the currently licenced capacity of Solina at approximately 200 MW is around 640 MWh [18], indicating further exploitable reserves. In this study, the total PSH capacity assumed for 2060 is 30 GWh.

Accelerated growth is also observed in the segment of large-scale BESS. The Polish Energy Group (PGE) plans to build storage facilities with a total capacity of approximately 8 GWh by 2035 [19], including BESS at Żarnowiec (260 MW, 900 MWh) [20] and Gryfino (40 MW, 800 MWh) [21]. The total planned capacity is expected to reach around 17.5 GWh [22–28]. In the capacity market main auctions for the 2029 delivery year, battery storage systems secured approximately 1.8 GW of capacity (with a 17-year capacity obligation) [29]. According to the national investment strategy, installed BESS capacity is projected to increase to approximately 8.7 GW by 2040 [30].

1.2. Hydrogen as an Energy Storage Medium

High capital expenditure [31], capacity degradation, calendar ageing [32], self-discharge behaviour [33], as well as issues related to raw material availability and environmental degradation from extraction [34], indicate that Poland's energy storage system cannot and should not rely solely on batteries. Owing to its capacity for long-term chemical energy storage, hydrogen should become an integral component of the permanent energy system infrastructure. This solution is particularly attractive due to its zero-emission production from RES electricity and the multifunctionality of its applications [35]. Poland's geological conditions allow hydrogen storage in three types of formations: salt caverns, porous formations such as depleted hydrocarbon reservoirs, and saline aquifers [36].

Salt caverns are considered the most efficient and well-established technology for large-scale hydrogen storage. They enable rapid injection and withdrawal cycles, making them ideal for peak demand balancing. They require less cushion gas compared to porous structures, reducing operational costs. Hydrogen stored in this way maintains high purity (no chemical reactions with salt). Depleted natural gas reservoirs are larger than salt caverns, making them suitable for seasonal storage. However, these formations pose risks of hydrogen contamination from residual natural gas and reduced recovery efficiency due to gas mixing and microbial activity [37]. Saline aquifers are the least explored but offer large capacities. They require the highest amount of cushion gas (up to 80% of volume) to maintain structure and pressure. Chemical interactions between hydrogen, brine, and rock may lead to resource losses and require monitoring [38].

https://doi.org/10.3390/en19040969

Currently in Poland, cavern storage facilities used for underground natural gas storage have a total working capacity of 3.23 billion Nm^3 (9.65 $TWhH_2$); after planned expansion, this will increase to 4.70 billion Nm^3 (14.0 $TWhH_2$) [39]. Regarding depleted gas reservoirs, the most significant region in Poland is the Polish Lowlands [40]. Uliasz-Misiak et al. analysed 114 natural gas fields in this area, identifying seven sites (Bogdaj-Uciechów, Kościan S, Pniewy, Radlin, Załęcze, Żuchlów, Brońsko) as the first candidates for hydrogen storage. The technical storage potential in these locations is estimated at 20.3–42.3 $TWhH_2$, enabling storage of over 200 $TWhH_2$ in total [41]. For deep saline aquifers, particularly promising areas include the Mid-Polish Mesozoic Basin. Among the analysed structures, 36 were previously identified for CO_2 storage [42]. For three aquifers—Konary, Suliszewo, and Sierpc—the hydrogen storage capacity was estimated at 5.1 TWh H_2. The total capacity of deep saline aquifers is estimated by Tarkowski et al. at 28.6–38.1 $TWhH_2$ [43]. The theoretical working gas capacity of the assessed salt caverns, depleted hydrocarbon reservoirs, and saline aquifers is 4.70 billion Nm^3 (14.0 $TWhH_2$) [39], 66.7 billion Nm^3 (200 $TWhH_2$) [41], and 11.4 billion Nm^3 (38.1 $TWhH_2$) [43], respectively.

Alternatives to large-scale physical hydrogen storage in geological formations include chemical methods. In many countries, governments and energy agencies recognise ammonia as the most promising hydrogen carrier and storage medium in the energy transition. Its synthesis and cracking are well-established processes, and Poland has 60 years of experience in ammonia production, ranking as the third-largest producer in the EU-27 [44]. Research is also focused on hydrogen storage in organic substances. Liquid Organic Hydrogen Carriers (LOHC) primarily involve cyclic organic compounds such as cyclohexane/benzene and decalin/naphthalene, which can undergo reversible hydrogenation and dehydrogenation reactions [45], with pilot projects confirming LOHC feasibility [46]. Simple organic compounds such as formic acid, methanol, and ethanol can also act as hydrogen carriers, facilitating liquid-phase storage [47]. Previous studies have shown that hydrogen can be released in a controlled manner from polyoxymethylene [48]. In this form, hydrogen stored as a formaldehyde polymer (solid, stable under ambient conditions, non-toxic) can be kept in standard silos.

In this study, hydrogen is considered as a chemical energy carrier within a Power-to-Power (P2P) cycle. In this approach, surplus energy is used for hydrogen production via water electrolysis. The produced hydrogen can then be stored and later reconverted into electricity in fuel cells during periods of high demand. Each stage of this cycle—electrolysis, compression, liquefaction, storage, distribution, and conversion in fuel cells—incurs energy losses. The overall efficiency of the process, referred to as round-trip efficiency (RTE), is a key parameter for assessing P2P system performance. For example, systems based on high-efficiency solid oxide electrolysers (SOEC), followed by storage in hydrides and combustion in micro gas turbines (mGT), achieve a maximum RTE of 29% [49]. Conversely, reversible fuel cells (RFCs) are attracting significant attention due to their potential to substantially improve efficiency. RFCs are electrochemical devices capable of operating both as fuel cells for electricity generation and as electrolysers for hydrogen production [50]. Literature reports various efficiency values for such devices, mainly from modelling studies. Examples include reversible solid oxide cells (rSOC), with declared efficiencies of 70% for SOEC and 60% for solid oxide fuel cells (SOFC) [51], or even 93% for SOEC and 77% for SOFC [52]. Santhanam et al. [53] estimated a theoretical maximum RTE of 98%, although current technology achieves up to 60% at 30 bar pressure. It should be noted that these efficiencies exclude intermediate stages such as hydrogen storage and distribution. Other RFC types include Proton Exchange Membrane (PEM)-based systems, which are currently popular. Comparative analyses of rSOC and PEM-RFCs systems address aspects such as lifetime, cost, and efficiency, indicating, for example, that the power ratio of electrolysers to

https://doi.org/10.3390/en19040969

fuel cells may be approximately 2:1 for PEM-RFCs and 6:1 for rSOC [51,54]. Nevertheless, predicting the parameters of such devices in 2060 remains challenging; therefore, this study adopts values consistent with those reported in the literature.

1.3. Challenges Associated with a Net-Zero Energy Model

Energy transitions aimed at achieving the net-zero target must not disregard the critical issue of energy security. This concept extends beyond the balancing of production, consumption, and storage. A fundamental aspect is ensuring the stability of the national power system, which becomes particularly challenging in systems with a high penetration of weather-dependent generation sources. Furthermore, compliance with power quality standards defined by relevant regulations (e.g., [55]) is essential.

Power quality encompasses multiple factors. Among the most significant are the control of voltage and frequency, as these are the two primary parameters for maintaining system stability. Any imbalance between energy generation and consumption can result in voltage and frequency deviations, which can potentially lead to grid instability. Short-term disturbances such as flicker, swells, sags, or transients not only reduce user comfort but may also cause equipment damage. Additionally, elevated levels of harmonic distortion, quantified by the Total Harmonic Distortion (THD) index, contribute to increased energy losses [56]. These phenomena disrupt the operation of electrical devices and, in severe cases, may trigger failures within the power network.

Recent examples of large-scale emergency events [57–59], classified at the highest OB3 level (blackout) according to the Incident Classification Scale (ICS) methodology [60], demonstrate how local disturbances in grid stability can propagate across entire regions. Published reports [57–59] emphasise the importance of compensation devices for mitigating network disturbances, as well as their adequate quantity and capacity. They also highlight the role of advanced control systems in limiting the consequences of such failures.

With the increasing share of weather-dependent RES, grid control issues have become the subject of intensive research efforts [61–63]. The gradual reduction in the role of conventional generation units based on rotating machines with significant inertia leads to a decrease in the overall system inertia. Consequently, the contribution of so-called 'spinning reserve', an inertial buffer that enables compensation for network imbalances, is also diminished. As a result, the power system becomes more vulnerable to destabilising effects, even from short-term disturbances. To counteract this, solutions referred to as virtual inertia [64] are being introduced. These are devices designed to emulate the inertial behaviour of synchronous generators by employing power electronic converters and advanced control techniques.

Currently, the most rapidly developing solutions for electrical energy storage are battery-based systems, often directly integrated with renewable energy sources such as photovoltaic and wind power. As previously mentioned, subsequent stages of power system modernisation advocate the gradual replacement of battery storage with technologies based on hydrogen. Regardless of the chosen technology, the complexity of conversion and storage systems utilising direct current necessitates the use of inverter-based resources (IBR). Presently, inverters can be broadly categorised into two groups: grid-following inverters (GFL) and grid-forming inverters (GFM) [65,66]. A grid-forming inverter is a device that actively regulates voltage and frequency, emulating the inertial response typically provided by synchronous generators. This capability enables grid stabilisation and facilitates the integration of renewable energy sources. The current predominance of GFL inverters, particularly in distributed prosumer photovoltaic systems, adversely affects grid stability. It is recommended that the share of GFM inverters should reach at least 30%.

https://doi.org/10.3390/en19040969

A key role in stabilising modern power systems is played by devices belonging to the Flexible AC Transmission Systems (FACTS) family [66,67]. Their essential function is to compensate for disturbances such as voltage fluctuations and frequency instability, thereby ensuring grid balance and reactive power management. The integration of FACTS devices into contemporary grids is therefore critical, especially given the growing penetration of renewable energy sources. Commonly deployed FACTS solutions include Static VAR Compensators (SVC), Thyristor-Controlled Series Capacitors (TCSC), Static Synchronous Compensators (STATCOM), and Static Synchronous Series Compensators (SSSC) [67]. A common feature of all FACTS devices is the use of fast-switching power electronic circuits controlled by purpose-specific algorithms. Some FACTS devices are connected in series with the transmission line, while others are installed in parallel.

It is noteworthy that specific electrical devices, although primarily designed for other purposes, can exhibit functionalities similar to FACTS systems. One such example is water electrolysers, provided that they operate under an appropriate control regime [68]. In ref. [69], the properties of electrolysers functioning as smart loads were analysed. The study demonstrated that, depending on the utilisation level of the electrolyser, it was possible to modulate the network load, indirectly contributing to voltage regulation. Moreover, improvements were observed in other aspects of power quality. These characteristics were further explained in ref. [70], where impedance spectroscopy was employed to determine the frequency response of electrolysers. The results indicate the potential for controlling the phase angle and, consequently, sourcing or sinking reactive power. Additionally, Valikhany et al. [64] investigated the feasibility of employing electrolysers and battery storage systems equipped with appropriate power electronics and control strategies as sources of virtual inertia. Attention is also drawn to the potential application of supercapacitor banks, which are significantly better suited for short-term buffering compared to conventional battery systems. These properties highlight an additional, beneficial aspect of hydrogen-based energy storage.

1.4. Demographic Impact on the Power System

As of 31 December 2024, the population of Poland was 37,489,087 [71]. The population forecast for 2023–2060, prepared by Statistics Poland [72], indicates the following, depending on the scenario:

- a decline in population to approximately 30.93 million in the baseline (medium) variant, 26.65 million in the low variant, and 34.80 million in the high variant,
- an advanced process of societal ageing—a dynamic increase in the share of the 65+ population,
- a reduction in the working-age population by 25–40%, depending on the scenario,
- changes in the structure of households and the spatial distribution of the population.

These factors may affect energy demand across various sectors, including residential and municipal, transport, and services. A decline in the number of inhabitants leads to reduced energy consumption for residential heating, cooling, lighting, and the services sector. At the same time, a growing share of elderly people alters the demand profile—energy is consumed differently than by a younger, economically active population. A shrinking working-age population results in lower economic activity and, thereby, reduced overall energy demand of the economy. Furthermore, anticipated population movements—including continued suburban expansion—will reshape future energy consumption profiles in transport and buildings. In light of the above, it is justified to apply a demographic correction factor in forecasting energy demand to 2060.

https://doi.org/10.3390/en19040969

1.5. A New Balancing Model for the Polish Power System

The net-zero balancing strategy established in our earlier work [73] serves as a baseline for this analysis. However, while the 2060 target remains unchanged, this paper introduces a pivotal shift in system architecture to evaluate the performance of a novel integrated model. In the previous approach [73], supply-side balancing relied on biomethane-fired gas turbines, serving as centrally dispatched generation units (CDGUs). Conversely, the demand side utilised a centralised electrolyser system, acting as centrally dispatched consumption units (CDCUs). These electrolysers were the final recipients of surplus RES energy (after electricity and heat storage), and the produced hydrogen was allocated to decarbonization of hard-to-abate industries, particularly heavy industry.

In view of the global trend recognising hydrogen as an energy storage medium, we simulate a system in which hydrogen directly functions as storage within a Power-to-Power configuration. This variant becomes feasible if electrolysers are replaced by reversible fuel cells (RFCs). In the proposed concept, RFCs are the first recipients of surplus RES energy, while simultaneously assuming the role of stabilising generation units (currently provided by coal-fired plants). Consequently, RFCs operate both as centrally dispatched demand units and centrally dispatched generation units, ensuring system stability on both the demand and supply sides.

The differences between the two models are summarised in Table 1. Simulations were conducted for four modelling variants: the reference model from [73] based on biomethane-fired gas turbines (Model G) and the RFCs-based model (Model H). Both models were analysed under two demographic scenarios: the current population level (100%) and the "low" population forecast (71%) [57].

Table 1. The comparison of approaches to the functioning of a net-zero energy system in Poland.

	Model G	**Model H**
System balancing	biomethane-fired gas turbines	reversible fuel cells
Industry decarbonization	hydrogen from electrolysis	biomethane
Priority of electrolysers	the final recipient of excess energy	the first recipient of excess energy
Electrolyser capacity	optimised to meet industrial hydrogen production needs	optimised to balance the system
Biomethane production and storage	connected to gas turbines	connected to industrial facilities
Hydrogen production and storage	connected to industrial facilities	connected to renewable energy sources
Centrally dispatched generation units	biomethane-fired gas turbines	reversible fuel cells
Centrally dispatched consumption units	electrolysers	reversible fuel cells

This study extends previous research by presenting a long-term comparative assessment of two fundamentally different net-zero balancing strategies for the Polish energy system. It anticipates the comprehensive electrification of transportation and heating sectors, complemented by the decarbonisation of industrial production. Both systems were evaluated under identical meteorological and demographic conditions. The key innovation lies in the combination of multi-year hourly simulations, demographic projections, cross-sector electrification and optimised storage sizing. This enables a comprehensive assessment of the capacity requirements of energy infrastructure components and decarbonisation pathways.

https://doi.org/10.3390/en19040969

2. The Methods of Simulation

2.1. Simulation Tools

The calculations were performed using an energy system simulator developed by the Polish governmental institution, the National Centre for Research and Development (NCBR) [74,75]. The parameters characterising the power system were computed using MS Excel 2021 and MATLAB R2025b. For optimisation of system parameters, the 'fmincon' function with the 'sqp' algorithm was applied. A detailed description of the simulator is provided in [73].

2.2. Input Data

Simulations were conducted with an hourly time step to align with the resolution of meteorological data (solar irradiance, wind speed, ambient temperature). These data were used to calculate RES generation, heating demand, and heat pump performance. Input datasets were sourced from publicly available tools and databases (PVGIS [76], ENTSO-E [77], Open-meteo [78], and Instrat [79]).

Input parameters for the net-zero energy system in 2060 were selected based on analyses presented in [73]. In summary, the minimum renewable capacities ensuring a positive energy balance under weather conditions from 2015 to 2023 were determined. For wind energy, the potential installed onshore capacity in Poland ranges from 63.4 GW to 118.3 GW, depending on the legally required minimum distance between wind farms and residential areas (700 m or 500 m, respectively). The potential offshore wind capacity in the Polish Baltic Sea region is 33 GW [80]. The photovoltaic potential in Poland has not been precisely defined, but estimates for agrophotovoltaics alone indicate a capacity of 118.8 GW [81]. In the model, installed capacities were assumed as follows: 64 GW onshore wind, 33 GW offshore wind, and 136 GW PV. An additional balancing element is nuclear power. According to estimates by the Ministry of Climate, the economic feasibility and rationale for developing nuclear generation with a total capacity of 9.9 GW by 2050 have been confirmed [82]. The current electrical capacity of heat pumps (1.5 GW_E) was estimated based on the following assumptions: 684,000 units with an average thermal capacity of 8 kW and a seasonal coefficient of performance (SCOP) of 3.5.

2.3. Energy Storage Modelling and the Role of RFCs

In the new model (Model H), surplus electricity is first directed to RFCs operating in electrolysis mode, followed by stationary battery storage, electric vehicles, PSH, and thermal storage systems. Energy deficits are compensated sequentially by RFCs (in fuel cells mode), residential thermal storage, industrial thermal storage, and electricity storage. This configuration positions RFCs as key stabilising units within the system.

It was assumed that the electrolysers and fuel cells' capacities within RFCs are equal. Electrolysis efficiency was set at 90%, fuel cells efficiency at 70%, and hydrogen storage and transmission efficiency at 97%, resulting in an overall Power-to-Power round-trip efficiency (RTE) of 61.1%. Additionally, it was assumed that 90% of waste heat from RFCs can be stored in industrial thermal storage systems, subject to sufficient storage capacity.

2.4. Assumptions for Electricity and Heat Storage

The total electricity storage capacity was preliminarily set at 400 GWh, including 30 GWh for PSH and 230 GWh in electric vehicles. The latter assumes 23 million EVs by 2060, each with an average battery capacity of 50 kWh, of which 20% of this capacity is utilised for energy storage purposes. Thermal storage capacities were set at 70 GWh for residential systems and 35 GWh for industrial systems.

https://doi.org/10.3390/en19040969

2.5. Energy Import as a Balancing Mechanism

A novel feature of the model is the inclusion of electricity imports as the final balancing mechanism. The annual average imports were assumed to be 2 TWh throughout the analysed period. Considering Poland's electricity production (166–180 TWh) and import level (15 TWh) during 2021–2023 [83], the adopted value is realistic and facilitates a reduction in the required generation capacities.

2.6. Biomethane and Industrial Decarbonisation

Biomethane plays a significant role in both models (G and H). The projected biomethane production of 7.8 billion Nm^3/year by 2060 is based on the current technical potential presented in national reports [84,85]. In contrast to Model G (described in [73], where green hydrogen was used to decarbonise industry), Model H adopts extensive biomethane utilisation. It could be used as: a feedstock for hydrogen production (fertilisers, refineries), a methane substitute (steel, cement), and a base for synthetic fuels (aviation, shipping). Assuming hydrogen production via steam methane reforming with Carbon Capture and Storage (CCS efficiency of 90%) and a consumption rate of 3.7 kg biomethane/kg H_2 [86], industrial hydrogen demand corresponds to 7.87 billion Nm^3 of biomethane annually. Only domestic sources of biomethane and hydrogen were considered.

2.7. Optimisation Procedure

Remaining system parameters for 2060 were determined using a sequential optimisation procedure. In the first step, the minimum heat pump capacity was calculated to ensure that 90% of the heating demand is met by heat pumps (10% by electric heaters). Next, the minimum RFCs capacity was determined under the constraint of $\leq$2 TWh/year electricity import. Subsequently, the minimum hydrogen storage capacity required to guarantee a continuous supply to fuel cells was established. The final step involved optimising the minimum capacities of electricity and thermal storage systems to maintain the assumed electricity import level. For Model G [73], an analogous optimisation approach was applied, but with a different sequence: minimum heat pump capacity, minimum gas turbine capacity, minimum electricity and thermal storage capacities, minimum electrolyser capacity, and minimum biomethane and hydrogen storage capacities. In all scenarios, PSH capacity was fixed at 30 GWh, and the minimum battery storage capacity was set at 17.5 GWh.

2.8. Simulation Scenarios

The simulations covered a nine-year period following 2060, during which: weather conditions corresponded to those observed in 2015–2023, electricity demand replicated the patterns observed in 2015–2023, and heating and transport demand remained constant at 2023 levels. For both models (G and H), simulations were conducted under demographic constraints, assuming a demographic index of 71% (low population scenario [72]). This index was applied to estimate future electricity production and demand, alongside transport, heating, and storage capacities relative to the baseline (100% population). An exception was made for the industrial sector, for which constant demand for heat and hydrogen was assumed at 70.8 TWh and 56.3 TWh, respectively (2024 levels). This approach reflects the fact that industrial production in Poland is strongly influenced by global market conditions, making the impact of domestic demographics difficult to quantify.

https://doi.org/10.3390/en19040969

3. Results and Discussion

3.1. Electricity Demand

The current average annual electricity demand in Poland is 168.1 TWh (Base 2025, Table 2). In the proposed climate-neutral models, this value increases significantly to 446.0 TWh (2.7× current demand), 352.2 TWh (2.1×), 541.5 TWh (3.2×), and 395.7 TWh (2.4×) for models G-100%, G-71%, H-100%, and H-71%, respectively. This indicates that achieving net-zero GHG emissions requires preparing the power system to transmit electricity volumes 2.7–3.2 times higher than today in models G-100% and H-100%, or 2.1–2.4 times higher in models G-71% and H-71%. This is a direct consequence of the complete electrification of sectors such as heating and transport, alongside the introduction of hydrogen as a chemical energy carrier. The higher energy demand in Model H stems from the operational characteristics of electrolysers integrated with RFCs. These units produce hydrogen not only for industrial decarbonisation but also as part of the power system balancing mechanism. While this mechanism increases total electricity consumption, it simultaneously enhances system flexibility under conditions of high variability in RES generation. In Model G, hydrogen production is limited to industrial demand, leading to lower overall electricity consumption.

3.2. Generation Structure and the Role of Balancing Resources

Currently, the Polish power system has limited RES capacities—approximately 21.6 GW of PV and 10.9 GW of onshore wind (Table 2). Based on weather conditions from 2015 to 2023, these capacities yield an average annual generation of around 51.3 TWh, covering one-third of national demand.

In the analysed scenarios, G-100% and H-100%, electricity is generated primarily by onshore and offshore wind farms and PV installations, yielding approximately 445 TWh annually. In models G-71% and H-71%, this output is around 314 TWh. The system is supported by nuclear power plants (10 GW) and balancing units, whose characteristics depend on the adopted model. In Model G, stabilisation is provided by biomethane-fired gas turbines with capacities of 37.9 GW (G 100%) and 26.9 GW (G-71%), and electrolysers rated at 12.1 GW and 14.5 GW, respectively. In Model H, this role is assumed by RFCs with capacities of 41.1 GW (H-100%) and 29.7 GW (H-71%).

A notable result is that the required electrolyser capacity in Model G-71% (14.5 GW) is higher than in Model G-100% (12.1 GW). This stems from the lower installed RES capacity; to meet constant industrial hydrogen demand, the system requires more intensive operation of electrolysers during narrower windows of surplus generation. This system behaviour is consistent with the findings of Liponi et al. [87], who show that reduced renewable energy penetration leads to lower electrolyser utilisation rates. Consequently, it requires higher installed electrolyser capacities to meet steady hydrogen demand with variable renewable supply.

To meet residential heating demand, heat pump capacities of 9.81 GW_E in 100% scenarios and 6.92 GW_E in 71% scenarios are required, reflecting the projected population decline.

https://doi.org/10.3390/en19040969

Table 2. The comparison of the electricity system parameter values used to carry out simulations for 2025 and 2060 (energy values are the result of simulations).

Model	Production						Heat Pumps	Storage					Demand				
	Wind Onshore	Wind Offshore	PV	Nuclear	Gas	Fuel Cells		Batteries	EV Batteries	PSH	Heat Storage (Industry)	Heat Storage (Household)	Transportation	Industry Heat	Household Heat	Electricity	Electrolysers
	GW						GW_E	GWh_E			GWh_{Th}		Electrification—Share of Current Demand				GW
Base (2025)	10.9	0.0	21.6	0.0	6.0	0.0	1.5	0.2	0.0	7.3	2.4	29.2	0%	0%	11%	100%	0.0
G-100% (2060)	64.0	33.0	136.0	10.0	37.9 *	0.0	9.81 *	17.5 *	0.0 *	30.0	20.2 *	48.2 *	100%	100%	100%	100%	12.1 *
G-71% (2060)	45.2	23.3	96.0	7.1	26.9 *	0.0	6.92 *	17.5 *	0.0 *	30.0	24.7 *	40.2 *	71%	100%	71%	71%	14.5 *
H-100% (2060)	64.0	33.0	136.0	10.0	0.0	41.1 *	9.81 *	140.0 *	230.0 *	30.0	35.0 *	70.0 *	100%	100%	100%	100%	41.1 *
H-71% (2060)	45.2	23.3	96.0	7.1	0.0	29.7 *	6.92 *	98.8 *	162.3 *	30.0	24.7 *	49.4 *	71%	100%	71%	71%	29.7 *
Energy, TWh																	
Base (2025)	31.1	0.0	20.2	0.0	116.8	0.0							0.0	0.0	21.3	146.8	0.0
G-100% (2060)	183.2	135.1	126.9	70.8	45.5	0.0							104.6	70.8	67.5	146.8	56.3
G-71% (2060)	129.2	95.3	89.5	49.9	36.4	0.0							73.8	70.8	47.6	103.6	56.3
H-100% (2060)	183.2	135.1	126.9	70.8	0.0	60.8							104.6	70.8	67.5	146.8	151.7
H-71% (2060)	129.2	95.3	89.5	49.9	0.0	50.1							73.8	70.8	47.6	103.6	99.9

* optimised values

https://doi.org/10.3390/en19040969

3.3. Energy Balance, Surplus Generation, and the Significance of Unused Energy

In models G-71% and H-71%, the average annual generation from RES and nuclear power amounts to 364.0 TWh, while the combined demand for electricity, heat, and transport totals 295.8 TWh (Tables 3 and 4). Despite this apparent surplus on an annual energy basis, limitations in electricity and thermal storage capacities necessitate the activation of supporting sources or energy imports during periods of low wind and solar availability. This illustrates a fundamental challenge of highly decarbonized energy systems: positive annual energy balances do not necessarily translate into real-time power adequacy or system reliability under unfavourable weather conditions as widely discussed in the literature [88,89].

Table 3. The average annual balance of the electric energy system in the model G-71%.

Supply	TWh	Demand	TWh
Wind onshore	129.2	Electricity	103.6
Wind offshore	95.3	Industry heat	70.8
PV	89.5	Household heat	47.6
Nuclear	49.9	Transportation	73.8
Gas	36.4	Electrolysers (Hydrogen)	56.4
Cogeneration	7.2	Unused	54.3
Batteries	7.7	Batteries—charging	9.0
Heat storage (industry)	1.2	Heat storage (industry)—charging	1.3
Heat storage (household)	8.0	Heat storage (household)—charging	7.8
Import	2.0	Transmission losses of gas	1.9
SUM	426.5	SUM	426.5

Table 4. The average annual balance of the electric energy system in the model H-71%.

Supply	TWh	Demand	TWh
Wind onshore	129.2	Electricity	103.6
Wind offshore	95.3	Industry heat	70.8
PV	89.5	Household heat	47.6
Nuclear	49.9	Transportation	73.8
Reversible Fuel Cells (Fuel Cells)	50.1	Reversible Fuel Cells (Electrolysers)	99.9
Heat from RFCs	0.2	Unused	20.6
Batteries	0.8	Batteries—charging	1.0
Heat storage (industry)	0.2	Heat storage (industry)—charging	0.2
Heat storage (household)	0.3	Heat storage (household)—charging	0.1
Import	2.0		
SUM	417.6	SUM	417.6

In Model G-71%, gas turbines operating in cogeneration deliver 36.4 TWh of electricity and 7.2 TWh of heat. A transmission loss of 5% for gas was assumed, resulting in 1.9 TWh on the demand side. Electrolysers utilise surplus energy in this scenario, producing 56.4 TWh of hydrogen for industrial use. In Model H-71%, RFCs stabilise the system on both the demand side (99.9 TWh) and the supply side (50.1 TWh).

Despite the application of advanced balancing mechanisms, both models exhibit significant amounts of unused energy, amounting to 20.6 TWh in H-71% and 54.3 TWh in G-71%. On an annual basis, this volume ranges from 20.6 TWh in H-71% to 123.4 TWh

https://doi.org/10.3390/en19040969

in G-100%. These values remain stable across subsequent simulation years (Figure 1) suggesting that inter-annual weather variability has a limited impact on the overall structural surplus. This indicates a structural overcapacity within the system, which converges with findings from other European energy transition models [90] suggesting that intentional overcapacity is often more cost-effective than long-term storage. Such surpluses represent significant potential for further development of the hydrogen sector or energy-intensive industrial branches—particularly those that can operate flexibly in response to grid conditions. Examples include energy-intensive artificial intelligence training processes with adjustable intensity or autonomous dark factories, which do not require continuous human-staffed operation.

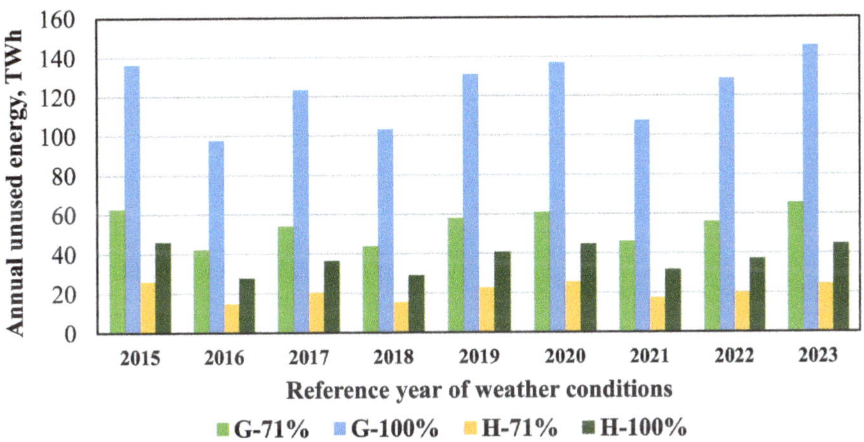

Figure 1. The comparison of the annual unused energy in individual years of the long-term simulation.

3.4. Electricity and Heat Storage

Electricity and thermal storage systems support the system's energy balance. In Model H-100%, storage capacities remained at the levels established at the beginning of the analysis; however, their contribution to the overall energy balance was minimal (Table 2). Optimisation performed for Model G-100% reduced the electricity storage capacity to 47.5 GWh_E, which is 352.5 GWh_E less than in Model H-100%. The capacities of industrial thermal storage and residential thermal storage were also reduced by 14.8 GWh_{Th} and 21.8 GWh_{Th}, respectively. In Model H-71%, it was not possible to reduce the capacities of electricity and thermal storage through optimisation. Conversely, in Model G-71%, the total battery storage capacity was 47.5 GWh_E, which is 243.5 GWh_E less compared to Model H-71%. Additionally, residential thermal storage capacity was reduced by 9.2 GWh_{Th} in Model G-71%.

The operation of electricity and thermal storage systems is detailed in Tables 5 and 6, which present the nine-year cumulative performance of each storage type. It was assumed that all storage units were initially charged to 50%. Electricity and thermal storage play a crucial role in short-term system balancing; however, their effectiveness is significantly constrained by technical parameters.

According to the simulator's logic, residential thermal storage accumulates heat from three sources: waste heat from gas turbines operating in cogeneration, heat available from heat pumps during periods of electricity surplus, and heat from electric heaters located adjacent to the storage units. Residential thermal storage, with a charging/discharging time of 50 h, achieves practical efficiencies of 66.3% in Model G-71% and 35.5% in Model H-71%, despite a theoretically assumed efficiency of 98%. Limited storage capacity results

https://doi.org/10.3390/en19040969

in significant dissipation of surplus heat to the environment. Even a 1000-fold increase in storage capacity in Model G-71% raises efficiency only to 73%, indicating the inherent limitations of this solution.

Table 5. The long-term balance of the storage systems in the model G-71%.

Type of Storage	Heat Household	Heat Industry	Electric Energy	Hydrogen	Biomethane
Unit	GWh_{Th}		GWh_E	Million Nm^3	
Storage capacity	40.2	24.7	47.5	13,014	9974
Capacity at the start	20.1	12.3	23.8	6507	4987
To storage (available)	108,835	45,653			
- heat from electrolysers		45,653			
- heat from heat pumps	5101				
- heat from electric heaters	4954				
- heat from cogeneration	98,780				
To storage (effective)	73,662	11,673	81,359	152,871	65,964
	67.7%	0.0%			
From storage	72,172	11,077	69,132	152,812	65,964
Capacity at the end	35.5	24.7	47.5	1980	3008
Efficiency (effective)	98.0%	95.0%	85.0%	97.0%	97.0%
Efficiency	66.3%	24.3%	85.0%	97.0%	97.0%

Table 6. The long-term balance of the storage systems in the model H-71%.

Type of Storage	Heat Household	Heat Industry	Electric Energy	Hydrogen
Unit	GWh_{Th}		GWh_E	Million Nm^3
Storage capacity	49.4	24.7	291.1	12,513
Capacity at the start	24.7	12.3	145.5	6256
To storage (available)	6711	202,733		270,911
- heat from electrolysers		80,904		
- heat from fuel cells		121,829		
- heat from heat pumps	4374			
- heat from electric heaters	2336			
- hydrogen production				270,911
To storage (effective)	2431	1628	8684	226,216
	36.2%	0.8%		83.5%
From storage	2359	1534	7236	215,846
Capacity at the end	48.7	24.7	291.1	9841.0
Efficiency (effective)	98.0%	95.0%	85.0%	97.0%
Efficiency	35.5%	0.8%	85.0%	81.0%

Industrial thermal storage exhibits even lower efficiency, at 24.3% in Model G-71% and 0.8% in Model H-71%. In the simulator, industrial thermal storage is assumed to use electric heaters that activate only during temporary electricity surpluses. The charging/discharging time for this storage type was set at 50 h. Additional heat sources for industrial storage include waste heat from electrolysers and, in Model H, waste heat from fuel cells. In practice, due to limited storage capacity and long charging times, waste heat from electrolysers and RFCs is almost entirely lost.

Electricity storage systems, with a charging/discharging duration of 6 h, achieve the assumed efficiency of 85%, making them the most effective tool for short-term power

https://doi.org/10.3390/en19040969

balancing. It should be noted that the simulator does not differentiate between battery storage, electric vehicles acting as storage units, and pumped-storage hydropower plants.

3.5. Hydrogen and Biomethane Storage

While battery storage systems handle short-term power balancing, hydrogen and biomethane storage play a crucial role in ensuring long-term supply stability.

In Model G, hydrogen storage serves as a safeguard for industrial supply, and its location should prioritise proximity to major industrial hubs to optimise delivery and minimise transport losses. The minimum required hydrogen storage capacity in Model G-71% is 6.5 billion Nm^3, which, assuming 50% initial filling, corresponds to a total capacity of 13.0 billion Nm^3 (Table 7). As evidenced by Figure 2, this facility reaches a peak of 7.0 billion Nm^3 without ever attaining full capacity. Nevertheless, this requirement represents a technological, logistical, and economic challenge, particularly since Model G allocates all available salt caverns and methane storage facilities exclusively for biomethane.

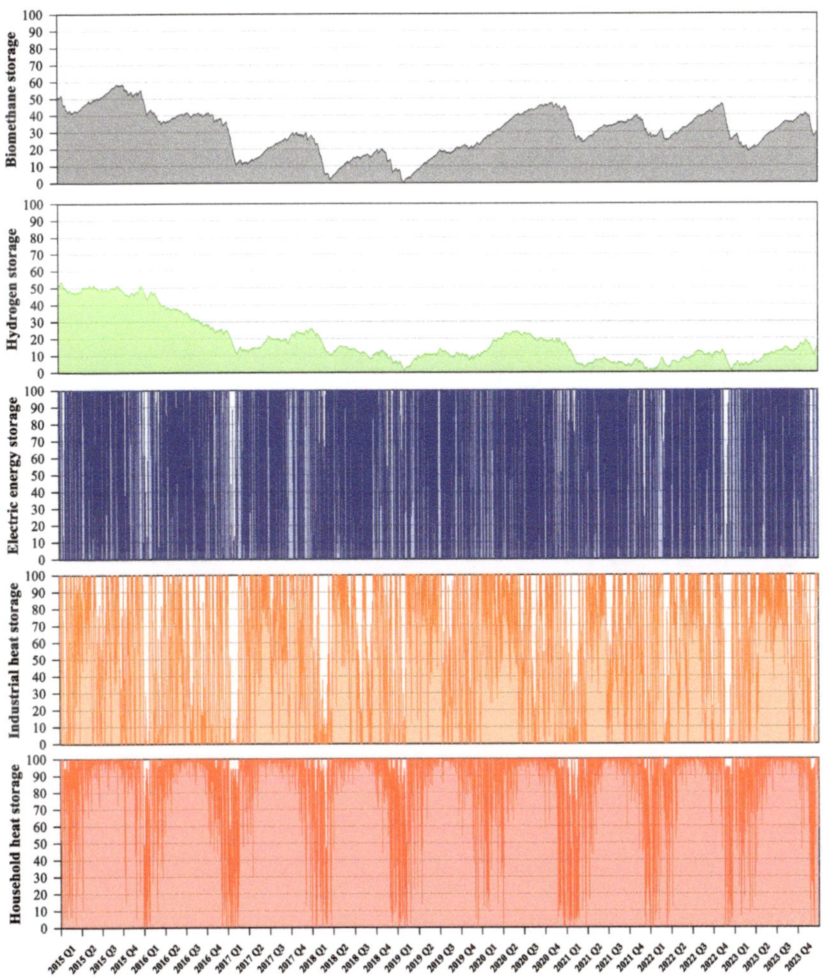

Figure 2. The state of charge of the storage systems over a nine-year simulation period for the Model G-71%.

https://doi.org/10.3390/en19040969

Table 7. The comparison of the models' performance.

Model	G 100%	G 71%	H 100%	H 71%
Gas turbines, GW	37.9	26.9	0.0	0.0
Electrolysers/Fuel cells, GW	12.1	14.5	41.1	29.7
Unused energy, TWh/year	123.4	54.3	37.5	20.6
Biomethane demand, billion m^3/year	9.2	7.3	7.9	7.9
Biomethane storage capacity, billion m^3	12.8 (7.5) *	10.0 (5.8) *	-	-
Hydrogen demand in industry, billion m^3/year	17.0	17.0	0.0	0.0
Hydrogen storage capacity, billion m^3	11.7 (6.2) *	13.0 (7.0) *	12.4	12.5
Hydrogen production, billion m^3/year	17.0	17.0	45.7	30.1
Unused hydrogen, billion m^3/year	0.0	0.0	15.3	5.0
Import, TWh/year	2.0	2.0	2.0	2.0

* in brackets: the minimum capacity when the initial filling exceeds 50%.

In Model H, the minimum total hydrogen storage capacity was determined to ensure a continuous supply to fuel cells and prevent storage depletion. For Model H-71%, this results in a capacity of 12.5 billion Nm^3 (Table 7). This value poses a major infrastructural challenge, considering that salt cavern storage capacity is estimated at only 4.7 billion Nm^3 (14.0 $TWhH_2$) after expansion [44]. While depleted hydrocarbon reservoirs, with a potential capacity of 66.7 billion Nm^3 [42], could theoretically meet this demand; however, they require a detailed geological assessment and the deployment of hydrogen conversion and off-take infrastructure. The theoretical hydrogen storage efficiency assumed in the simulation is 97%; however, due to the limited storage capacity, the actual efficiency for Model H-71% was 81%.

Biomethane demand in the analysed scenarios is less than 7.9 billion Nm^3/year, remaining below Poland's theoretical production potential—except in Model G-100% (Table 7). In the latter case, the shortfall would need to be met by importing biomethane or by increasing the capacity of other generation elements within the system. For Model G-71%, biomethane production stands at a slightly lower level of 5.8 billion Nm^3, resulting in a lower required storage capacity.

For biomethane storage in Model G-71%, the initial filling level was set at 5.0 billion Nm^3. Assuming 50% initial filling, the total storage capacity should be 10 billion Nm^3, although this could be reduced to 5.8 billion Nm^3 with a higher initial filling level (Table 7).

Model H exhibits a different characteristic, where biomethane production is closely correlated with industrial activity. It is therefore assumed that during periods of increased demand for methane or hydrogen, biomethane production is intensified. Theoretically, this approach eliminates the need for long-term biomethane storage and the use of existing methane storage facilities. Instead, this strategy shifts the requirement toward the storage of biomass feedstocks intended for biomethane production—a challenge that is less problematic than constructing gas storage infrastructure. In practice, however, ensuring the stable operation of a biomethane plant may still require a biomethane buffer to maintain a continuous supply to industry.

In both models, the combined storage capacity for biomethane and hydrogen is comparable, although Model G requires 0.3–1.3 billion Nm^3 more (Table 7). However, the hydrogen balance profiles differ significantly. In Model G, hydrogen losses due to

https://doi.org/10.3390/en19040969

overproduction are avoided, as production is limited to quantities required for industrial decarbonisation. In contrast, Model H exhibits surplus hydrogen production, stemming from the assumption of equal capacities for electrolysers and fuel cells. If literature-based ratios for RFCs were applied (electrolyser-to-fuel cell power ratios of 2:1 to 6:1), even greater hydrogen overproduction would occur in the absence of centralised electrolyser capacity control. Thus, similar to surplus electricity, this excess hydrogen could be considered an opportunity for the development of certain industrial sectors or for widely discussed applications in transport. However, if surplus hydrogen were assumed to be lost, the round-trip efficiency (RTE) of the electricity–hydrogen–electricity cycle would decrease from the projected 61.1% to 40.1% in Model H-100% or 50.2% in Model H-71%.

3.6. State of Charge of Storage Systems

An analysis of the state of charge across different storage types reveals significant differences between Models G and H (Figures 2 and 3). Regardless of storage type, both models exhibit periods of low charge levels due to adverse weather conditions and increased electricity and heat demand.

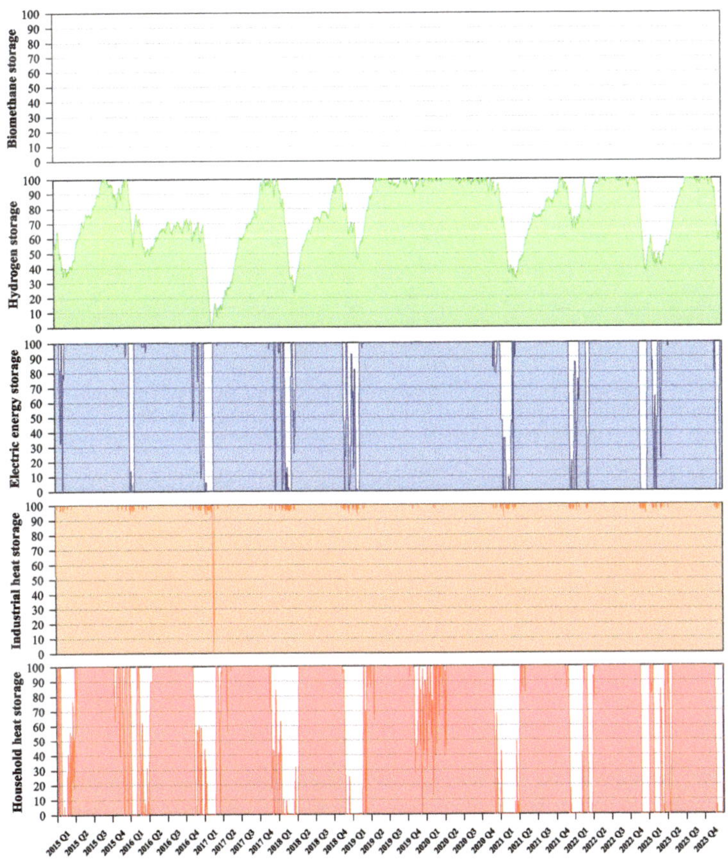

Figure 3. The state of charge of the storage systems over a nine-year simulation period for the Model H-71%.

In Model G, electricity and thermal storage systems serve as the primary recipients of surplus energy, leading to intensive utilisation year-round, particularly from November to March (Figure 2). Energy flows to and from storage in Model G-71% were 7–30 times higher than in Model H-71% (Tables 5 and 6). Such continuous operation and a high number of cycles may accelerate wear and degradation, increasing the frequency of replacements.

In Model H-71%, storage systems played a secondary role, as electrolysers absorbed the majority of the surplus RES generation. In this model, electricity and residential thermal storage remained fully charged for most of the analysed period, while also experiencing extended periods of complete discharge (Figure 3). Such extreme states can be detrimental to durability—particularly for batteries, which should not operate below a defined threshold (typically 10%). Industrial thermal storage remained nearly fully charged throughout the entire period, preventing the absorption of waste heat from RFCs and rendering its contribution to system performance negligible. Consequently, while storage systems in Model H have large nominal capacities (Table 2), their actual contribution to the energy balance is minimal. This suggests that their capacity could be significantly reduced with only a modest increase in RFCs' capacity.

The situation differs for compressed gas storage, which is designed to avoid full depletion. In Model G-71%, the biomethane storage reached its minimum level during the period corresponding to Q1 2019 weather conditions, while hydrogen storage reached its lowest levels during Q1 2019, Q1 2022, and Q1 2023. In Model H-71%, the most critical period occurred under Q1 2017 conditions, during which the hydrogen storage was nearly exhausted.

3.7. Operating Modes of Storage Systems

In Model G-71% (Figure 4A), storage systems remained in active operation (charging/discharging) for the majority of the period, with utilisation rates ranging 51% and 66%. Full charge durations varied significantly, from 17% for industrial thermal storage to 31% for residential thermal storage. In Model H-71% (Figure 4B), electricity and thermal storage systems operated only during short intervals (4–8%), remaining fully charged for the majority of the time (60–92%), which confirms their limited role in this scenario. The simulations did not account for self-discharge effects, which are particularly relevant for thermal storage. In practice, maintaining a full charge for such prolonged periods would lead to substantial energy losses in this model.

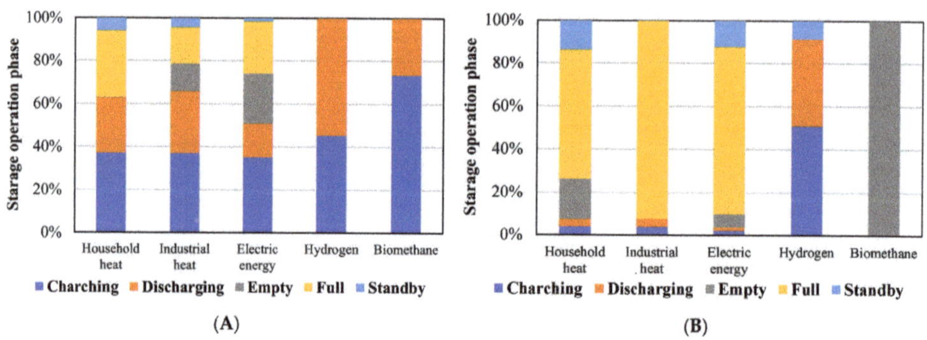

Figure 4. The operation profiles of storage systems over a nine-year simulation period: (**A**) Model G-71%; (**B**) Model H-71%.

Analysis of battery storage revealed that in Model G-71%, batteries remained fully discharged for 23% of the time, compared to only 6% in Model H-71%. This indicates that despite intensive cycling in Model G, cumulative downtime in a fully discharged state is significant, potentially constraining system flexibility during critical moments. A possible solution is to increase battery capacity to maintain a minimum state of charge, ensuring balancing capability during periods of low RES generation.

https://doi.org/10.3390/en19040969

In Model G-71%, gas storage systems operated continuously throughout the analysed period. The biomethane storage remained in charging mode for 73.4% of the time, while the hydrogen storage was being charged for 45.3% of the time. In Model H-71% (Figure 4B), hydrogen storage was being charged for 50.9% of the time, but also exhibited standby states (8.4%).

3.8. Operation of Electrolysers, Gas Turbines, and RFCs

Using hourly data, Figures 5–7 illustrate the operational profiles of the main centrally dispatched units in both models under 2018 weather conditions. Figure 8 provides a quantitative summary of the operating modes for these components over the entire nine-year simulation period.

In Model G-71%, the average electrolyser power reached 12.1 GW (84% of available capacity), compared to 11.0 GW (91%) in Model G-100%. Electrolysers operated at maximum power for 35–48% of the time and at partial load for 11–18% (Figure 8A). Gas turbines ran at maximum power for 3% of the time and at reduced power for 27%, with an average output of 13.3 GW (49% of available capacity) in G-71% and 17.4 GW (46%) in G-100%.

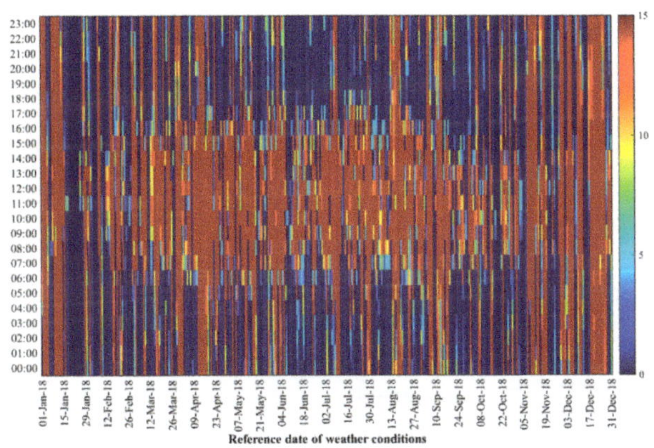

Figure 5. The operation of electrolysers in the model G-71% under 2018 weather conditions.

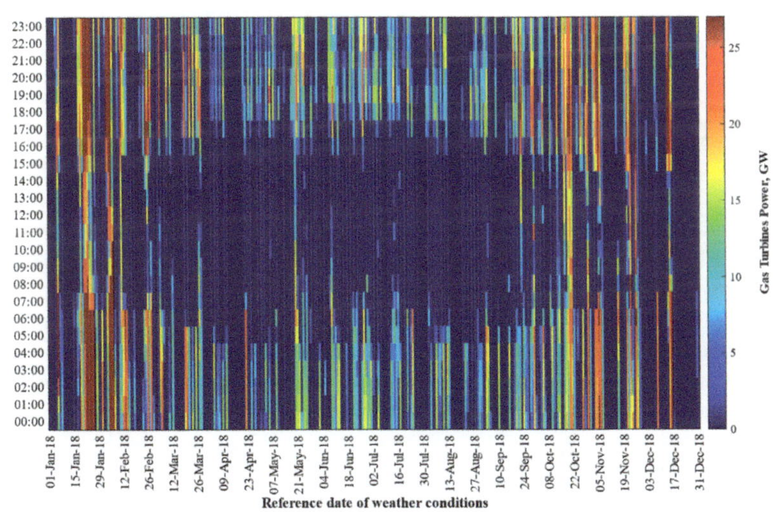

Figure 6. The operation of gas turbines in the model G-71% under 2018 weather conditions.

https://doi.org/10.3390/en19040969

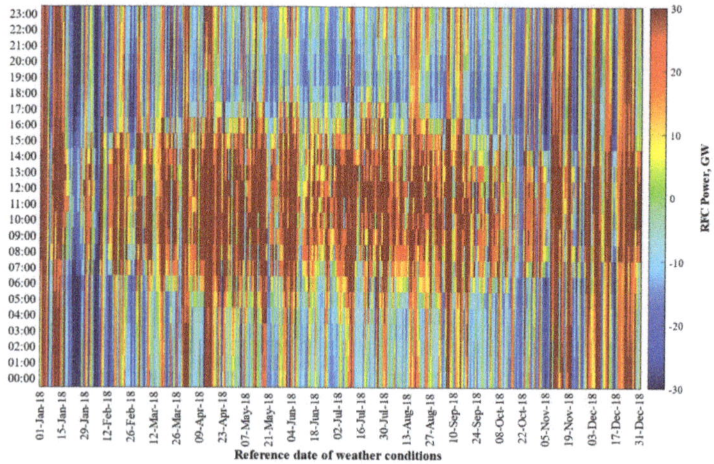

Figure 7. The operation of RFCs in the model H-71% under 2018 weather conditions (Electrolysers—positive values, Fuel cells—negative values).

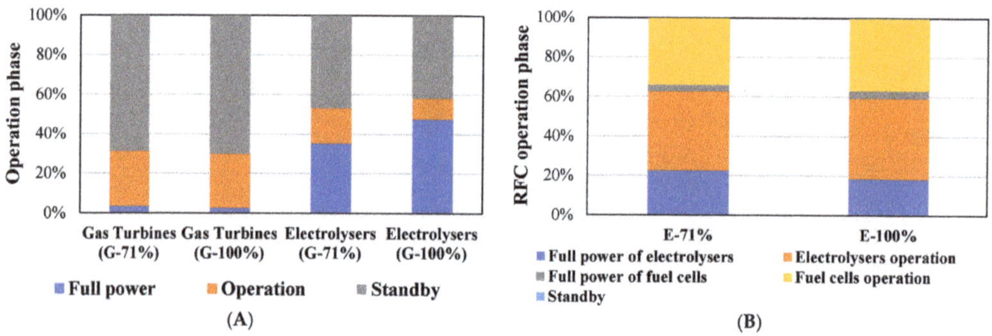

Figure 8. The operation profiles of the major energy consumers and producers over a nine-year simulation period: (**A**) Model G-71%; (**B**) Model H-71%.

Hourly power profiles (G-71%, 2018, Figure 5) reveal a strong correlation between electrolyser activity and PV generation. Electrolysers predominantly reached maximum power between 07:00 and 15:00. Extended periods of continuous peak operation—lasting entire days or several consecutive days—occurred during wind generation surplus. Conversely, prolonged periods of inactivity were also observed. These intervals provide opportunities for necessary maintenance and servicing. However, the system must account for start-up losses and ramp-up times. Furthermore, such periods result in heat losses associated with reheating the equipment to required operating temperatures.

Gas turbines operated predominantly during morning and evening peaks throughout the year (G-71%, 2018, Figure 6), with maximum output coinciding with adverse weather conditions—typically January–March and October–December. These units serve as balancing and reserve capacity; therefore, their siting and capacity must align with demand characteristics and RES variability to ensure reliability during peak hours and unfavourable conditions.

The operational profile of RFCs in Model H differs markedly. In this scenario, RFCs predominantly functioned in electrolyser mode, operating at maximum power for 19–23% and at partial load for 40–41% (Figure 8B). The average electrolyser power reached 19.3 GW (65% of available capacity) in H-71% and 27.7 GW (67%) in H-100%. Conversely, RFCs

https://doi.org/10.3390/en19040969

operation in fuel cells mode accounted for 3–4% of the time at maximum power and 34–37% at partial load, with average fuel cells output at 14.1 GW (47%) in H-71% and 18.6 GW (45%) in H-100%.

A key finding is the minimal share of standby periods for RFCs, ranging from 0.12 to 0.19%. This high utilisation rate supports the validity of the proposed balancing approach in Model H. Hourly power profiles (H-71%, 2018, Figure 7) confirm near-continuous RFCs operation at loads near maximum capacity. However, such intensive operation may have significant implications for device durability and lifetime. Although the simulations did not account for maintenance downtime, all equipment requires periodic servicing in practice. Therefore, for the solution proposed in Model H to be feasible, it is necessary to provide redundant RFCs capacity to compensate for maintenance activities and other typical operational constraints.

3.9. Energy Imports and Their Role

The average annual electricity import assumed in the simulations was 2 TWh. However, import levels varied across simulation years (Figure 9). The analysis shows that electricity imports were low or zero under weather conditions similar to those in 2020, whereas the most adverse conditions for most scenarios corresponded to those observed in 2018. An exception was Model H-71%, where the maximum annual electricity import reached 5.67 TWh/year under weather conditions observed in 2017.

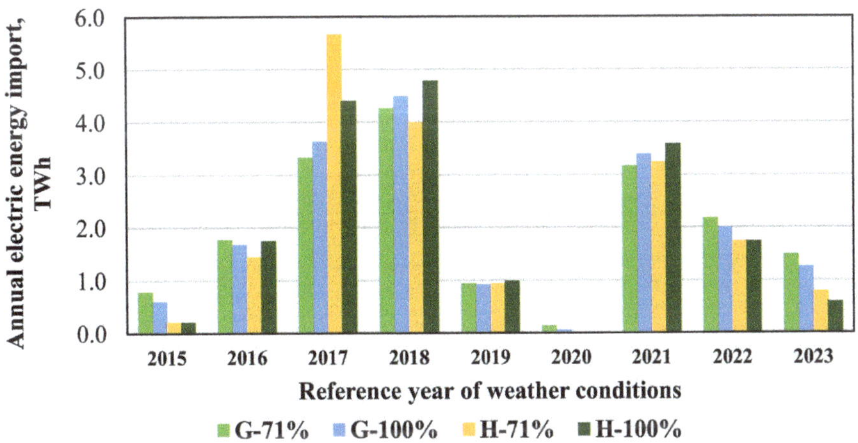

Figure 9. The comparison of annual electricity imports in a long-term simulation.

Using this model as an example—where substantial periodic imports were required—specific time intervals exhibiting power shortages were identified based on simulation results (Table 8). These periods coincided with months characterised by particularly unfavourable weather conditions combined with increased electricity and heat demand. Under Polish climatic conditions, these occurred primarily November–March, with January and February being the most prevalent.

Hourly analysis of Model H-71% indicates that, despite the small proportion of imports over the entire nine-year simulation period (2.1% of the time, across 152 days or 23 months), imports played a critical balancing role during short, high-stress intervals. Maximum hourly import values—34 GWh in Model G-71% and 57 GWh in Model H-71%—highlight the need to ensure adequate transmission capacity and access to external sources during periods of peak energy deficit.

https://doi.org/10.3390/en19040969

Table 8. The long-term electricity imports on a monthly basis in the model H-71%, TWh.

02.2015	**01.2016**	**01.2017**	**02.2017**	**12.2017**	**01.2018**	**02.2018**	**03.2018**	**11.2018**	**12.2018**	**01.2019**	**02.2019**	**01.2021**	**02.2021**	**03.2021**	**12.2021**	**01.2022**	**03.2022**	**11.2022**	**12.2022**	**02.2023**	**11.2023**	**12.2023**
0.21	1.44	2.21	3.38	0.08	0.01	2.53	1.26	0.03	0.18	0.93	0.01	1.07	1.46	0.04	0.67	0.14	0.02	0.21	1.36	0.26	0.04	0.49

These findings suggest that the complete elimination of imports could lead to excessive installed capacity oversizing. Considering current cross-border flow levels (<2 GW [91]), the import volumes determined in the simulations represent an additional challenge for both models.

From a technical perspective, the two system variants represent fundamentally different approaches to energy balancing and RES integration (Table 9), as reflected in the differences in generation, storage, and conversion technologies. In Model G, the role of the central balancing unit is performed by biomethane-fired gas turbines, whereas in Model H, this function is assumed by reversible fuel cells. This distinction translates into markedly different scales of electrolyser capacity requirements: approximately 12–15 GW in Model G, compared to 30–41 GW in Model H, where RFCs also serve as balancing units.

Table 9. The comparative overview of the features of the two proposed models for Poland's net-zero energy system.

	Model G	**Model H**
Generation units	biomethane-fired gas turbines	RFC
Electrolyser power	low (12–15 GW)	high (30–41 GW)
Battery storage capacity	low (approx. 20 GWh)	high (approx. 260–370 GWh)
Heat storage capacity	smaller than in the Model H	higher than in the Model G
Hydrogen storage	large storage facilities, competing with biomethane	large storage facilities, possibility of using salt caverns
Biomethane storage	large storage facilities, competing with hydrogen	none or small
Industrial decarbonisation	focused on hydrogen with a difficult transformation	easy transformation based on biomethane
Centrally controlled units	gas turbines and electrolysers	reversible fuel cells

Distinct differences are also evident in storage infrastructure. In Model G, battery storage capacity is relatively low, remaining at around 20 GWh, whereas Model H requires significantly larger short-term storage capacities, reaching 260–370 GWh. A similar pattern applies to thermal storage: capacities in Model G are smaller than in Model H, where thermal storage is considered as an integral element for stabilising the heating system.

The most pronounced differences occur in the scale and function of hydrogen and biomethane storage. Model G necessitates the construction of large hydrogen storage facilities, which compete for available capacity and infrastructure with biomethane storage, essential for industrial decarbonisation and maintaining gaseous fuels in the system. Model H also requires substantial hydrogen storage; however, due to the different role of biomethane, there is no competition for resources. In Model H, biomethane storage is either negligible or limited to small volumes.

These differences also influence the pathways for industrial decarbonisation. In Model G, the transition relies primarily on hydrogen, requiring deep technological and infrastructural changes, particularly in energy-intensive sectors. In contrast, Model H

https://doi.org/10.3390/en19040969

enables a simpler decarbonisation pathway through the use of biomethane in existing installations or with minimal technological adjustments.

The expansion of large scale systems is subject to several practical constraints. The deployment of wind and PV infrastructure is contingent upon land availability that meets a series of technical, environmental and spatial criteria, which may significantly restrict potential development areas. Furthermore, permitting procedures often introduce significant lead times, as projects must undergo comprehensive environmental impact assessments, public consultations and multi-level administrative reviews. Grid reinforcement is equally critical to accommodate escalating energy flows, the construction of new transmission lines and substations is essential for integrating renewable sources while maintaining system stability. Additionally cross border transfer limits may constrain the effectiveness of imports, as interconnector capacity might be insufficient to support system balancing during periods of peak stress. Seasonal variability exacerbates these bottlenecks, particularly when interconnectors reach congestion limits. Collectively, these factors may impede the projected build out and increase the total cost of the energy transition. Consequently, the establishment of a robust policy framework and the integration of planning processes are imperative to facilitate this transformation.

4. Conclusions

This study evaluated four variants of a national power system for a net-zero Poland, comparing Model G (biomethane-fired gas turbines) with Model H (reversible fuel cells, RFCs). A comparative assessment reveals a fundamental trade-off between technological complexity and resource dependency. Model G offers a lower technical threshold, requiring 12.1–14.5 GW of electrolyser capacity. However, it remains critically dependent on large-scale biomethane production and constant biomass availability, which pose logistical challenges and introduce risks to the long-term stability of biomass supply chains. In contrast, Model H represents a more technologically advanced paradigm; by employing 29.7–41.1 GW of RFCs, it obviates the need for gas turbines and streamlines industrial decarbonisation. It must be noted, however, that RFCs are currently at an early stage of development and are not yet available at the required technical scale, making their implementation dependent on rapid technological breakthroughs and high capital investment.

The analysis indicates that while Model H relies on an average annual electricity import of 2 TWh, this dependency is a strategic choice to enhance economic efficiency. By leveraging cross-border exchange, the system avoids the excessive over-dimensioning of domestic generation and storage infrastructure that would otherwise be required during rare periods of low RES availability. Regardless of the chosen model, the transition requires an extensive investment programme, as the combined storage demand for hydrogen and biomethane could reach 13.7 billion Nm^3 (approx. 41 $TWhH_2$). The current Polish infrastructure is insufficient to meet these requirements, necessitating the urgent development of salt caverns and the modernisation of existing gas storage facilities.

From a policy perspective, the findings highlight the need for a synchronised approach that integrates RES growth with the expansion of regional grid interconnectors and the R&D-driven deployment of emerging technologies such as RFCs. Furthermore, certain limitations regarding the fluctuating capital costs and the efficiency of long-term geological storage remain. Future research should therefore focus on scaling up of reversible fuel cells from pilot projects to industrial-scale applications, their operational stability under dynamic weather conditions, and the optimisation of surplus energy use across varying demographic scenarios, where required RES capacities may range from 164.5 GW to 233 GW. The presented calculations provide an essential framework for methodologies shaping long-

https://doi.org/10.3390/en19040969

term national investment policies aimed at eliminating anthropogenic environmental impacts associated with GHG emissions.

Author Contributions: Conceptualisation, W.Ż., G.B.-P. and J.P.; methodology, software, and validation, D.B., M.O. and G.B.-P.; formal analysis, investigation, and data curation, W.Ż., D.B. and M.O.; writing—original draft preparation, D.B., M.O., J.P. and G.B.-P.; writing—review and editing, D.B., W.Ż. and M.O. All authors have read and agreed to the published version of the manuscript.

Funding: This research received no external funding.

Data Availability Statement: The raw data supporting the conclusions of this article will be made available by the authors on request.

Conflicts of Interest: The authors declare no conflicts of interest.

References

1. The Climate Action Tracker (CAT). CAT Net Zero Target Evaluations. Available online: https://climateactiontracker.org/global/cat-net-zero-target-evaluations/ (accessed on 20 December 2025).
2. International Energy Agency (IEA). CO_2 Emissions. Available online: https://www.iea.org/reports/global-energy-review-2025/co2-emissions (accessed on 20 December 2025).
3. Chen, Z.Y.; Zhao, L.T.; Cheng, L.; Qiu, R.X. How does China respond to the Carbon Border Adjustment Mechanism? An approach of global trade analysis. *Energy Policy* **2025**, *198*, 114486. [CrossRef]
4. Fetting, C. *The European Green Deal*; ESDN Report; ESDN Office: Vienna, Austria, 2020.
5. Polish Transmission System Operator (Polskie Sieci Elektroenergetyczne S.A.). Types of Generation Sources and Definitions, How the National Power System Operates. Available online: https://www.pse.pl/jak-funkcjonuje-krajowy-system-elektroenergetyczny/rodzaje-zrodel-wytworczych-i-definicje (accessed on 20 December 2025). (In Polish).
6. International Energy Agency (IEA). Energy System of Poland. Available online: https://www.iea.org/countries/poland (accessed on 20 December 2025).
7. PORT PC. The Polish Heat Pump Market Is Returning to Growth. Available online: https://portpc.pl/polski-rynek-pomp-ciepla-wraca-do-wzrostow-pozytywne-sygnaly-w-i-polroczu-2025/ (accessed on 20 December 2025). (In Polish)
8. Association of Heating Equipment Manufacturers and Importers (Stowarzyszenie Producentów i Importerów Urządzeń Grzewczych). Report on the Heating Equipment Market in Poland 2024. Available online: https://spiug.pl/app/uploads/2025/06/Raport_Rynek_urzadzen_grzewczych_w_Polsce_2024.pdf (accessed on 20 December 2025). (In Polish)
9. Statistics Poland (Główny Urząd Statystyczny). Housing Conditions in Poland According to the Results of the National Population and Housing Census 2021. Available online: https://stat.gov.pl/download/gfx/portalinformacyjny/pl/defaultaktualnosci/6536/5/2/1/warunki_mieszkaniowe_w_polsce_w_swietle_wynikow_narodowego_spisu_powszechnego_ludnosci_i_mieszkan_2021.pdf (accessed on 20 December 2025). (In Polish)
10. Polish Energy Storage Association (Polskie Stowarzyszenie Magazynowania Energii). The Impact of Expanding Energy Storage Infrastructure on Economic Development in Poland—Forecast Until 2040. Available online: https://psme.org.pl/wp-content/uploads/2024/03/RAPORT_PSME_AGH.pdf (accessed on 20 December 2025). (In Polish)
11. Singh, H.; Saini, R.P.; Saini, J.S. A review on packed bed solar energy storage systems. *Renew. Sustain. Energy Rev.* **2010**, *14*, 1059–1069. [CrossRef]
12. Nunes, V.M.B.; Lourenço, M.J.V.; Santos, F.J.V.; Nieto de Castro, C.A. Molten alkali carbonates as alternative engineering fluids for high temperature applications. *Appl. Energy* **2019**, *242*, 1626–1633. [CrossRef]
13. Sharma, A.; Tyagi, V.V.; Chen, C.R.; Buddhi, D. Review on thermal energy storage with phase change materials and applications. *Renew. Sustain. Energy Rev.* **2009**, *13*, 318–345. [CrossRef]
14. Lizana, J.; Chacartegui, R.; Barrios-Padura, A.; Valverde, J.M. Advances in thermal energy storage materials and their applications towards zero energy buildings: A critical review. *Appl. Energy* **2017**, *203*, 219–239. [CrossRef]
15. Zwierzchowski, R.; Niemyjski, O.; Wołowicz, M. Energy Savings Analysis for Operation of Steam Cushion System for Sensible Thermal Energy Storages. *Energies* **2022**, *15*, 286. [CrossRef]
16. Institute for Renewable Energy (Instytut Energetyki Odnawialnej). Roadmap for the Heat Storage Market in Poland. Available online: https://ieo.pl/dokumenty/aktualnosci/06092024/raport mapa drogowa dla rynku magazynów ciepła w polsce.pdf (accessed on 20 December 2025). (In Polish)

https://doi.org/10.3390/en19040969

17. Expert Team for the Construction of Pumped Storage Power Plants (Zespół Ekspercki do spraw Budowy Elektrowni Szczytowo-Pompowych). The Role of Pumped Storage Power Plants in the National Power System: Conditions and Directions of Development. Available online: https://www.gov.pl/attachment/5560c5ac-fdc9-46e0-bf06-eb0dabd113dd (accessed on 20 December 2025). (In Polish)
18. Energy Regulatory Office (Urząd Regulacji Energetyki). Licence Nr MEE/3/2876/W/OŁO/2022/MWi. Available online: https://api.ure.gov.pl/api/ConcessionFuel/download/55842189-C402-402D-BC22-3257F9B6B728 (accessed on 20 December 2025).
19. PGE Polska Grupa Energetyczna S.A. PGE Group's 2035 Strategy. Available online: https://www.gkpge.pl/content/download/61ab3879bd5772c1ab88ace3f7b99f13/file/20250612_Strategia_PGE_2035_EN_final_share.pdf (accessed on 20 December 2025).
20. Polish Ministry of Climate and Environment. Poland's Largest Battery-Based Energy Storage Facility Is Developed in Żarnowiec. Available online: https://www.gov.pl/web/climate/polands-largest-battery-based-energy-storage-facility-is-developed-in-zarnowiec (accessed on 20 December 2025).
21. Raymond, X. Poland's PGE Launches Tender for 400 MW/800 MWh Battery Storage System. Available online: https://www.blackridgeresearch.com/news-releases/polands-pge-launches-tender-battery-storage-system (accessed on 20 December 2025).
22. Hynfra Energy Storage Webpage. Available online: https://www.hynfraenergystorage.pl/ (accessed on 20 December 2025).
23. OX2 Webpage. Available online: https://www.ox2.com/newsroom/press-releases-news/2024/ox2-secures-200-mw-of-battery-storage-at-government-auctions-in-poland/ (accessed on 20 December 2025).
24. E-Magazyny Webpage. Available online: https://e-magazyny.pl/aktualnosci/rekordowe-zainteresowanie-magazynami-energii-na-najnowszej-aukcji-rynku-mocy/ (accessed on 20 December 2025). (In Polish)
25. Markosyan, M. DTEK, Columbus Energy Plan to Add 398 MW of Energy Storage. Available online: https://renewablesnow.com/news/dtek-columbus-energy-plan-to-add-398-mw-of-energy-storage-837221/ (accessed on 20 December 2025).
26. DRI-ENERGY Webpage. Available online: https://dri-energy.com/en/dri-acquires-112mw-battery-storage-project-poland-from-greenvolt/ (accessed on 20 December 2025).
27. N2OFF Inc. N2OFF via Solterra Expands European Footprint with Entry into Fourth Project—A Battery Storage Venture in Poland. Available online: https://www.globenewswire.com/news-release/2025/05/09/3078159/0/en/N2OFF-via-Solterra-Expands-European-Footprint-with-Entry-into-Fourth-Project-a-Battery-Storage-Venture-in-Poland.html (accessed on 20 December 2025).
28. Axpo Holding AG Webpage. Available online: https://www.axpo.com/group/en/news-and-stories/media-releases/2025/axpo-and-edp-sign-landmark-agreement-to-optimise-energy-storage-.html (accessed on 20 December 2025).
29. Energy Regulatory Office (Urząd Regulacji Energetyki). Information from the Chairman of the Energy Regulatory Office No. 2/2025 of 3 January 2025 on the Announcement of the Final Results of the Main Auction for the 2029 Delivery Year. Available online: https://www.ure.gov.pl/download/9/15112/Ostateczneogloszeniewynikowaukcjiglownejna2029.pdf (accessed on 20 December 2025). (In Polish)
30. Ministry of Climate and Environment of the Republic of Poland. Appendix 3 to aKPEiK: Forecast Assumptions and Forecasting Methodology. Available online: https://bip.mos.gov.pl/fileadmin/user_upload/bip/dostep_do_informacji/2024-10-11_Zal_3_do_proj_aKPEiK_Zalozenia_prognostyczne.pdf (accessed on 20 December 2025). (In Polish)
31. Chatzigeorgiou, N.G.; Theocharides, S.; Makrides, G.; Georghiou, G.E. A review on battery energy storage systems: Applications, developments, and research trends of hybrid installations in the end-user sector. *J. Energy Storage* **2024**, *86*, 111192. [CrossRef]
32. Garttan, G.; Alahakoon, S.; Emami, K.; Jayasinghe, S.G. Battery Energy Storage Systems: Energy Market Review, Challenges, and Opportunities in Frequency Control Ancillary Services. *Energies* **2025**, *18*, 4174. [CrossRef]
33. Holze, R. Self-discharge of Batteries: Causes, Mechanisms and Remedies. *Adv. Mater. Sci. Technol.* **2022**, *4*, 32–47. [CrossRef]
34. Ngoy, K.R.; Lukong, V.T.; Yoro, K.O.; Makambo, J.B.; Chukwuati, N.C.; Ibegbulam, C.; Eterigho-Ikelegbe, O.; Ukoba, K.; Jen, T.C. Lithium-ion batteries and the future of sustainable energy: A comprehensive review. *Renew. Sustain. Energy Rev.* **2025**, *223*, 115971. [CrossRef]
35. Al Kareem, S.S.A.; Hassan, Q.; Fakhruldeen, H.F.; Hanoon, T.M.; Jabbar, F.I.; Algburi, S.; Khalaf, D.H. A review on physical and chemical hydrogen storage methods for sustainable energy applications. *Unconv. Resour.* **2025**, *8*, 235. [CrossRef]
36. Ligęza, K.; Narloch, P. Prospects for storing hydrogen from renewable energy sources in Poland, part 1. *Nowa Energ.* **2023**, *1*, 39–45. (In Polish)
37. Maniglio, M.; Rivolta, G.; Elgendy, A.; Panfili, P.; Cominelli, A. Evaluating the Impact of Biochemical Reactions on H_2 Storage in Depleted Gas Fields. In Proceedings of the SPE Annual Technical Conference and Exhibition, San Antonio, TX, USA, 16–18 October 2023. [CrossRef]
38. Delshad, M.; Umurzakov, Y.; Sepehrnoori, K.; Eichhubl, P.; Batista Fernandes, B.R. Hydrogen Storage Assessment in Depleted Oil Reservoir and Saline Aquifer. *Energies* **2022**, *15*, 8132. [CrossRef]
39. Filar, B.; Miziołek, M.; Cicha-Szot, R.; Moska, A.; Kwilosz, T.; Szpunar, T. Analysis of the possibility of increasing the working capacity of underground gas storage facilities in Poland. *Naft. Gaz* **2024**, *80*, 571–580. (In Polish) [CrossRef]
40. State Geological Service (Państwowa Służba Geologiczna). Balance Sheet of Mineral Resources In Poland as at 31 December 2024. Available online: https://www.pgi.gov.pl/images/surowce/2024/bilans_2024.pdf (accessed on 20 December 2025). (In Polish)

https://doi.org/10.3390/en19040969

41. Uliasz-Misiak, B.; Lewandowska-Śmierzchalska, J.; Matuła, R. Hydrogen Storage Potential in Natural Gas Deposits in the Polish Lowlands. *Energies* **2024**, *17*, 374. [CrossRef]
42. Tarkowski, R. Perspectives of using the geological subsurface for hydrogen storage in Poland. *Int. J. Hydrogen Energy* **2017**, *42*, 347–355. [CrossRef]
43. Tarkowski, R.; Lankof, L.; Luboń, K.; Michalski, J. Hydrogen storage capacity of salt caverns and deep aquifers versus demand for hydrogen storage: A case study of Poland. *Appl. Energy* **2024**, *355*, 122268. [CrossRef]
44. Bonnet-Cantalloube, B.; Espitalier-Noël, M.; Ferrari De Carvalho, P.; Fonseca, J.; Pawelec, G. Clean Ammonia in the Future Energy Systems. Available online: https://hydrogeneurope.eu/wp-content/uploads/2023/03/2023.03_H2Europe_Clean_Ammonia_Report_DIGITAL_FINAL.pdf (accessed on 20 December 2025).
45. Agyekum, E.B.; Tarawneh, B.; Rashid, F.L. State-of-the-Art review of liquid organic hydrogen carriers for hydrogen storage and transport: Challenges and perspectives. *Appl. Energy* **2025**, *400*, 126555. [CrossRef]
46. Li, H.; Zhang, X.; Zhang, C.; Ding, Z.; Jin, X. Application and Analysis of Liquid Organic Hydrogen Carrier (LOHC) Technology in Practical Projects. *Energies* **2024**, *17*, 1940. [CrossRef]
47. Żukowski, W.; Berkowicz-Płatek, G.; Wencel, K.; Migas, P.; Wrona, J.; Reyna-Bowen, L. Hydrogen production from liquid organic hydrogen carriers in a fluidized-bed made out of Co_3O_4-cenosphere catalyst. *Int. J. Hydrogen Energy* **2025**, *136*, 358–370. [CrossRef]
48. Berkowicz-Płatek, G.; Żukowski, W.; Leski, K. Production of hydrogen from polyoxymethylene in a binary fluidized bed. *Appl. Energy* **2024**, *360*, 122833. [CrossRef]
49. Escamilla, A.; Sánchez, D.; García-Rodríguez, L. Assessment of power-to-power renewable energy storage based on the smart integration of hydrogen and micro gas turbine technologies. *Int. J. Hydrogen Energy* **2022**, *47*, 17505–17525. [CrossRef]
50. Nguyen, V.N.; Blum, L. Reversible fuel cells. *Compend. Hydrog. Energy Hydrog. Energy Convers.* **2016**, *3*, 115–145. [CrossRef]
51. Lamagna, M.; Ferrario, A.M.; Astiaso Garcia, D.; Mcphail, S.; Comodi, G. Reversible solid oxide cell coupled to an offshore wind turbine as a poly-generation energy system for auxiliary backup generation and hydrogen production. *Energy Rep.* **2022**, *8*, 14259–14273. [CrossRef]
52. Yang, Y.; Lei, J.; Huang, X.; Liao, Z.; Liu, Y.; Tu, Z. Recent Development in Reversible Solid Oxide Fuel Cells: Theory, Integration and Prospective. *ChemElectroChem* **2024**, *11*, e202300593. [CrossRef]
53. Santhanam, S.; Heddrich, M.P.; Riedel, M.; Friedrich, K.A. Theoretical and experimental study of Reversible Solid Oxide Cell (r-SOC) systems for energy storage. *Energy* **2017**, *141*, 202–214. [CrossRef]
54. Wei, M.; Levis, G.; Mayyas, A. Reversible Fuel Cell Cost Analysis. Available online: https://www.hydrogen.energy.gov/docs/hydrogenprogramlibraries/pdfs/review20/fc332_wei_2020_o.pdf (accessed on 20 December 2025).
55. The European Parliament and the Council of the European Union. Regulation (EU) 2019/943 of the European Parliament and of the Council of 5 June 2019 on the Internal Market for Electricity. Available online: https://eur-lex.europa.eu/legal-content/EN/TXT/PDF/?uri=CELEX:32019R0943 (accessed on 20 December 2025).
56. Daniel, K.; Kütt, L.; Naveed Iqbal, M.; Shabbir, N.; Ashraf Raja, H.; Sardar, M.U. A Review of Harmonic Detection, Suppression, Aggregation, and Estimation Techniques. *Appl. Sci.* **2024**, *14*, 966. [CrossRef]
57. ICS Investigation Expert Panel Grid Incident in South-East Europe on 21 June 2024. Available online: https://eepublicdownloads.blob.core.windows.net/public-cdn-container/clean-documents/Publications/2024/entso-e_incident_report_240621_250225_02.pdf (accessed on 20 December 2025).
58. ICS Investigation Expert Panel Grid Incident in North Macedonia on 18 May 2025. Available online: https://eepublicdownloads.blob.core.windows.net/public-cdn-container/clean-documents/Publications/2025/251106_factual_report_on_MEPSO_incident_18_May_2025.pdf (accessed on 20 December 2025).
59. ICS Investigation Expert Panel Grid Incident in Spain and Portugal on 28 April 2025. Available online: https://eepublicdownloads.blob.core.windows.net/public-cdn-container/clean-documents/Publications/2025/entso-e_incident_report_ES-PT_April_2025_06.pdf (accessed on 20 December 2025).
60. ENTSO-E Incident Classification Scale Methodology. Available online: https://eepublicdownloads.entsoe.eu/clean-documents/SOC%documents/Incident_Classification_Scale/200629_Incident_Classification_Scale_Methodology_revised_and_in_ use_as_of_2020.pdf (accessed on 20 December 2025).
61. Rashid, S.M. Employing advanced control, energy storage, and renewable technologies to enhance power system stability. *Energy Rep.* **2024**, *11*, 3202–3223. [CrossRef]
62. Ugwuagbo, E.; Ray, B.; Sul, J.; Nabi, N.; Datta, U.; Emami, K. Optimal allocation of grid-connected converter-based energy resources for securing power system stability: A comprehensive review. *Electr. Power Syst. Res.* **2025**, *248*, 111978. [CrossRef]
63. Jeyaraj, T.; Ponnusamy, A.; Selvaraj, D. Hybrid renewable energy systems stability analysis through future advancement technique: A review. *Appl. Energy* **2025**, *383*, 125355. [CrossRef]
64. Valikhany, M.; Astero, P.; Lehtonen, M.; Peltoniemi, P. A review on the suitability of virtual inertia strategies for the next generation of low-inertia power systems. *e-Prime Adv. Electr. Eng. Electron. Energy* **2025**, *14*, 101131. [CrossRef]

https://doi.org/10.3390/en19040969

65. Bahrani, B.; Ravanji, M.H.; Kroposki, B.; Ramasubramanian, D.; Guillaud, X.; Prevost, T.; Cutululis, N.A. Grid-Forming Inverter-Based Resource Research Landscape: Understanding the Key Assets for Renewable-Rich Power Systems. *IEEE Power Energy Mag.* **2024**, *22*, 18–29. [CrossRef]
66. Moore, P.; Alimi, O.A.; Abu-Siada, A. A Review of System Strength and Inertia in Renewable-Energy-Dominated Grids: Challenges, Sustainability, and Solutions. *Challenges* **2025**, *16*, 12.
67. Alajrash, B.H.; Salem, M.; Swadi, M.; Senjyu, T.; Kamarol, M.; Motahhir, S. A comprehensive review of FACTS devices in modern power systems: Addressing power quality, optimal placement, and stability with renewable energy penetration. *Energy Rep.* **2024**, *11*, 5350–5371. [CrossRef]
68. Flamm, B.; Peter, C.; Büchi, F.N.; Lygeros, J. Electrolyzer modeling and real-time control for optimized production of hydrogen gas. *Appl. Energy* **2021**, *281*, 116031. [CrossRef]
69. Fernández, R.D.; Martínez, L.A.; Peña, R.R.; Mantz, R.J. Electrolyzers as smart loads, preserving the lifetime. *Int. J. Hydrogen Energy* **2024**, *52*, 72–79. [CrossRef]
70. Sanchez-Ruiz, A.; de Groot, M.T. High-power electrolyzer characterization via smart power converters. *Int. J. Hydrogen Energy* **2024**, *96*, 1243–1250. [CrossRef]
71. Statistics Poland (Główny Urząd Statystyczny). Demographic Yearbook of Poland. Available online: https://stat.gov.pl/download/gfx/portalinformacyjny/pl/defaultaktualnosci/5515/3/19/1/rocznik_demograficzny_2025_pl.pdf (accessed on 20 December 2025). (In Polish)
72. Statistics Poland (Główny Urząd Statystyczny). Population Projection 2023–2060. Available online: https://stat.gov.pl/download/gfx/portalinformacyjny/pl/defaultaktualnosci/5469/11/1/1/1-prognoza_ludnosci_na_lata_2023-2060_.pdf (accessed on 20 December 2025). (In Polish)
73. Bradło, D.; Żukowski, W.; Porzuczek, J.; Olek, M.; Berkowicz-Płatek, G. Towards Net Zero in Poland: A Novel Approach to Power Grid Balance with Centralized Hydrogen Production Units. *Energies* **2025**, *18*, 1576. [CrossRef]
74. National Center of Research and Development Polish Power System Simulator. Available online: https://symulatorsystemuenergetycznego.ncbr.gov.pl/en/ (accessed on 20 December 2025).
75. Popkiewicz, M. NCBR's Zero-emission Energy System Symulator. *Energetyka* **2024**, *5*, 181–198. (In Polish)
76. Photovoltaic Geographical Information System (PVGIS) Webpage. Available online: https://re.jrc.ec.europa.eu/pvg_tools/en/ (accessed on 20 December 2025).
77. ENTSO-E Transparency Platform. Available online: https://transparency.entsoe.eu/ (accessed on 20 December 2025).
78. Open-Meteo Webpage. Available online: https://open-meteo.com/ (accessed on 20 December 2025).
79. Instrat Foundation Webpage. Available online: https://energy.instrat.pl/en/electrical-system/generation-capacity-are/ (accessed on 20 December 2025).
80. Polish Wind Energy Association (Polskie Stowarzyszenie Energetyki Wiatrowej). Rethinking Wind Energy: Necessary Legislative Changes and Onshore Potential Analysis. Available online: https://www.psew.pl/wp-content/uploads/2024/09/Psew_2024_Energia_na_nowo_poprawione.pdf (accessed on 20 December 2025). (In Polish)
81. Czyżak, P.; Mindeková, T. Empowering Farmers in Central Europe: The Case for agri-PV. Available online: https://ember-energy.org/app/uploads/2024/09/Empowering-farmers-in-Central-Europe-1.pdf (accessed on 20 December 2025).
82. Ministry of the Climate of the Republic of Poland. *The Polish Nuclear Power Programme*; Annex to Resolution No. 141 of the Cabinet of Ministers of 2 October 2020; Ministry of the Climate of the Republic of Poland: Warsaw, Poland, 2020; p. 946. (In Polish)
83. Statistics Poland (Główny Urząd Statystyczny). Energy. 2024. Available online: https://stat.gov.pl/download/gfx/portalinformacyjny/pl/defaultaktualnosci/5485/1/12/1/energia_2024.pdf (accessed on 20 December 2025).
84. Polish Biomethane Association Biogas and Biomethane in Poland Report. Available online: https://magazynbiomasa.pl/biogas-and-biomethane-in-poland-report-pobierz-za-darmo/ (accessed on 20 December 2025).
85. Krasuska, E.; Waliszewska, H.; Krzymiński, M.; Lenart, K.; Popkiewicz, M.; Racięcki, W. The Real Potential for Biomethane Production in Poland. Available online: https://www.gov.pl/attachment/1d27ccd0-14be-46ee-9535-68cef31dd0b0 (accessed on 20 December 2025). (In Polish)
86. Clean Air Task Force Hydrogen Production Calculator. Available online: https://www.catf.us/hydrogen-converter/#smr (accessed on 20 December 2025).
87. Liponi, A.; Pasini, G.; Baccioli, A.; Ferrari, L. Evaluating the Potential and Limits of Green Electrolysis in Future Energy Scenarios with High Renewable Share. *Energies* **2025**, *18*, 1654. [CrossRef]
88. Sepulveda, N.A.; Jenkins, J.D.; de Sisternes, F.J.; Lester, R.K. The Role of Firm Low-Carbon Electricity Resources in Deep Decarbonization of Power Generation. *Joule* **2018**, *2*, 2403–2420. [CrossRef]
89. Brown, T.; Schlachtberger, D.; Kies, A.; Schramm, S.; Greiner, M. Synergies of sector coupling and transmission reinforcement in a cost-optimised, highly renewable European energy system. *Energy* **2018**, *160*, 720–739. [CrossRef]

https://doi.org/10.3390/en19040969

90. Hasan, M.H.; Keiner, D.; Breyer, C. Techno-economic analysis of inter-annual energy storage and overcapacity in 100% renewable energy systems for 145 regions globally. *Appl. Energy* **2025**, *401*, 126736. [CrossRef]
91. Forum Energii Energy Transition in Poland. 2024 Edition. Available online: www.forum-energii.eu/en/download/download/transformacja-edycja-2024 (accessed on 20 December 2025).

https://doi.org/10.3390/en19040969

MDPI AG
Grosspeteranlage 5
4052 Basel
Switzerland
Tel.: +41 61 683 77 34

Energies Editorial Office
E-mail: energies@mdpi.com
www.mdpi.com/journal/energies

www.ingramcontent.com/pod-product-compliance
Lightning Source LLC
Chambersburg PA
CBHW061836210926
51850CB00006B/903
9783725870707